Springer
Tokyo
Berlin
Heidelberg
New York
Hong Kong
London
Milan
Paris

Takamitsu Sawa (Ed.)

International Frameworks and Technological Strategies to Prevent Climate Change

With 123 Figures

Springer

Takamitsu Sawa, Ph.D.
Director
Institute of Economic Research
Kyoto University
Yoshidahonmachi, Sakyo-ku
Kyoto 606-8501, Japan

ISBN 4-431-00001-1 Springer-Verlag Tokyo Berlin Heidelberg New York

Library of Congress Cataloging-in-Publication data applied for.

Printed on acid-free paper

Typesetting: SNP Best-set Typesetter Ltd., Hong Kong
Printing and binding: Nikkei Printing Inc., Japan
SPIN: 10897294

Preface: Strategies to Prevent Global Warming After the Possible Effectuation of the Kyoto Protocol

Will It Be Possible to Achieve the Reduction Target of Greenhouse Gas Emissions as Set Down in the Kyoto Protocol?

Measures Taken Against Global Warming Based on a Market Economic System

Three different types of measures are being taken in order to reduce greenhouse gas (GHG) emissions: voluntary measures, regulatory measures, and economic measures. As long as one believes in the market, priority should be given to economic measures such as carbon taxation, or making sales and ownership taxation on vehicles proportional to their fuel-efficiency. In particular, the recent trend of market liberalization and internationalization should be taken into consideration when we consider preferences among the three measures.

Japan has an open market economic system, i.e., not a centrally planned and command-and-control one. The appropriate approach against global warming for a free economy is to take predominantly economic measures and to supplement these with regulatory measures, i.e., prohibitions and obligations.

In March 2002 the Japanese government published *Fundamental Principles of Promoting Measures Against Global Warming*, in which numerous concrete measures were proposed and their effects were numerically estimated. It should be noted, however, that nothing was mentioned about how much these measures would cost, and how the costs would be borne, and by whom. Furthermore, policies to motivate households and firms to implement these measures were totally overlooked.

Past Trends and Future Forecasts of Carbon Dioxide (CO_2) Emissions

From 1986 to 1996, the levels of CO_2 emissions increased at an annual rate of 2.8%, but in 1997 emissions fell by 0.2% and again by 3.6% in 1998, reversing the past trend. This was followed by 3.4% and 1% increases in 1999 and 2000, respectively. One of the causes for the leveling off of emissions since 1997 is the negative economic growth rate during this period. However, during the so-called Heisei recession that started in May 1991 and ended in October 1993, in spite of the nearly 0% economic growth rate, there was a significant increase in CO_2 emissions. This suggests that a significant structural change occurred in energy consumption patterns in the late 1990s.

The significant increase in energy consumption and hence CO_2 emissions by the household sector from the late 1980s to the early 1990s was due to the rapid spread of domestic appliances such as air conditioners that consume large amounts of electricity daily, the propagation of domestic electrical goods with a standby function, and the rapid development of large-scale domestic electrical appliances during that period.

Furthermore, the significant increase in CO_2 emissions within the transportation sector during the same period was due to the transition to large-size cars, the continuing spread of recreation vehicles (RVs), and the decrease in the average fuel-efficiency of vehicles.

The spread of domestic appliances that consume large amounts of electricity and the decrease in fuel-efficiency of vehicles within Japan is reaching what may be called a state of saturation. At the same time the Energy Conservation Law has been in force. The law requires companies that manufacture electrical appliances and cars to improve energy-efficiency to the level determined by the government for each type of appliance and vehicle. Moreover, the recent tendency has been to move from the CIMA phenomenon to the VITZ phenomenon, i.e., smaller cars have become preferred to larger ones by the majority of consumers.

According to the government's forecast of the BAU (business-as-usual) CO_2 emissions, in 2010 total CO_2 emissions will have increased by 18% compared with 1990 unless effective measures are taken. I am afraid, however, that the government's forecast is exaggerated.

In particular, it is necessary to take into account possible technological innovations in the near future. For instance, will the anticipated IT innovations increase or decrease energy consumption? If one supposes that they will decrease energy consumption, the expected reduction in CO_2 emissions must be included in the BAU forecast.

Possible Change in Industrial Structure and Globalization

The possibility of a change in industrial structure should not be overlooked. In 1985 the manufacturing industry accounted for 29.5% of Japanese GDP, but in 1999 this figure fell to 23.1%. Furthermore, the share of energy-intensive heavy and chemical industry sectors, including steel, nonferrous metals, ceramics, and metal goods, fell from 24.3% to 14.5% of all manufacturing sectors in terms of GDP. These trends are expected to continue in the near future. Such changes in industrial structure could well cause a gradual decrease in CO_2 emissions per unit of the GDP.

One of the factors that have caused a significant fall in the share of GDP held by the manufacturing industry is overseas transfer of production bases. The change in industrial structure, namely, what is called "hollowing of the industrial structure," should be considered a necessary consequence of economic development. In fact, William Petty rightly predicted such a tendency at the end of the seventeenth century. Developed Western countries have gone through the same hardships.

In other words, the transition from an industrialized society to a post-industrial society is inevitable, where the latter is a society with an economic structure based mainly on the high-tech manufacturing and software industries. Those who are afraid that measures against global warming will cause hollowing of the industrial structure are ignoring an inevitable consequence of economic development.

Those who argue against globalization-driven change in the industrial structure actually are playing an unintentional role in the possible decline of the international competitive power of Japanese industries. However, it is an appropriate policy to change the industrial structure so that it may adequately adjust to such historical trends.

If you re-examine the economic history of Japan, you will immediately realize that many cases can be found in which restrictions and shortages served as the driving force for economic development and growth. Surely, CO_2 emission reduction based on the Kyoto Protocol could work as a springboard for economic growth. It is the government's responsibility to publicize and take advantage of a potential springboard for economic growth as quickly as possible.

Conclusion

As we have seen, it is by no means impossible to achieve a 6% reduction in annual GHG emissions during the period from 2008 to 2012 compared to those in 1990. As long as the appropriate measures are implemented without delay, the goal set out in the Kyoto Protocol might be fully attainable through domestic policies alone.

By effectively utilizing the clean development mechanism (CDM), credits for CO_2 emissions can be obtained at a relatively low cost, and hence we can avoid paying unnecessarily high costs to reduce CO_2 emissions domestically. An appropriate combination of domestic policies and CDM will enable us to achieve the greatest effect at the lowest cost.

Because our economic system is a market-oriented one, it is quite likely that we could fail to comply with the target. In such a noncompliance case the Kyoto Protocol permits us to make up for the deficiency by transferring emissions rights to us from other countries who have surpassed their reduction target.

The Effect on the Economy of Measures Taken Against Global Warming

Differences in Macroeconomic Effects Between Developed and Developing Countries

The three Scandinavian countries, Holland, and Denmark introduced carbon taxation in the early 1990s. Germany introduced somewhat irregular carbon taxation in 1999. France introduced carbon taxation in 2001, and Britain, a climate change levy from April 2001. Italy's policy is to implement taxation on fossil fuels in the form of a mineral oil tax from 2005. In principle, these countries are remaining revenue neutral, including reducing the burden of social security to compensate for the burden imposed by carbon taxation.

One of the most controversial issues is whether or not CO_2 emission reduction due to carbon taxation necessarily restrains economic growth. My conclusion is that it certainly does in developing countries, but does not necessarily do so in developed countries.

It is a matter of course that the CO_2 emission reduction requires firms and consumers to share a part of the necessary costs, but this does not necessarily mean that the macroeconomic growth rate is certain to decrease.

The investment in equipment to effectively reduce CO_2 emissions and the investment to expand production capacity are mutually linked to a trade-off in developing countries. Regulations and taxation to reduce CO_2 emissions reinforce the former and hence restrain the latter as a result of limited availability of funds. Consequently, CO_2 emission reduction restrains the potential growth rate.

On the other hand, in fully mature developed countries such as Japan, almost all industries have excess production capacity, and hence quite low incentive and necessity to invest in production equipment. Furthermore, because investment in plants and equipment has become highly electronic nowadays, its multiplier effect is relatively small compared with that of investment that consumes huge amounts of steel and cement. Relatively speaking, investment in equipment to reduce CO_2 emissions makes use of more steel and cement and hence has a stronger multiplier effect; it contributes to economic growth to a larger extent than does investment in electronic equipment.

The introduction of carbon taxation brings about an income transfer from consumers and firms to the government. Unless the government misuses the transferred income, a slowdown in the economic growth rate may be avoidable. In fact, personal consumption expenditure decreases as a result of such an income transfer.

In cases where the personal income tax is reduced so that tax revenue is kept neutral, the personal disposable income increases, and hence personal consumption expenditure will increase. The question is whether an offset account is positive or negative. It is difficult to give a definite answer to this question *a priori*, because it depends on so many macroeconomic structural parameters, such as the propensity to consume, in a very complex manner.

Will Carbon Taxation Really Reduce CO_2 Emissions?

Those who oppose the introduction of carbon taxation pose the question about its effectiveness of reducing CO_2 emissions. Indeed, electricity, gas, and gasoline are all necessities, and hence demand for them is quite inelastic with regard to price variation, i.e., demand is largely independent of price. However, this is only in the short term; the mid-term effect is another story. For example, when the price of gasoline rises, demand for gasoline will not decrease very significantly. However, after 3 to 6 years, when it comes to purchasing a new car the tendency to prefer more fuel-efficient cars ought to increase. Therefore, in the mid term, demand for gasoline is elastic enough with regard to price variation. This implies that the effectiveness of carbon taxation is significant.

By presenting a time series line graph of gasoline prices and consumption as well as the electricity prices and consumption, some economists argue that the graph shows that consumption of energy does not fall at all in response to a price hike, i.e., the price elasticity is nearly zero. This is a simple example of an elementary mistake of statistics. In short, this misleading argument is confusing partial correlation and simple correlation between demand and a price. The details are as follows.

Apparently, price is not the sole factor that determines the demand for gasoline and electricity: the demand for gasoline depends on many other factors including income, increase in size of appliances, improvement of energy- or fuel-efficiency, changes in lifestyle, and so on. If we can find no significant negative correlation between the residuals obtained by removing the effect of the other factors from the prices and consumption of gasoline and electricity, then we should conclude that the price elasticity of energy demand is zero.

Problematic Points and Side Effects of Carbon Taxation

One of the controversial aspects of carbon taxation concerns the choice among the following three alternatives: first, the carbon tax revenue should be included in the general fiscal account; second, it should be included in a special account for implementation of measures against global warming; third, based on the revenue-neutral principle, the personal income tax should be reduced by the amount of the carbon tax revenue. The financial authorities tend to support the first, while authorities concerned with taking measures against global warming support the second. The third alternative is most often supported by economists.

Since carbon taxation causes a rise in manufacturing costs for export industries such as steel, nonferrous metals, ceramics, and metal goods that consume large amounts of energy in their production processes, their international competitiveness is likely to be harmed. As a compensation, customs measures can be taken to lessen the negative effect. For example, when steel is exported, the carbon tax already paid by the steel company to produce steel to be exported is refunded at the port, while the carbon tax is charged on steel imports at the port. In short, taxing domestically consumed steel but not taxing steel consumed abroad would solve the problem. Another solution would be to follow Sweden's example: a carbon tax is exempted for some selected industries that consume larger amounts of fossil fuels in the production process.

Generally speaking, the promotion of measures against global warming, such as the introduction of carbon taxation, sets up an incentive for technical innovation. The competition in research and development of alternative fuels that emit low levels of CO_2 and of fuel-efficient cars will be intensified. Therefore, the Kyoto Protocol offers an opportunity for promotion of new research and development within the industrial world.

The Microeconomic Effects of Carbon Taxation

In the implementation of measures against global warming, such as the introduction of carbon taxation, it is hard to avoid a division into "winner industries" that make a profit and "loser industries" that make a loss.

The biggest loser industry is the coal industry. For this reason countries with a large coal industry, such as Australia, have reacted negatively to taking measures against global warming. On the other hand, as Japan's coal industry is close to extinction, Japan is one of the rare countries that can implement measures against global warming without any significant resistance by the coal industry.

As for the oil industry, given the increase in demand for natural gas which is a by-product of crude oil processing, difficulties in developing a liquid fuel to replace oil, and the fact that oil reserves will last only for the next 40 years, we cannot necessarily say for sure that it will be a loser.

Similarly, it is also hard to avoid a division of winner companies and loser companies in the same industry. Winners include car manufacturers that are developing fuel-efficient cars, and electrical equipment manufacturers that are developing appliances with electricity-saving features.

Given that sorting of winners and losers is developed not only on a national scale but also on an international scale, it is preferable for Japan to take the initiative in implementing measures against global warming. Early action is indispensable if Japanese automakers are to be winners in forthcoming tough competition in developing more fuel-efficient cars, such as those powered by fuel cells.

The Kyoto Protocol offers an opportunity for worldwide reorganization of the automobile industry. The merger of Daimler-Benz with Chrysler in 1998 can be seen as a forerunner to the reorganization of the industry triggered by the Kyoto Protocol. Fierce competition in research and development of fuel-efficient cars has already begun. The Kyoto Protocol is predicted to come into effect at latest by the end of 2003, which is considered to be the deadline of putting fuel-cell cars on the market.

The Kyoto Mechanism

The Kyoto Protocol introduced an international system of emissions trading, joint implementation, and the clean development mechanism, the common name of which is the Kyoto Mechanism.

The Kyoto Protocol requires Annex I countries to reduce their average GHG emissions for the period from 2008 to 2012 by at least 5% compared to total emissions in 1990. The assigned reduction rates varied among countries: 8% for the EU countries, 7% for the United States, 6% for Japan, and 0% for Russia and other East European countries.

Japan's assigned reduction rate of 6% suggests that Japan has acquired emissions rights of the amount equal to 5 times 94% of the 1990 GHG emissions, which are effective during the above time period.

The total cost of reductions would be minimized if the reduction rate of GHG emissions were assigned to each Annex I country so that the marginal reduction cost of GHG emissions could be distributed equally among all Annex I countries. However, as the marginal reduction cost curves for Annex I countries are unknown, the assigned reduction rates do not satisfy the above optimality condition. The Kyoto Mechanism was introduced in order to compensate existing unfairness of differentiation of assigned reduction rates.

The marginal reduction cost, i.e., the cost to reduce one more unit of GHG emissions, is largely different among countries. Generally speaking, it is cheaper in developing countries and more expensive in developed countries. In order to attain the assigned reduction rate we can try to minimize the total cost by adequately combining opportunities provided by the clean development mechanism with domestic measures.

The introduction of domestic emissions trading is often recommended by some economists and climate change experts. Dealers in sales of fossil fuels are obliged to attach emissions rights issued by the government depending on the carbon content of the fossil fuels being sold. The total amount of issued emissions rights is equal to the total GHG emissions planned by the government. It is expected that a market will be created to buy and sell available emissions rights.

A domestic market of emissions rights is certain to follow the creation of an international market. The creation of the market is indispensable to make private sector investment in energy-saving equipment in developing countries economically viable.

Takamitsu Sawa
Institute of Economic Research
Kyoto University
September 3, 2002

Contents

Section II: China and International Cooperation in Global Warming

Section III: Assessments of Technology Strategies Toward Energy, Economic, and Environmental Issues

List of Authors

Aiba, T., Development Bank of Japan (DBJ); Climate Design, ISER, Osaka University, Japan

Akita, J., Graduate School of Economics and Management, Tohoku University, Japan

Ezaki, M., Graduate School of International Development, Nagoya University, Japan

Fujii, Y., Department of Electrical Engineering, University of Tokyo, Japan

Furuse, T., Norin Chukin Securities Co., Ltd., Japan

Goto, N., Graduate School of Arts and Sciences, University of Tokyo, Japan

Hayashi, T., Department of Electrical Engineering, University of Tokyo, Japan

Imai, H., Kyoto Institute of Economic Research, Kyoto University, Japan

Kinjo, M., School of Political Science and Economics, Tokai University, Japan

Kusakawa, T., Graduate School of Economics, Osaka University, Japan

Li, Z.D., Department of Management and Information System Science, Nagaoka University of Technology, Japan

Mathur, R., Tata Energy Research Institute, India

Matsuhashi, R., Institute of Environmental Studies, Graduate School of Frontier Sciences, University of Tokyo, Japan

Matsuoka, I., Graduate School of Energy Science, Kyoto University, Japan

Mori, S., Department of Industrial Administration, Tokyo University of Science, Japan

Niizawa, H., Institute of Economic Research, Kobe University of Commerce, Japan

Okada, A., Institute of Economic Research, Kyoto University, Japan

Saijo, T., Institute of Social and Economic Research and Climate Design, ISER, Osaka University; Research Institute of Economy, Trade and Industry, Japan

Sawa, T., Institute of Economic Research, Kyoto University, Japan

Sun, L., Graduate School of International Development, Nagoya University, Japan

Tezuka, T., Graduate School of Energy Science, Kyoto University, Japan

Section I: Functioning of the Kyoto Mechanism

A Market Game Analysis of International CO_2 Emissions Trading: Evaluating Initial Allocation Rules

Akira Okada

Summary. We present a market game model of international CO_2 emissions trading, and evaluate the widely discussed allocation rules for emissions permits. We first show the general difficulty of an international agreement on emissions permit allocations by using a voting game before emissions trading. The calibration of emissions trading shows that three major emitting countries, the United States (USA), the former Soviet Union (FSU), and Japan, have diverse preferences over allocation rules. The USA prefers a per capita reduction rule, the FSU prefers a per GDP reduction rule, and Japan prefers an equalizing net cost per capita rule. Finally, we discuss the proposition that the reduction commitments in the Kyoto Protocol approximately equalize the allocations of total saving costs ($530 million) by emissions trading among the three countries.

Key words. CO_2 emissions trading, Competitive equilibrium, Initial allocation rules, Kyoto Protocol, JEL classification: C71, D41, Q25

Introduction

The third Conference of the Parties (COP-3) to the UN Framework Convention on Climate Change was held in Kyoto on December 1–10, 1997. It was expected to adopt a "legally binding protocol or other legal instrument" committing developed countries in Annex I of the convention to reducing their greenhouse gas (GHG) emissions after the year 2000. Before the Kyoto Conference, the 1992 Convention on Climate Change had committed developed countries to stabilizing their emissions to 1990 levels by the year 2000 without any legal obligation. Recognizing that this voluntary approach had not been successful, the parties to the convention agreed, under the "Berlin Mandate" in 1995, to strengthen the commitments of Annex I parties, and in particular to set quantified limitation and reduction objectives within specified time-frames as early as possible in 1997.

Institute of Economic Research, Kyoto University, Yoshidahonmachi Sakyo-ku, Kyoto 606-8501, Japan

After intense debates among more than 150 developed and developing countries, the Kyoto Protocol to the Framework Convention was finally agreed. The main contents of the Kyoto Protocol are summarized below.

(1) Annex I countries (OECD countries and countries in the former Soviet Union (FSU) and Eastern Europe) as a whole will reduce emissions by 5.2% below 1990 levels between 2008 and 2012 (Article 3). (2) A quantified emissions limitation or reduction commitment (QELRC) is assigned to each Annex I country (Article 4 and Annex B). The reduction rates for the major emitting countries are 0% for Russia and Ukraine, 6% for Japan, 7% for the United States (USA), and 8% for the European Union (EU). (3) The protocol includes three "flexible" mechanisms for international emission transfer: joint implementation (Article 6), the clean development mechanism (Article 12), and emissions trading (Article 17).

The aim of this chapter is to present a market game model of international CO_2 emissions trading, and to evaluate numerically the economic outcomes of emissions trading under various initial allocation rules, including the reduction commitments in the Kyoto Protocol.[1] The Kyoto Protocol is a set of agreements made in long and complex negotiation processes before and during the Kyoto meeting. Although some publications have appeared since the meeting which explain the negotiation process (e.g., Grubb et al. 1999), it is not clear on what kind of rational basis the reduction commitments in the Kyoto Protocol were determined. Furthermore, all the details of international emissions trading remain open for future negotiations. In order to move toward the establishment of an international emissions trading mechanism, we believe that it is important to scrutinize various aspects of the structure of conflict and cooperation in the international negotiations underlying the Kyoto Protocol. Using a theoretical game model and its calibration, we investigate the negotiations which allocated emission permits when permits are tradeable.[2]

We consider six widely discussed allocation rules for emissions permits. The first two rules, the per capita rule and the per GDP rule, require that an equal amount of emissions permits, either per capita or per GDP, should be allocated to each country. Both rules focus on the "zero emissions" level as an "ideal" point in the negotiations. The next two rules, the per capita reduction rule and the per GDP reduction rule, require that every country should reduce emissions by an equal amount, either per capita or per GDP, from the level of some agreed base year (e.g., 1990). None of these rules take into account a country's net reduction

---

[1] The Kyoto Protocol states that international emissions trading shall be supplementary to domestic actions for the purpose of meeting QELRC (Article 17). An analysis of such a binding supplementarity mechanism is beyond the scope of this chapter

[2] In the literature, there are several publications which analyze the global economic impacts of the Kyoto Protocol by using large-scale models. For example, for an analysis using a dynamic general equilibrium model with multisectors and multiregions, see Bernstein et al. (1999). As far as we know, however, there are few works which investigate the negotiations on emissions permit allocations in the Kyoto Protocol in the framework of game theory

costs after emissions trading. The last two rules, the equalizing net costs per capita rule and the equalizing net costs per GDP rule, require that every country should be burdened with an equal amount of net reduction costs after emissions trading, either per capita or per GDP.

We calibrate the competitive equilibrium of international emissions trading among three major emitting countries, the USA, the FSU, and Japan. Our analysis follows the work of Barrett (1992) and Bohm and Larsen (1994). Barrett (1992) investigated whether an "acceptable" allocation of emissions permits exists in a global warming treaty, based on the numerical results of emissions trading among the USA, the FSU, and China. Bohm and Larsen (1994) evaluated the distributional implications of several permit allocations in a tradeable permit regime for the region of Europe and the FSU. As in this previous work, we adopt the marginal cost functions for CO_2 emissions reductions estimated by Nordhaus (1991). This enables us to derive an explicit formula for the competitive equilibrium price of the emissions trading, and to evaluate the economic implications of different allocation rules of emissions. For a recent study of the marginal costs of CO_2 emissions reductions, see Tol (1999).

To consider the negotiations for permits allocations, we formulated a voting game in which countries collectively decide a permit allocation before emissions trading begins. It is shown that a stable outcome (defined by the "core" solution) of the voting game exists if and only if the voting rule allows at least one country to have the power of veto. This result implies the general difficulty of reaching an agreement on emissions permits allocations in international negotiations when no power of veto is allowed. On the other hand, the unanimous voting rule under which every country has the power of veto leads to the vacuous outcome that all allocations of emissions permits are stable. Consistent with this theoretical prediction, there was a strong conflict among participating countries in negotiations on emissions permits allocations at COP-3. Given this general insight to international negotiations on emissions permits allocations, the calibration of emissions trading is a useful way for us to consider how countries evaluate various allocation rules for emissions permits.

The main findings of our numerical analysis are as follows. First, about 95% of the total emissions reductions should be done by the USA and the FSU in any efficient reduction scheme. Owing to the high marginal costs of emissions reductions, Japan needs to reduce only 3.4% of the total target. In terms of saving reduction costs, Japan benefits the most from emissions trading, the FSU is second, and the USA benefits the least.

Second, neither the per capita rule nor the per GDP rule is practically feasible in international negotiations because each of these rules imposes an excessively severe limitation of emissions on a single country. The per capita rule is not acceptable to the USA, and the per GDP rule is not acceptable to the FSU. The preferences of these three countries over the four remaining rules are very diverse. The USA prefers the per capita reduction rule, the FSU prefers the per GDP reduction rule, and Japan prefers the equalizing net cost per capita rule.

Third, the reduction rates (USA 7%, FSU 0%, Japan 6%) agreed in the Kyoto Protocol are estimated to give net reduction costs of $431 million (0.01% of GDP) to the USA and $97 million (0.003% of GDP) to Japan, and net profits of $193 million (0.03% of GDP) to the FSU after emissions trading. Emissions trading saves a total of about $530 million for the group. The reduction cost allocation imposed by the Kyoto Protocol is close to equal allocations of this total saving costs among the three countries.

The chapter is organized as follows. The next section presents a market game model for international CO_2 emissions trading, and this is followed by a section giving an estimated marginal cost function for reducing emissions, and deriving a formula for competitive equilibrium in the market game. There is then a discussion of six allocation rules of permits, which shows the general difficulty of international agreement on emissions permits allocations. The penultimate section numerically evaluates various allocation rules for permits, including the reduction commitments in the Kyoto Protocol, and the chapter ends with our conclusions.

The Model

Let $N = \{1, \ldots, n\}$ be the set of countries. For every $i = 1, \ldots, n$, we denote country i's current level of CO_2 emissions by E_i. The total level of CO_2 emitted by n countries is given by $E = \Sigma_{i \in N} E_i$. Let ω_i denote the amount of initial emissions permits allocated to country i, and let $\omega = (\omega_1, \ldots, \omega_n)$ be a vector of initial emissions permits. The total amount of CO_2 emissions permits is given by $\overline{\omega} = \Sigma_{i=1}^{n} \omega_i$. We assume that $0 < \overline{\omega} < E$. In this section, the initial emissions permits allocation $\omega = (\omega_1, \ldots, \omega_n)$ is fixed. We discuss later how an emissions permits allocation ω should be determined, given the total permits $\overline{\omega}$. For a subset S of N, notation R^S means the set of all vectors $x = (x_i : i \in S)$ of real numbers. R_+ denotes the set of all nonnegative real numbers.

Let x_i denote country i's reduction of CO_2 emissions where $0 \leq x_i \leq E_i$. Country i's CO_2 abatement cost function is denoted by $C_i(x_i)$, which is assumed to be a differentiable, strictly convex, and monotonically increasing function of R_+.

We present a market game model for international emissions trading. Every subset S of N is called a coalition of countries. All member countries in coalition S can minimize their total costs of CO_2 reduction by trading emissions permits within the coalition. The total cost of reducing CO_2 emissions for coalition S is given by

$$\min_{x \in R^S} \sum_{i \in S} C_i(x_i) \quad s.t. \quad \sum_{i \in S} x_i \geq \sum_{i \in S} (E_i - \omega_i), 0 \leq x_i \leq E_i, \text{for any } i \in S \tag{1}$$

assuming that $\Sigma_{i \in S}(E_i - \omega_i) \geq 0$. The first constraint means that coalition S as a whole should not emit more CO_2 than the total emissions permits assigned to all member countries. Note that equality $\Sigma_{i \in S} x_i = \Sigma_{i \in S} (E_i - \omega_i)$ must hold at an optimal solution of Eq. 1.

Let $x^{*S} = (x_i^{*S} : i \in S)$ denote an optimal solution of Eq. 1, and define

$$C^{\omega}(S) = \sum_{i \in S} C_i(x_i^{*S})$$

The optimal value $C^{\omega}(S)$ is the minimum cost of CO_2 emissions reductions by coalition S where emissions are freely tradeable among its member countries. Notation $C^{\omega}(S)$ emphasizes that the total reduction costs for coalition S may depend on the initial allocation $\omega = (\omega_1, \ldots, \omega_n)$ of emissions permits. It can easily be seen that the optimal cost $C^{\omega}(\cdot)$ is subadditive: $C^{\omega}(S \cup T) \leq C^{\omega}(S) + C^{\omega}(T)$ for any S and T with $S \cap T = \varnothing$. This fact implies that, given the predetermined level $\overline{\omega}$ of emissions, the total costs of CO_2 emissions abatement across n countries can be minimized if emissions permits are tradeable in the largest coalition N. Formally, a *market game of international CO_2 emissions trading* is defined by the pair (N, C^{ω}).[3]

We now define a competitive equilibrium in a market game (N, C^{ω}) for emissions trading. Let p be the market price of emissions, and let x_i be the level of emissions reduction by country i after emissions trading.

Definition 1: A *competitive equilibrium* for a market game (N, C^{ω}) of international CO_2 emissions trading is defined to be a vector $(p^*, x_1^*, \ldots, x_n^*) \in R_+^{n+1}$ satisfying

(i) $x_i^* \in \operatorname{argmin}\{C_i(x_i) + p^*(E_i - x_i - \omega_i) \mid 0 \leq x_i \leq E_i\}$ for any $i \in N$

(ii) $\sum_{i \in N} (E_i - x_i^*) = \sum_{i \in N} \omega_i$

When country i reduces CO_2 emissions by x_i from E_i, it must possess $E_i - x_i$ amounts of emissions permits. If the initial allocation of emissions, ω_i, is less than this level, country i has to purchase $E_i - x_i - \omega_i$ amounts of emissions permits from other countries. Equation (i) means that every country i minimizes its CO_2 abatement net costs, given the equilibrium price p^* of emissions. Equation (ii) is the balanced condition of demand and supply for CO_2 emissions permits. If x_i^* is an interior solution of country i's cost-minimizing problem in (i), then it must hold that $p^* = MC_i(x_i^*)$ for all $i \in N$, where $MC_i(x_i)$ is the marginal reduction cost function of country i. This is a well-known principle of *marginal cost pricing*. Then, a competitive equilibrium $(p^*, x_1^*, \ldots, x_n^*) \in R_+^{n+1}$ of a market game (N, C^{ω}) can be characterized as a solution of the system

$$\begin{aligned} p - MC_1(x_1) &= 0 \\ &\vdots \\ p - MC_n(x_n) &= 0 \\ E - \overline{\omega} - \sum_{i \in N} x_i &= 0 \end{aligned}$$

[3] For detailed discussions of market games, see Shubik (1983)

Note that the competitive equilibrium of emissions trading is determined solely by the total reduction $r = E - \overline{\omega}$, given the marginal abatement cost functions.

The *competitive equilibrium cost* for country i is defined by

$$c_i^e(\omega_i) \equiv C_i(x_i^*) + p^*(E_i - x_i^* - \omega_i) \tag{2}$$

Since countries can anticipate the equilibrium outcome of international emissions trading, it is natural to assume that countries evaluate various initial allocations of emissions permits by their competitive equilibrium costs. Negotiations on initial allocations will be discussed later. In the next section, we derive a formula for the competitive equilibrium of emissions trading by employing the marginal cost function for CO_2 emissions abatement from Nordhaus (1991).

The Estimated Marginal Cost Function for CO_2 Emissions Abatement

A country's marginal costs of carbon reductions depend on various factors such as domestic fossil fuel prices and taxes, and energy-use patterns. These factors affect the country's carbon intensity (i.e., carbon emissions per GDP). High consumer prices for fossil fuel decrease carbon emissions per unit of output, which increases marginal costs of emissions reductions. In our analysis, following Bohm and Larsen (1994), we assume that the marginal cost function of CO_2 emissions reductions for different countries are mainly determined by their carbon intensities.

Based on a survey of nine different estimates of marginal costs of CO_2 emissions in the USA and Western Europe, Nordhaus (1991) derived a logarithmic form of the marginal cost function

$$\mathrm{MC}(R) = -185.2 \cdot \ln(1 - R)$$

where R is the rate of emissions reductions (x/E), and MC is the marginal cost in units of 1989 US dollars per ton of carbon. In the above form, we assume a zero intercept (Nordhaus estimates an intercept of –4.13). Since most studies surveyed by Nordhaus (1991) were based on data in the USA, we can assume that the estimated marginal cost function is applicable to a country with the same carbon intensity (E/GDP) as the USA. The carbon intensity of the USA is 0.26 kg/US\$. The USA was chosen as the "reference country" in the following analysis.

To derive marginal cost functions for other countries, we employ the method used by Bohm and Larsen (1994). Let $\bar{e}$ be the carbon intensity of the USA. It is assumed that a country l with carbon intensity e_l lower than $\bar{e}$ has already taken measures to reduce its carbon intensity, and thus that its marginal cost function starts at some higher level R_l^0 along the MC-curve of the reference country. Figure 1 illustrates the marginal cost function of the reference country (USA), and the

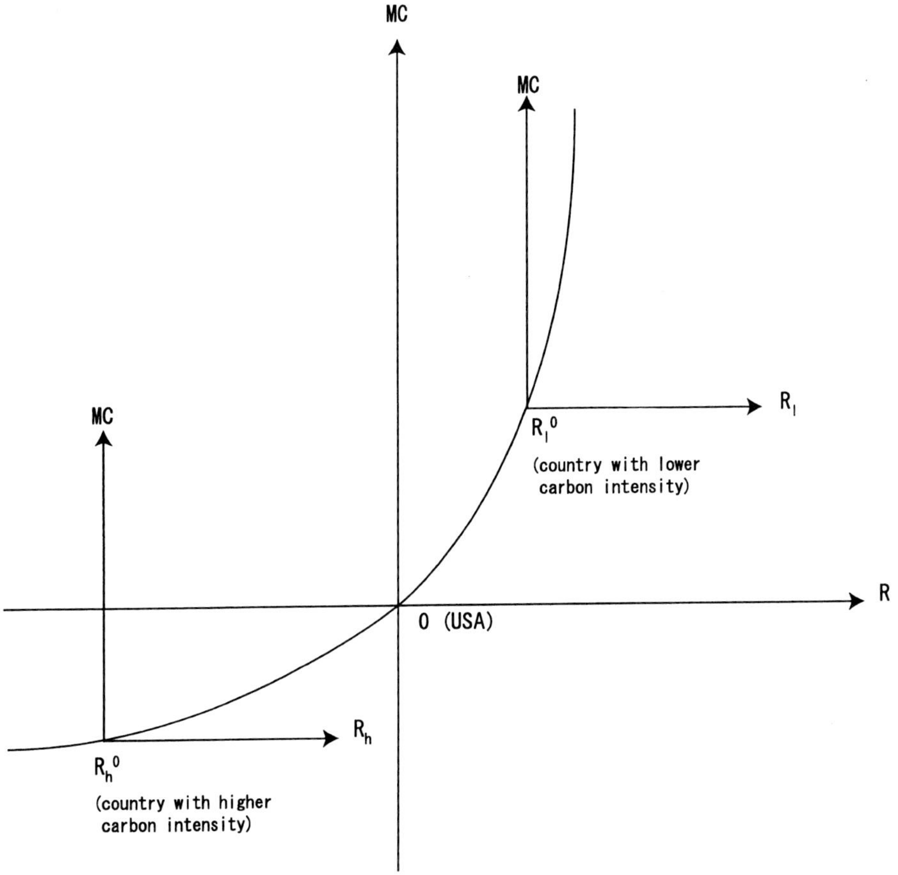

FIG. 1. Marginal cost functions for different countries

starting point R_l^0 for country l with a lower carbon intensity. The curve is steeper at R_l^0. The starting level R_l^0 is assumed to be

$$R_l^0 = 1 - \frac{e_l}{\bar{e}}$$

The distance R_l^0 is interpreted as indicating the percent age by which the reference country would have to reduce its carbon intensity if it were to reach the same marginal costs curve as country l. Then the marginal cost function of country l is given by

$$\begin{aligned} \mathrm{MC}_l(R_l) &= 185.2 \cdot \ln(1 - R_l^0) - 185.2 \cdot \ln(1 - R_l^0 - R_l) \\ &= -185.2 \cdot \ln\left(1 - \frac{R_l}{1 - R_l^0}\right) \end{aligned}$$

where R_l is the rate of carbon reductions by country l.

In the same way, a country h with carbon intensity e_h higher than $\bar{e}$ has a marginal cost function which starts at some lower level R_h^0 along the MC-curve of the reference country (see Fig. 1). The distance which R_h^0 shifts to the left is assumed to be

$$R_h^0 = 1 - \frac{\bar{e}}{e_h}$$

which can be interpreted in the same way as R_l^0. Then the marginal cost function of country h is given by

$$\begin{aligned} \mathrm{MC}_h(R_h) &= -185.2 \cdot \ln(1 + R_h^0 - R_h) + 185.2 \cdot \ln(1 + R_h^0) \\ &= -185.2 \cdot \ln\left(1 - \frac{R_h}{1 + R_h^0}\right) \end{aligned}$$

where R_h is the rate of carbon reductions by country h. The curve is flatter at R_h^0.

We note that in the estimations above, each country's marginal cost function for CO_2 emissions reductions has a zero intercept. This implies that countries' net marginal costs of emissions reductions are zero, regardless of their carbon intensities. It is assumed here that each country will optimize its energy consumption by some future year (say 2010) before the carbon reductions enforced by the Kyoto Protocol start, and thus that each country wil have chosen a carbon intensity so that the marginal cost is equal to the marginal (non-GHG-related) benefits of further reductions in fossil fuel consumption. For more detailed arguments on this point, see Bohm and Larsen (1994).

To summarize, the marginal cost function of country i is given by

$$\mathrm{MC}_i(x_i) = -185.2 \cdot \ln\left[1 - \frac{x_i}{E_i(1 - r_i)}\right] \tag{3}$$

where $r_i = 1 - e_i/\bar{e}$ if $e_i < \bar{e}$, and $r_i = -1 + \bar{e}/e_i$ if $e_i > \bar{e}$. The CO_2 abatement cost function of country i is given by

$$\begin{aligned} C(x_i) &= \int_0^{x_i} -185.2 \cdot \ln\left[1 - \frac{t}{E_i(1 - r_i)}\right] \mathrm{d}t \\ &= 185.2[E_i(1 - r_i) - x_i] \cdot \ln\left[1 - \frac{x_i}{E_i(1 - r_i)}\right] + 185.2 x_i \end{aligned}$$

We are now ready to compute the competitive equilibrium of CO_2 emissions trading under the estimated marginal cost functions. The optimal emission reduction x_i^{*S} of country i within coalition S satisfies

$$-185.2 \cdot \ln\left[1 - \frac{x_i^S}{E_i(1 - r_i)}\right] = \lambda, \qquad \forall i \in S$$

$$\sum_{i \in S} x_i^S = \sum_{i \in S} (E_i - \omega_i)$$

where λ is the Lagrange multiplier. By solving this equation, we obtain

$$\lambda = -185.2 \cdot \ln\left[1 - \frac{\sum_{i \in S}(E_i - \omega_i)}{\sum_{i \in S} E_i(1 - r_i)}\right]$$

$$x_i^{*S} = \frac{E_i(1 - r_i)}{\sum_{j \in S} E_j(1 - r_j)} \sum_{j \in S}(E_j - \omega_j)$$

By putting $S = N$, we can prove the following theorem.

Theorem 1: *Under the marginal cost functions of emissions abatement in Eq. 3, the competitive equilibrium of a market game (N, C^{ω}) for international CO_2 emissions trading is given by*

$$p^* = -185.2 \cdot \ln\left[1 - \frac{E - \overline{\omega}}{\sum_{i=1}^{n} E_i(1 - r_i)}\right] \tag{4}$$

$$x_i^* = \frac{E_i(1 - r_i)}{\sum_{j=1}^{n} E_j(1 - r_j)}(E - \overline{\omega}) \tag{5}$$

$$c_i^e = 185.2 x_i^* + p^*(E_i r_i - \omega_i) \tag{6}$$

where p^* is the market price of emissions, x_i^* is the emissions reduction of country i, and c_i^e is the competitive equilibrium costs of country i.

Proof: Equations 4 and 5 are easily proved. Eq. 6 is proved by Eq. 2 and

$$\begin{aligned} C_i(x_i^*) &= 185.2[E_i(1 - r_i) - x_i^*] \cdot \ln\left(1 - \frac{x_i^*}{E_i(1 - r_i)}\right) + 185.2 x_i^* \\ &= p^*[x_i^* - E_i(1 - r_i)] + 185.2 x_i^* \end{aligned}$$

■

Theorem 1 shows that the terms $E_i(1 - r_i)$, $i = 1, \ldots, n$, crucially determine the competitive equilibrium of the CO_2 emissions permits market. By definition, $E_i(1 - r_i)$ is equal to $E_i(e_i/\bar{e})$ if country i's carbon intensity (e_i) is smaller than that ($\bar{e}$) of the USA, and is equal to $E_i(2 - \bar{e}/e_i)$ otherwise. We may interpret $E_i(1 - r_i)$ as a *compensated level* of country i's CO_2 emissions by carbon intensity. When country i's carbon intensity is smaller (greater) than that of the reference country, the compensated level of emissions is smaller (greater) than the actual level of emissions. The equilibrium ratio of emissions reductions between country i and country j is equal to that of their compensated levels of emissions, namely, $x_i^*/x_j^* = E_i(1 - r_i)/E_j(1 - r_j)$. This equality gives us a convenient measure to examine the efficiency of a reduction scheme.

Allocation Rules of Emissions Permits

The analysis so far has assumed that an initial allocation $\omega = (\omega_1, \ldots, \omega_n)$ of CO_2 emissions permits is given exogenously. In this section, we show the general difficulty of an international agreement on emissions permits allocations, and consider six different allocation rules which are widely discussed in the literature.

As we have seen, the competitive equilibrium cost for country i is decreasing in its initial permits ω_i, given the total reduction of emissions. This fact highlights a strong conflict on emissions permits allocations in international negotiations. The collective decision problem of allocating emission permits is formulated by a voting game model. A *voting game* of emissions permits allocations is given by $V = (N, W, A)$, where $N = \{1, 2, \ldots, n\}$ is the set of countries, W is the class of winning coalitions, and $A = \{\omega = (\omega_1, \ldots, \omega_n) \in R^n_+ \mid \sum_{i=1}^{n} \omega_i = \overline{\omega}\}$ is the set of alternatives (emissions permits allocations).

Definition 2: *(1)* A coalition S is said to have an *objection* to $\omega \in A$ if $S \in W$ and there exists some $\omega' \in A$ such that $c_i^e(\omega_i') < c_i^e(\omega_i)$ for all $i \in S$. *(2)* The *core* of a voting game $V = (N, W, A)$ is the set of all $\omega \in A$ such that no coalition has an objection to ω. *(3)* Country i is said to have the *power of veto* if $i \in \bigcap_{S \in W} S$.

Proposition 1: *The core of a voting game $V = (N, W, A)$ of emissions permits allocations is nonempty if and only if $\bigcap_{S \in W} S \neq \emptyset$.*

Proof: A voting game $V = (N, W, A)$ corresponds to a simple game (N, v) with transferable utility such that $v(S) = 1$ if $S \in W$, and $v(S) = 0$ otherwise. It can easily be seen that a permit allocation $\omega = (\omega_1, \ldots, \omega_n) \in A$ is in the core of the voting game V if and only if the corresponding payoff vector $\omega/\overline{\omega} = (\omega_1/\overline{\omega}, \ldots, \omega_n/\overline{\omega})$ of the simple game (N, v) is in the core of (N, v). Then the theorem is derived by a well-known result in cooperative game theory (Owen 1995, p. 223). ■

The proposition implies that if no country is allowed to have the power of veto in an international negotiation on emissions permits allocations, then the negotiation is unstable in the sense that for every allocation of emissions permits, there exists some winning coalition of countries which object to it.

Given this general insight to international negotiations on emissions permits allocations, we consider six allocation rules for emissions permits, and evaluate their economic implications in the next section. In what follows, let N_i denote the population of country i, and let GDP_i be the gross domestic product of country i.

Per Capita Rule (Rule I)

One of the most frequently discussed rules about allocating emissions permits is the *per capita emissions equalization rule* (known as the *per capita rule*). This rule requires that "an equal number of permits should be allocated to each individual independent of nationality and income level" (Bohm and Larsen 1994; Akita and Sawa 1997). Formally, this rule imposes the condition that

$$\frac{\omega_1}{N_1} = \ldots = \frac{\omega_n}{N_n}$$

By solving this equation, we obtain the initial emissions permits of country i:

$$\omega_i = \frac{N_i}{N}\bar{\omega}, \qquad i = 1, \ldots, n$$

where $N = \sum_{i=1}^{n} N_i$ is the total population.

Per GDP Rule (Rule II)

It is often argued that the economic conditions of each country should be taken into account as a key factor in determining the initial allocation of emissions permits. A typical rule of this kind is the *per GDP emissions equalization rule* (known as the *per GDP rule*), which requires that an equal amount of emissions permits should be allocated to each country per GDP. Formally, this rule imposes the condition that

$$\frac{\omega_1}{\mathrm{GDP}_1} = \ldots = \frac{\omega_n}{\mathrm{GDP}_n}$$

which implies that

$$\omega_i = \frac{\mathrm{GDP}_i}{\Sigma_{i=1}^{n}\mathrm{GDP}_i}\bar{\omega}, \qquad i = 1, \ldots, n$$

Both the per capita rule and the per GDP rule are independent of the current CO_2 emission levels $E = (E_1, \ldots, E_n)$ of each country. These two rules focus on the "zero emissions" level as an "ideal" in the negotiations. The arguments of a fair allocation rule may depend on the "status quo" as well as the ideal point in negotiations. Although it may be considered that the per capita rule is fair at least in the long run, it is probable that this rule is not acceptable to a country such as the USA which emits a large amount of CO_2 relative to its population. By a similar reasoning, the per GDP rule is not acceptable to countries in the FSU which emit a large amount of CO_2 relative to their GDP. The next two rules are based on the current emission levels of each country.

Per Capita Reduction Rule (Rule III)

This rule requires that every country should reduce its CO_2 emissions per capita by an equal amount. Formally, this rule imposes the condition that

$$\frac{E_1 - \omega_1}{N_1} = \ldots = \frac{E_n - \omega_n}{N_n}$$

which yields

$$\omega_i = E_i - \frac{N_i}{N}(E - \overline{\omega}), \qquad i = 1, \ldots, n$$

where $N = \sum_{i=1}^{n} N_i$ is the total population.

Per GDP Reduction Rule (Rule IV)

If we employ the criterion of GDP instead of population, the per capita reduction rule changes to the *per GDP reduction rule*, which requires that every country should reduce its CO_2 emissions per GDP by an equal amount. This rule imposes the condition that

$$\frac{E_1 - \omega_1}{\text{GDP}_1} = \ldots = \frac{E_n - \omega_n}{\text{GDP}_n}$$

which yields

$$\omega_i = E_i - \frac{\text{GDP}_i}{\Sigma_{i=1}^{n}\text{GDP}_i}(E - \overline{\omega}), \qquad i = 1, \ldots, n$$

The four allocation rules of emissions permits described above do not take into account emissions trading. Since countries may anticipate the outcome of emissions trading, it can be argued that a fair allocation of emissions permits should take into account the costs of purchasing permits and the profits of selling them in the market. A natural condition based on this idea is that every country should incur an equal amount of reduction costs after emissions trading, either per capita or per GDP. We introduce these two rules below.

Equalizing Net Costs Per Capita Rule (Rule V)

This rule requires that every country should be burdened equally with the competitive equilibrium cost per capita in the emissions trading market. Formally, this rule imposes the condition that

$$\frac{c_1^e(\omega_1)}{N_1} = \ldots = \frac{c_n^e(\omega_n)}{N_n}$$

where $c_i^e(\omega_i) = C_i(x_i^*) + p^*(E_i - x_i^* - \omega_i)$ is the competitive equilibrium cost for country i $(= 1, \ldots, n)$. It is convenient to decompose $c_i^e(\omega_i)$ as

$$c_i^e(\omega_i) = T_i - p^* \omega_i$$

where $T_i = C_i(x_i^*) + \text{p}^*(E_i - x_i^*)$ is the total costs of country i when it buys $(E_i - x_i^*)$ amount of emissions. Note that T_i does not include any profits from selling initial emissions permits ω_i. This rule implies that

$$\omega_i = \frac{1}{p^*}\left(T_i - \frac{N_i}{N}c^{\omega}(N)\right), \qquad i = 1, \ldots, n$$

where $c^{\omega}(N) = \sum_{i=1}^{n} C_i(x_i^*)$.

Equalizing Net Costs Per GDP Rule (Rule VI)

If we employ the criterion of GDP instead of population, the previous rule changes to the *equalizing net costs per GDP rule*, which requires that every country should bear an equal competitive equilibrium cost per GDP in emissions trading. This rule imposes the condition that

$$\frac{c_1^e(\omega_1)}{\text{GDP}_1} = \ldots = \frac{c_n^e(\omega_n)}{\text{GDP}_n}$$

In a similar way to the equalizing net costs per capita rule, this rule yields

$$\omega_i = \frac{1}{p^*}\left(T_i - \frac{\text{GDP}_i}{\Sigma_{i=1}^{n}\text{GDP}_i}c^{\omega}(N)\right), \qquad i = 1, \ldots, n$$

Numerical Results

To evaluate the six allocation rules for CO_2 emissions permits discussed in the previous section, we calibrate a market game model of emissions trading for three major emitting countries, the USA, the FSU, and Japan. Table 1 shows all the relevant data on these countries, i.e., carbon emissions E_i (million tons), GDP (billion 1987 \$), population N_i (million), carbon intensity e_i (ton/billion), r_i, compensated emissions $E_i(1 - r_i)$. The data sources were Marland et al. (1999) on carbon emissions, and EDMC (1997) on GDP and populations.

Table 2 shows the equilibrium prices of emissions permits at four reduction rates: 5%, 10%, 15%, and 20%. In the literature, Bohm and Larsen (1994) and Rose and Stevens (1993) estimated the equilibrium prices of emissions at a 20%

TABLE 1. Data on the USA, the FSU, and Japan

Country	Carbon emissions: E_i (million tons)	GDP (billion 1987 US\$)	Population: N_i (million)	Carbon intensity: e_i (ton/billion)	r_i	Compensated emissions: $F_i = E_i\ (1 - r_i)$	F_i/F
USA	1293	4861	250	0.266	0	1293	0.4
FSU	1013	729	287	1.3896	−0.8086	1832	0.566
Japan	288	2814	124	0.1023	0.6154	111	0.034
Total	2594	8404	661			3236	

FSU, Former Soviet Union
Sources: Marland et al. 1999; EDMC 1997

TABLE 2. Emissions permits price

Reduction rates (%)	Total permits (million tons)	Total reductions (million tons)	Permit price (US$/ton)
5	2464.3	129.7	7.58
10	2334.6	259.4	15.47
15	2204.9	389.1	23.73
20	2075.2	518.8	32.36

TABLE 3. Efficient CO_2 emissions reductions

Reduction rates as a group	Country	Efficient reduction x_i (million tons)	Reduction costs (million US$)	Reduction rates (%)
	USA	51.88	195.4	4
5%	FSU	73.41	276.1	7.2
	Japan	4.41	16.4	1.5
	Total	129.7	487.9	
	USA	103.76	792.5	8
10%	FSU	146.82	1119.9	14.5
	Japan	8.82	66.7	3.1
	Total	259.4	1979.1	
	USA	155.64	1808.9	12
15%	FSU	220.23	2556.1	21.7
	Japan	13.23	152.2	4.6
	Total	389.1	4517.2	
	USA	207.52	3263.8	16
20%	FSU	293.64	4611.8	29
	Japan	17.64	274.5	6.1
	Total	518.8	8150.1	

reduction rate for different groups of countries. Bohm and Larsen (1994) estimated the emissions price to be $33.5 for a region consisting of Western and Eastern Europe and the FSU. Rose and Stevens (1993) estimated the emissions price to be $38.35 for a region consisting of the USA, Canada, Western Europe, the Commonwealth of Independent States (CIS), Brazil, Central Africa, Indonesia, and China. The estimated price in our model is $32.36 at a 20% reduction rate. We have not found a large difference of the emissions prices for these different groups of countries.

Table 3 shows the reduction levels x_i (million tons) of CO_2 emissions for three countries, and their reduction costs $C_i(x_i)$ in an efficient reduction scheme at each reduction rate. A remarkable observation is that the 3.4% reduction share of Japan is much lower than those of the USA or the FSU at all reduction rates. This result is implied by Theorem 1 because the proportions of the compensated emissions levels of these three countries are 0.4 for the USA, 0.566 for the FSU, and 0.034 for Japan (see Table 1).

The marginal reduction costs for Japan are estimated to be much higher than those of the other two countries, and therefore an efficient reduction scheme for the three countries would be to reduce the total emissions by the USA and the FSU by about 95%. The FSU would be expected to reduce its CO_2 emissions by about 40% more than the USA in terms of reduction share.

The distributional results under the six allocation rules for emissions permits at a 5% reduction rate are given in Table 4. Since the results are almost the same as for other reduction rates, we only present the calibration at a 5% reduction rate. Under both the per capita rule and the per GDP rule, there exist some countries which are allocated emissions permits in excess of their current emissions levels. In such cases, we assume that no country is allowed to increase its emissions levels by using excessive permits. Under the per capita rule, only the USA bears positive net reduction costs, although its reduction costs decrease significantly by emissions trading. For this reason, it is argued that the USA will reject the per capita rule. This result reflects the fact that the USA currently emits too much CO_2 relative to its population. Under the per GDP rule, the FSU is in the same position as the USA under the per capita rule, and the FSU is known to reject the per GDP rule.

Under the last four allocation rules, no countries are allocated emissions permits in excess of their current emissions levels. All three countries must reduce CO_2 emissions, regardless of whether emissions trading is allowed or not. Under all these allocation rules, the FSU sells permits and Japan buys them. The USA becomes a seller of permits under the per capita reduction rule (III) and the equalizing net costs per capita rule (V), and becomes a buyer of permits under the other two rules. Comparing reduction costs before and after emissions trading, we can see that Japan benefits most from the scheme of tradeable permits, the FSU is next, and the USA benefits least. For example, under the per capita reduction rule, Japan can save about \$365 million by emissions trading, the FSU saves about \$16 million, and the USA saves only about \$1 million (see Table 4).

Comparing reduction costs after emissions trading, we can identify the preference order of each country for the four feasible allocation rules. The USA's preference is III–V–VI–IV, the FSU's preference is IV–VI–III–V, and Japan's preference is V–VI–III–IV. The preferences of the three countries over allocation rules are quite different. For example, the FSU can earn a positive net profit from emissions trading under the per GDP reduction rule (IV), but the USA and Japan evaluate this rule as the worst. The USA and Japan rank three rules, the equalizing net cost per capita rule (V), the equalizing net cost per GDP rule (VI), and the per GDP reduction rule (IV), in the same order, but each one has a different first choice.

In addition to preferences about the allocation rules for emissions permits, each country has a different incentive to participate in the scheme of tradeable permits. Judging from the present condition of its economy, the FSU may be reluctant to accept the scheme if it incurs any reduction costs. For the USA, the

Table 4. Reduction rate 5%

Allocation rule	Country	Initial permits: w_i (million tons)	Reductions: $\max(E_i - w_i, 0)$ (million tons)	Cost without trade (million US$)	Net cost with trade: c_e (million US$)	c_e/GDP (%)	c_e/N (US$/person)	Reduction rate without trade (%)
I: per capita	US	932	361	10348	2538.4	0.0522	10.15	28
	FSU	1070	0	0	−712.6	−0.0978	−2.48	0
	Japan	462.3	0	0	−1338.2	−0.0476	−10.79	0
	Total	2464.3	361	10348	487.6			
II: per GDP	US	1425.4	0	0	−1201.5	−0.0247	−4.81	0
	FSU	213.8	799.2	38386	5777	0.7925	20.13	79
	Japan	825.1	0	0	−4088.3	−0.1453	−32.97	0
	Total	2464.3	799.2	38386	487.2			
III: per capita reduction	US	1243.9	49.1	175	174.2	0.0076	1.48	3.8
	FSU	956.7	56.3	162	146.2	0.0201	0.51	5.6
	Japan	263.7	24.3	533	167.2	0.0059	1.35	8.4
	Total	2464.3	129.7	870	487.6			
IV: per GDP reduction	US	1218	75	411	370.6	0.0076	1.48	5.8
	FSU	1001.7	11.3	6	−194.9	−0.0267	−0.68	1.1
	Japan	244.6	43.4	1829	311.9	0.0111	2.52	15.1
	Total	2464.3	129.7	2246	487.6			
V: equal net cost per capita	US	1242.5	50.5	185.1	184.5	0.0038	0.74	3.9
	FSU	948	65	216	211.8	0.0291	0.74	6.4
	Japan	273.7	14.3	178	91.5	0.0033	0.74	5
	Total	2464.2	129.8	579.1	487.8			
VI: equal net cost per GDP	US	1229.7	63.3	292	282.2	0.0058	1.13	4.9
	FSU	970.4	42.6	92	42.3	0.0058	0.15	4.2
	Japan	264.2	23.8	510	163.4	0.0058	1.32	8.3
	Total	2464.3	129.7	894	487.9			

TABLE 5. Kyoto Protocol

Reduction rate as a group	Permit price (US$/ton)	Country	Initial permits (million tons)	Reduction x_i (million tons)	Cost without trade (million US$)	Net cost with trade: c_e (million US$)	c_e/GDP (%)	c_e/N (US$/person)
		USA	1202.5	90.5	601	431.33	0.0089	1.73
4.16%	6.27	FSU	1013	0	0	–192.51	–0.0264	–0.67
		Japan	270.7	17.3	264	96.71	0.0034	0.78
		Total	2486.2	107.8	865	335.53		

savings in reduction costs from tradeable permits are not very high, regardless of the allocation rule used for emissions permits. If the costs of creating an emissions trading market are high, the USA may not have any good reason to support the introduction of such a market. As far as our numerical analysis is concerned, Japan benefits the most from the creation of an emissions permits market. The cost savings of emissions trading by Japan are the highest under all four allocation rules. Even if the worst allocation rule, the per GDP reduction rule (IV), is introduced, Japan's cost saving by emissions trading is greater than that of the USA under all allocation rules.

Table 5 gives our evaluation of the reduction commitments for the USA, the FSU, and Japan agreed in the Kyoto Protocol. As stated in the introduction, the reduction rates for these three countries are 7% for the USA, 0% for the FSU (Russia and Ukraine), and 6% for Japan. The reduction rate of the three countries as a group is 4.16%. The abatement costs without trade are $601 million for the USA, and $264 million for Japan. The FSU is not constrained to any reduction, and thus it bears no costs. The competitive equilibrium price of emissions in the trading among the three countries is $6.27 per carbon ton. It can be seen from Table 5 that emissions trading can significantly reduce the total reduction costs, i.e., by about $530 million (61%). The FSU can earn a profit of $192.51 million from trading. It is interesting to observe that the net cost vector (431.33, –192.51, 96.71) of the three countries is close to (425, –176, 88), in which the total surplus of $530 million is equally distributed among the three countries. Formally, this means that the reduction commitment in the Kyoto Protocol is approximately equal to the emissions permits allocation $\omega = (\omega_1, \ldots, \omega_n)$, which is a solution of the equation

$$c_i^e(\omega_i) = C_i(\omega_i) - \frac{1}{n}\left(\sum_{i \in N} C_i(\omega_i) - c^{\omega}(N)\right), \qquad i \in N$$

where $c_i^e(\omega_i)$ is country i's competitive equilibrium costs in emissions trading. Under the initial allocation of emissions permits in the Kyoto Protocol, the competitive equilibrium of emissions trading can attain an almost equal allocation of saving costs among the USA, the FSU and Japan.

Finally, we compare the Kyoto Protocol with the four allocation rules evaluated above. The Kyoto agreement gives the FSU almost the best result among the four rules ($194.9 million profits with the per GDP reduction rule), and it also gives Japan very good reduction costs ($91.5 million with the equalizing net costs per capita rule). In this sense, the reduction commitments in the Kyoto Protocol may be acceptable to the FSU and Japan. The USA incurs higher reduction costs under the Kyoto Protocol than under all four feasible allocation rules discussed here. One good aspect of the Kyoto Protocol for the USA seems to be that its reduction costs are less than 0.01% of GDP. Taking into account the facts that the USA emits the largest amount of CO_2, i.e., about 20% of the total world emissions, and that the USA, the FSU, and Japan as a group emit about 44% of the CO_2, it may be argued that the reduction rates of the USA (and possibly of Japan) should be more than in the Kyoto agreement.

Conclusions

We have presented a market game model of international CO_2 emissions trading. Employing the marginal cost function of CO_2 emissions abatements estimated by Nordhaus (1991), we have numerically analyzed the economic implications of emissions trading among three major countries, the USA, the FSU, and Japan, under various allocation rules for emission permits, including the reduction commitments made in the Kyoto Protocol. We have shown by a voting game model that an agreement on emissions permits allocation is very difficult in international negotiations if no country is allowed to have the power of veto. In addition to this theoretical result, a numerical analysis showed that the USA, the FSU, and Japan have different preferences for allocation rules.

The model calibration evaluates the Kyoto Protocol as follows. Emissions trading can significantly mitigate the total costs of emissions reductions. The cost saving by emissions trading is about $530 million (61%). Each of the three countries can evaluate the Kyoto Protocol positively in at least one aspect. The USA, which is the largest emitting country in the world, can keep its reduction costs below 0.01% of GDP ($1.73 per person). The FSU can earn net profits of $193 million. Japan can achieve reduction costs close to the best possible outcome under the four feasible allocation rules studied here.

The most important event that happened after the Kyoto meeting is that the USA, led by the new Republican regime, changed its policy to oppose the Kyoto Protocol (March 2001). This deviation from the Kyoto agreement seems partly consistent with our numerical results, which show that the saving in reduction costs for the USA by trading permits is not high, and that it incurs higher reduction costs under the Kyoto Protocol than under all four feasible allocation rules discussed here. It is currently expected that the international trading of emissions will start without the USA.

Finally, we admit that the calibration of reduction costs in this chapter is simpler than in other detailed models in the literature. In our view, it is sensible to study the nature of conflicts in international negotiations on global warming without having to depend on many model parameters which cannot be estimated precisely in uncertain situations. The simple game model in this chapter is well suited to this purpose. We hope that the game theoretical insights obtained will be incorporated into more detailed models in future research.

Acknowledgments. I am grateful to an anonymous referee for very useful comments and suggestions. This research was supported by CREST (Core Research for Evolutional Science and Technology) of Japan Science and Technology Corporation, and the Asahi Glass Foundation.

References

Akita J, Sawa T (1997) On the determination of QELROs at the 3rd FCCC Conference. WP No. 5, CREST, Japan Science and Technology Corporation

Barrett S (1992) "Acceptable" allocations of tradeable carbon emission entitlements in a global warming treaty. Combating Global Warming. UNCTAD, UN, New York

Bernstein PM, Montgomery WD, Rutherford TF (1999) Global impacts of the Kyoto agreement: results from the MS–MRT model. Resource Energy Econ 21:375–413

Bohm P, Larsen B (1994) Fairness in a tradeable-permit treaty for carbon emissions reductions in Europe and the former Soviet Union. Environ Resource Econ 4:219–239

EDMC (Energy Data and Modelling Center) (1997) Handbook of energy and economic statistics in Japan. Energy Conservation Center, Tokyo

Grubb M, Christiaan V, Brack D (1999) The Kyoto Protocol: a guide and assessment. Royal Institute of International Affairs, London

Marland G, Andres RJ, Boden TA, Johnston C, Brenkert A (1999) Global, regional and national CO_2 emission estimates from fossil fuel burning, cement production, and gas flaring: 1751–1996. NDP-030, CDIAC, Oak Ridge National Laboratory

Nordhaus WD (1991) The cost of slowing climate change: a survey. Energy J 12:37–65

Owen G (1995) Game theory, 3rd edn. Academic Press, New York

Rose A, Stevens B (1993) The efficiency and equity of marketable permits for CO_2 emissions. Resource Energy Econ 15:117–146

Shubik M (1983) Game theory in the social sciences: concepts and solutions. MIT Press, Cambridge

Tol RSJ (1999) The marginal costs of greenhouse gas emissions. Energy J 20:61–81

The Kyoto Protocol and Global Environmental Strategies of the EU, the USA, and Japan: A Perspective from Japan

Takao Aiba*,§ and Tatsuyoshi Saijo†,‡,§

Summary. Despite the objections of the Bush administration to the USA's ratification of the Kyoto Protocol, its entry into force has come to be realistic as a result of the Bonn Agreement in July 2001. The purpose of this chapter is first to survey the framework of the Protocol, and then to analyze the strategic positions of the EU, the USA, and Japan in negotiations to design the details of the Protocol. Finally, we consider the strategy of Japan, and identify some of the problems of the Protocol which will help to design a new framework that will allow the participation of developing countries.

Key words. Climate change, Commons, Emission trading, Global warming, Global public goods, Kyoto Protocol, Kyoto mechanism, Post-Kyoto scheme

Introduction

Global warming caused by the emissions of greenhouse gases (GHGs), including CO_2, is accompanied by a double "exploitation." One is the exploitation of future generations by generations that have emitted a significant portion of these GHGs. GHGs that are being emitted now do not immediately contribute to current global warming. Instead, future generations will be affected by a temperature rise caused by accumulated GHGs. The other exploitation appears within the same generation. Those who live in countries that are enjoying rich lifestyles by emitting substantial amounts of GHGs are exploiting those who live

*Development Bank of Japan (DBJ), 1-9-1 Otemachi, Chiyoda-ku, Tokyo 100-0004, Japan
†Institute of Social and Economic Research, Osaka University, Ibaraki, Osaka 567-0047, Japan
‡Research Institute of Economy, Trade and Industry, 1-3-1 Kasumigaseki, Chiyoda-ku, Tokyo 100-8901, Japan
§Climate Design, ISER, Osaka University, Ibaraki, Osaka 567-0047, Japan

in countries that are not emitting GHGs by not paying for their GHG emissions. Thus, the global warming problem is a complicated problem which involves intra-generational and intergenerational "commons."

The Kyoto Protocol[1]

The UN Framework Convention on Climate Change (UNFCCC) was adopted in 1992 to address global warming, and was brought into force in 1994. The third session of the Conference of the Parties to UNFCCC (COP-3) was held in Kyoto in 1997, where the Kyoto Protocol was adopted. The Protocol stipulates that 38 countries, including developed countries and economies in transition, will reduce their GHGs, including CO_2, to total emissions 5.2% below the 1990 level during a period from 2008 to 2012. For example, a reduction of 8% below 1990 levels is required for the EU, 7% for the USA, 6% for Japan, and 0% for Russia.

In order to achieve this target, the Protocol has employed three mechanisms, known as the Kyoto mechanisms, while urging each country to make domestic reductions. Emissions trading is one of the Kyoto mechanisms. Suppose the reduction cost per unit of GHG in Japan is 10, while that in Russia is only 1. If both countries are to reduce 1 unit each by themselves it will cost 11. However, if Russia is to reduce 2 units by itself, the reduction cost will be 2. In other words, emissions trading allows Japan to let Russia reduce Japan's emissions by 1 unit by paying a cost ranging from 1 to 10. Emissions trading sets positive prices on GHG emissions that have previously been free.

Another of the Kyoto mechanisms is the clean development mechanism (CDM). Suppose a country with an emissions cap, for example, Japan, builds power plants in a country without an emissions cap, for example, China. In this system, the difference between GHG emissions that would otherwise be caused by power generation using conventional Chinese technology in Chinese plants and those which would otherwise be caused by power generation using Japanese technology will be considered as joint reductions by Japan and China. The last of the Kyoto mechanisms is called joint implementation (JI), which allows the transfer of emissions reductions through technological transfer, etc., among countries with an emissions cap.

Joint fulfilment could perhaps be called the fourth Kyoto mechanism. For example, 15 countries in the EU have made a unified commitment of 8% reduction below 1990 levels across the board, but a reallocation among themselves is approved.

The economic interpretation of the Kyoto mechanisms is to protect the earth through putting a price on GHG emissions. There is some criticism about emis-

[1] For the Kyoto Protocol, see the home page of the United Nations Framework Convention on ClimateChange (http://www.unfccc.de/index.html)

sions trading, e.g., "it entrusts the fate of the earth to the speculating elite,"[2] or "it allows those who cannot make actual reductions to buy virtual emission reductions."[3] However, these criticisms are mere misunderstandings.

We can expect to minimize the total global abatement cost using the Kyoto mechanisms by equating the marginal GHG abatement costs of each country. At the same time, the mechanisms may mitigate the inequalities among countries in achieving their respective abatement targets, which differ country by country in the Kyoto Protocol. A country with a high marginal abatement cost can achieve its target by obtaining emission credits from the market, rather than by reducing emissions domestically. On the other hand, a country with a low marginal abatement cost can sell surplus emission credits if it over shoots its target. As a result, the differences in economic burden among countries become smaller. In this respect, the Kyoto mechanisms are a relief measure that equalizes the economic loads of each country, which were unfairly allocated through political negotiations.

Strategies of the EU, the USA, and Japan

We now look at the strategies of the EU, the USA, and Japan regarding the Kyoto Protocol. In Fig. 1, the horizontal axis shows the amount of CO_2[4] emitted to produce a GDP of US$ 1 million evaluated in purchasing power parity (CO_2 ton), and the vertical axis shows CO_2 emissions per capita (CO_2 kg) in major countries in 1996. In this figure, the least efficient countries will be in the upper-right corner. The slope of a line between the origin and each point shows the GDP per capita (US$10000). Therefore, the steeper the slope, the larger the GDP per capita. Countries, which have no points on the graph which are higher or further to the right than they are, can be considered to be inefficient countries. The USA, Australia, Russia, and Ukraine are considered to be such countries. They can be called "inefficient countries" in terms of CO_2 emissions. These four countries form a group called the "umbrella," with some noninefficient countries, such as Japan, in order to negotiate the Protocol with the EU and developing countries.[5] The efficiency of the major EU countries, such as the UK and Germany, is relatively high. As for Japan, CO_2 emissions in terms of both GDP and per capita are a little lower than those of Germany and the UK. However, it can be considered to be comparable to typical EU countries when it is evaluated in terms of both

[2] See Shohei Yonemoto (1997) CO_2 international emissions trading: risky American proposal (in Japanese). Ronza, p. 112–115

[3] NHK TV program, "Greedy society: how far do markets prevail?" December 2000

[4] Emission trend analyses of CO_2, the data for which are easily available, can be regarded as an analysis of all GHGs because CO_2 occupies 80%–90% of the total GHG emissions

[5] The Umbrella is a negotiating group of countries consisting of Australia, Canada, Iceland, Japan, New Zealand, Norway, Russia, Ukraine, and the USA, but it is not a group of joint fulfilment such as the EU

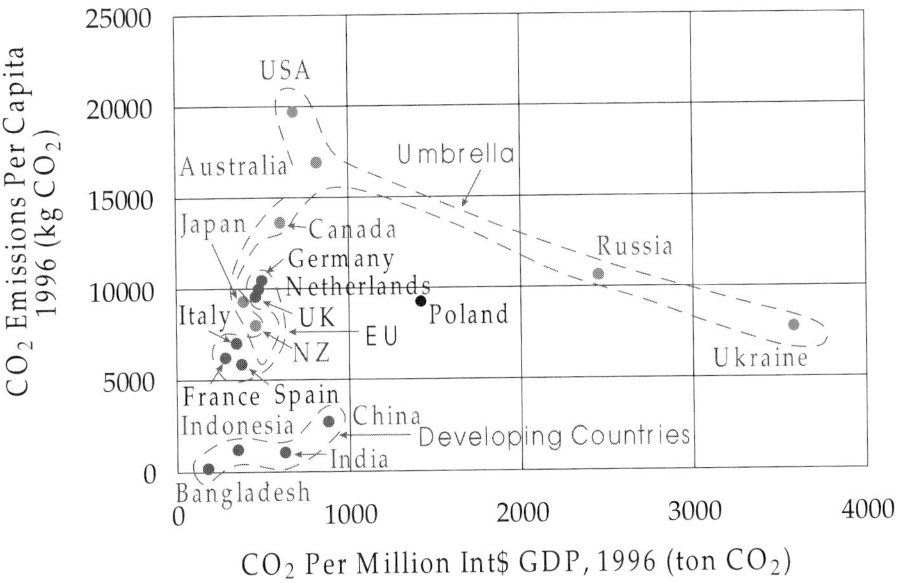

FIG. 1. CO_2 emissions per capita and per GDP. Data Source http://earthtrends.wri.org/

of these two indicators. Located near the origin along the horizontal axis are developing countries such as China, India, Indonesia, and Bangladesh. It should be noted that India and Bangladesh are more efficient than the USA in terms of emissions per GDP.

Although we tend to look at per capita emissions, what should really receive attention may be the amount of GHGs directly/indirectly input into goods and services that are used by any given individual. Suppose one country produces goods that would produce a large amount of GHGs, and another country produces services that would produce a relatively smaller amount. Per capita emissions might be larger in the former country and smaller in the latter, but they each use the other country's goods and services through bilateral trading. This would narrow the difference in GHG emissions per capita that have been input directly/indirectly into those goods and services. Therefore, it can be said that the entity which should take responsibility is the consumer. In this respect, it is highly likely that the USA, as shown in Fig. 1, is using more GHGs than the data show when viewed from the perspective of the consumers' responsibility because it is buying goods and services from developing countries.

We must pay attention to an asymmetric information problem here in order to make consumer responsibility work efficiently. Only producers of goods can efficiently and effectively know the amount of GHG emissions which were caused in the production of one good or service. We should also examine the producer's responsibility, because only producers can directly control and reduce GHG emissions relating to production. Only after adequate information about

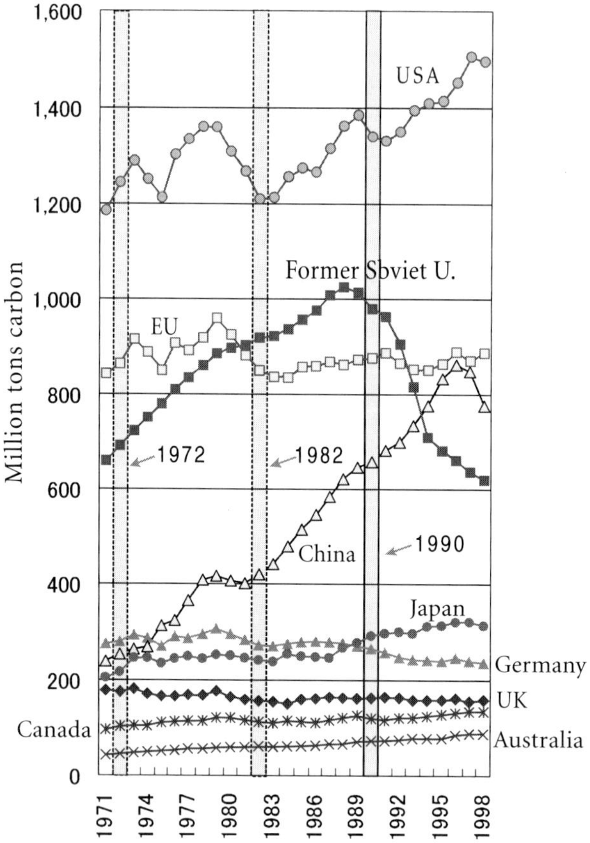

FIG. 2. CO_2 emissions from 1971 to 1998

GHG emissions is obtained from producers do we have the necessary base to consider consumer responsibility. Then, if consumers choose goods and services made by production processes with fewer GHG emissions in terms of life-cycle assessment, the consumers can lead the producers in the desired direction.

Figure 2 shows the total emissions of CO_2 in major countries from 1971 to 1998.[6] The base year for the reductions stipulated in the Protocol is 1990. The emissions from the former USSR, i.e., Russia, Ukraine, etc., were reduced by approximately 40% below 1990 levels in 1998 owing to an economic recession. As the reduction target of Russia and Ukraine is 0% below 1990 levels, they can sell these 40% excess reductions as emissions permits as long as their emission level remains below the 1990 level in the first commitment period. However, this portion, called "hot air," is criticized by NGOs because these countries can sell such portions as emissions permits without any reduction effort. Hot air is the

[6] Data are based on the *Handbook of Energy and Economy*, Econometric Analysis Division of the Institute of Energy Economics, Japan

result of problematic target-setting in the Kyoto Protocol, which decided on each country's target after political compromises based on the emissions in the single year of 1990, and ignoring the subsequent economic situation of each country. At least target setting during and after the second commitment period should avoid such problems. However, the problem may not be the hot air itself, but the fact that hot air is in the hands of inefficient countries. A characteristic of the USA and Australia, which are inefficient countries, is that their emissions have increased compared with 1990 levels. On the other hand, in Germany and the UK, emissions are declining. In Japan, emissions in 1999 had increased by 6.8% above the 1990 level, which requires a reduction of 12.8% as of 1999 to achieve the Protocol target. This is one of the reasons that Japan is included in the umbrella group, along with the USA and other countries.

We now look at the position of the EU.[7] The total GHG emissions of the EU in 1990 was about 1144 million tC, with the main emitting countries of Germany and the UK emitting 341 million tC and 199 million tC, respectively (Table 1). The reallocated targets for EU countries, decided by the above-mentioned Joint Fulfillment (the so-called "bubble") mechanism, vary widely from a 27% increase to a 28% reduction from the baseline of 1990 emissions. We can see the true abatement margins required by comparing these reallocated internal EU targets with each country's emission increase/decrease projections without measures (business as usual (BAU) projections) around 2010.[8] For example, although Germany's 21% reduction seems rather large, it is only a 5% reduction from its BAU projection, and therefore is not a very wide reduction margin. The reduction margin of the EU as a whole is 6% below BAU projections. We can therefore understand that the 26% reduction margin required of Japan is quite large compared with that of major countries of the EU.

For the abatement amounts required from individual countries below the 1990 GHG emissions baseline after reallocation, see Germany will meet 77% of the total EU reduction, and the UK will account for 27%. Although the reduction amounts for these two countries surpass the total EU reduction level, they both have special circumstances, such as the reunification of Germany and the drastic change in energy supply structure.

We now consider these special situations in Germany and the UK. Firstly, Germany and the UK are taking advantage of the base year being set as 1990. It was in 1990 that the unification of East and West Germany—and the

---

[7] See also J. Gummer and R. Moreland (2000) The European Union and global climate change: a review of five national programmes. Pew Center on Global Climate Change (http://www.pewclimate.org/projects/pol_review.cfm)

[8] Emission projections without measures are called business as usual (BAU) projections. BAU projections are merely projections and should be treated to some discounts, although other countries have reviewed these figures. However, BAU should be the baseline when we consider necessary reduction margins taking into account each country's economic growth projections, etc

TABLE 1. The real state of the EU (compiled by the authors) (units, million carbon equivalent tons)

Country	Kyoto targets (*a*) (%)	Internal EU bubble targets (*b*) (%)	BAU projection in 2010; % change from 1990 (*c*) (%)	Net targets ($b - c$) (%)	GHG emissions amount in 1990 (*d*)	Target reduction amount within EU ($e = d \times b$)	Share (%)	Special factor in amount (*f*)	Real reduction targets ($(e - f)/d$) (%)	Emission trends: 1998/1990 (*g*) (%)	Gaps ($g - b$) (%)
Austria	–8.0	–13.0	–6.0	–7.0	21	–3	3.0			6	19
Belgium	–8.0	–7.5	10.1	–17.6	38	–3	3.1			7	15
Denmark	–8.0	–21.0	–26.0	5.0	20	–4	4.4			9	30
Finland	–8.0	0.0	15.7	–15.7	18	0	0.0			1	1
France	–8.0	0.0	8.0	–8.0	138	0	0.0			1	1
Germany	–8.0	–21.0	–16.0	–5.0	341	–72	**76.7**	36	–10.4	–16	5
Greece	–8.0	25.0	27.0	–2.0	27	7	–7.2			18	–7
Ireland	–8.0	13.0	33.0	–20.0	16	2	–2.2			19	6
Italy	–8.0	–6.5	–4.0	–2.5	145	–9	10.1			4	11
Luxembourg	–8.0	–28.0	–43.0	15.0	4	–1	1.1			–24	4
The Netherlands	–8.0	–6.0	9.0	–15.0	59	–4	3.8			8	14
Portugal	–8.0	27.0	68.0	–41.0	19	5	–5.4			17	–10
Spain	–8.0	15.0	25.0	–10.0	82	12	–13.2			21	6
Sweden	–8.0	4.0	10.0	–6.0	18	1	–0.8			6	2
UK	–8.0	–12.5	–14.9	2.4	199	–25	26.6	25	0.1	–8	5
Total		–8.0	–2.4	–5.6	1144	–93	100.0	61	–2.8	–2	6
Reference: Japan	–6.0		20.0	–26.0	325	–19				10	16

Notes:

1. BAU projection in 2010; % change from 1990 (*c*) is the increase or decrease ratio of projected GHG emissions without measures in 2010 based on the 1990 emission level. The basic data are from material by the UNFCCC. The UK figure is from the latest governmental projection published in November 2001, which includes some policy effects. Here we treat it as BAU because policy effects, except energy market liberalization, seem to be very small. Data on Belgium and Finland are from their Second National Communications
2. Net target ($b - c$) is the margin between *b* and *c*. These are necessary reduction rates to achieve internal EU targets, which virtually replace the Kyoto targets at the EU's disposal. Figures without minus signs refer to a surplus that can be sold to other countries without taking any policy measures. The Japanese figure, as a reference, is a margin between the Kyoto target and the BAU projection
3. GHG emission amount in 1990 (*d*) is the total GHG emissions in 1990 before subtracting carbon sequestration or emissions from land use, land-use change, and forestry. Data are from the table compiled by the UNFCCC secretariat based on data in the second national communication. CO_2 tons were converted into carbon tons by multiplying by 3/11
4. Target reduction amount within EU is ($e = d \times b$), assuming that emission amounts are stable between 1990 and 2010
5. Share is a percentage of each country's target reduction amount when the total EU target reduction amount is 100%
6. The special factor for Germany, 36 million carbon tons, is the materialized CO_2 reduction amount in former East Germany until 1995 through economic rearrangements. The former East Germany is still inefficient in terms of energy use, and Germany is likely to be able to reduce more in the former East Germany quite easily

7 The special factor of the UK, 25 million carbon tons, is the projected reduction until 2010 from energy market liberalization that has been facilitating an energy changeover from coal to natural gas. By the end of 2000, 17 million carbon tons reduction seems to have been materialized. Parallel to energy market liberalization, subsidies and protection for the UK's coal industry has been reduced and oil and gas wells in the North Sea have been developed

8. Emission trends: 1998/1990 (*g*) means actual GHG emission trends expressed as a ratio of the 1990 emission level
9. Gaps ($g - b$) means the difference between the actual emissions in 1998 and internal EU targets

Data sources

(1) Secretariat of UNFCCC, FCCC/CP/1998/11/Add.2, Oct. 1998. (2) Secretariat of UNFCCC, FCCC/SBI/2000/INF.13, Oct. 2000. (3) Government of the Federal Republic of Germany, Second Report of the Government of the Federal Republic of Germany Pursuant to the UNFCCC, Apr. 1997. (4) Secretaries of State for the Environment and the Foreign and Commonwealth Office, The United Kingdom's Second Report under the Framework Convention on Climate Change, Feb. 1997. (5) UK Energy Paper 68: Energy Projections for UK, Nov. 2000. (6) WS Atkinsons Environment, Projections of Non-CO_2 Greenhouse Gas Emissions for the UK and Constituent Countries, Final Report, Nov. 2000. (7) Royaume de Belgique, Deuxieme Communication Nationale Conformement Aux Articles 4 Et 12 De La Convention, Aug. 1997. (8) Finland, Second National Communication of Finland, Apr. 1997

BAU, business as usual; GHG, greenhouse gas; UNFCCC, United Nations Framework Convention on Climate Change

emergence of the largest CO_2-emitting country in the EU—was realized. In other words, East Germany, which was an inefficient country with a lot of hot air, joined this group. Looking at the regional breakdown of CO_2 emissions in Germany, those in the former East Germany showed a 44% decrease over the 1990 level in 1995, while those in the former West Germany showed a 2% increase. Germany as a whole showed a 12% decrease. The CO_2 emissions reduction achieved in East Germany by 1995 accounts for about a half of the German emissions reduction target.[9] Thus, the real reduction target for Germany can be considered not as 21%, but as less than 10%. According to the UNFCCC data, Germany realized large emissions reductions by 1998, when the level was about 84% of the 1990 figure. A significant part of this reduction seems to have materialized in East Germany. As a reference, we note that German per capita CO_2 emissions in 1998, which is after the large reduction, are still 17% larger than those of Japan.[10] Secondly, both Germany and the UK had used a lot of coal, which causes greater CO_2 emissions. In both countries, a switchover from coal to natural gas took place mainly after 1990. Germany and the UK are being supplied with natural gas from Russia and from the oil fields in the North Sea, respectively. They changed to natural gas because it was cheaper. For example, in the the UK, the energy switch from coal to gas took place because of liberalization of the energy market, which was unrelated to the Kyoto Protocol. If we take into account the foreseeable progress of energy market liberalization until 2010, we can say that the UK can achieve their reallocated internal EU target, which surpasses the UK's Kyoto target, only with this liberalization, which has a negative cost. In this context, the UK's real reduction target can be seen to be zero.[11] In fact, the latest UK BAU projection in 2010 is below its internal EU target.[12]

It could be said that the EU position in negotiations over the Protocol is determined by the unification of Germany, as well as by the oil fields in the North Sea and the natural gas pipelines. Furthermore, ten Eastern European countries with hot air are to join the EU in the near future, which could facilitate the achievement of their reduction target despite the absence of any actual reductions.

---

[9] We make this argument by assuming that GHG emissions after 1990 would have remained at the 1990 level

[10] International Energy Agency, Key world energy statistics from the IEA—2000 edition, November 2000

[11] Japanese energy market liberalization, which is still at a very early stage, probably will bring about a completely different outcome. This is because the gas available in Japan costs nearly three times the international market price, and the only cheap energy source available in Japan is coal. Thus, simple energy market liberalization in Japan is likely to increase the use not of gas, but of coal, which has a high carbon content

[12] Although the UK does not have to take any measures to achieve the Kyoto target, it has set its own target of a 20% reduction from the 1990 level, and is considering taking additional action. The UK can sell surplus credits from the Kyoto baseline to other countries

Within the EU, some countries are relying on the bubble and do not attempt to reduce GHG emissions significantly within their territories. The EU bubble virtually allows countries such as Greece, Spain, and Sweden to hand the major portion of their reduction targets on to other EU countries, such as Germany and the UK.[13] In addition, there are countries such as The Netherlands which have stated clearly that they will utilize cheap foreign reductions by the Kyoto mechanisms for half of their reduction amount. The spirit of the Kyoto Protocol is to try to make the best use of cost-effective reduction opportunities all over the world. Japan should comprehensively reconsider its domestic-dominant emissions reduction plan, paying serious attention to cross-border options in the spirit of the Protocol, as other countries are doing.

How much will it cost the EU to reduce GHGs? According to a survey by the EU, the marginal abatement cost (MAC) for reducing emissions by 8% below the 1990 level is about 70 euro (about 8000 yen) per ton of carbon.[14] This figure is estimated to increase to about 150 euro (about 16000 yen) per ton of carbon if the EU does not use the EU bubble. On the other hand, the MAC to Japan of achieving the Protocol target is believed to be around 30000 yen per ton of carbon in many cases, with extreme case reaching over 100000 yen per ton of carbon.[15] If the Kyoto mechanisms function fully, and the USA participates in the Protocol, the international price per ton of carbon could be around US$20–70. This means that the EU would be able to achieve its Protocol target efficiently within its boundaries with the EU bubble. This can be considered to be reliable and unlimited emissions trading, without depending on the Kyoto mechanisms.

If this is so, the diplomatic strategy of the EU, which acts in the EU's interests, would be to acquire a relatively favorable position compared with non-EU countries by making the Kyoto mechanisms inconvenient for the umbrella group countries. In other words, their negotiation cards would include a limitation on emissions trading as far as possible, no forest management sinks, and strict operation of the CDM. Furthermore, environment-oriented political powers within and outside the EU which responded to that situation would have been supporting a limitation on the use of the Kyoto mechanisms. On the other hand, the EU is trying not to limit emissions trading within the EU in order to minimize the total costs all over the region, while also trying to set a limit on international

---

[13] According to the authors' estimation based on BAU projections, Greece's 94% reduction, Spain's 70%, and Sweden's 67% will not be achieved within those countries, but in such countries as Germany and the UK owing to the EU bubble. The EU adamantly insisted that other countries outside the EU should not rely on reductions of more than 50%, utilizing the fact that it is not widely known that some countries within the EU are dependent on non-domestic reductions of more than 50%

[14] http://europa.eu.int/comm/environment/enveco/climate_change/sectoral_objectives.htm

[15] See http://www.env.go.jp/council/06earth/y062-08/mat02.pdf

emissions trading.[16] In other words, the EU is intentionally trying to apply different policies inside and outside their region.

However, in response to President Bush's statement, "I oppose the Kyoto Protocol because it was fatally flawed in fundamental ways, although the process used to bring nations together to discuss our joint response to climate change is important. The Protocol is not effective in preventing global warming since it exempts 80% of the world, including major population centers such as China and India, from compliance. In addition, the Protocol may cause serious harm to the USA and the world economy owing to its exceedingly stringent targets,"[17] the diplomatic strategy of the EU has entirely changed, and it made great concessions to the umbrella group in the resumed session of COP-6 in Bonn in July 2001. This is because the EU considered that there would be no other choice to maintain the force of the Protocol, as stated below, as well as because the international price of emissions permits would be significantly lowered by using the Kyoto mechanisms without the USA, and this would force the EU themselves to depend on those mechanisms. The EU may have tried to keep its relatively advantageous position in the Kyoto framework compared with the USA, Japan, etc. Rejecting the Kyoto framework meets the USA's interests, while preserving the Kyoto framework seems to meet the interests of the EU, which has to take some kind of domestic action because of domestic political pressures. For the EU, even making some compromises with the non-USA umbrella group in order to save the Protocol may have been a better option than killing the Protocol.

CO_2 emissions per capita in Japan continued to rise until the early 1970s, when the increase almost stopped owing to the oil shock in 1973–1974. It remained at around 8 CO_2t per capita thereafter, but has increased again since 1987. From 1990 to 1999, the increase in the industrial sector was 0.8%, while that in the transportation sector was 23%, and that in the commercial and residential buildings sector was 15%. In order to achieve the Protocol target based on 1990 levels, a reduction of 12.8% is required. However, this cannot easily be achieved, even if the economic recession continues, owing to the structural reforms of the Koizumi administration. Just after the adoption of the Protocol, the Global Warming Prevention Headquarters[18] of the Japanese government decided on the Climate Change Policy Program[19] in June 1998.[20] According to this Program, energy-derived CO_2 emissions control must be 0%, emissions control of methane, etc., must be –0.5%, reduction through technical innovation, etc., must be –2.0%, reduction by sinks, mainly from forest management, must be –3.7%, emissions

[16] See the EU green paper published in March 2000, http://europa.eu.int/comm/environment/docum/0087_en.htm
[17] These are not exact quotations from President Bush's statement, but a summary of his claims
[18] The headquarters is headed by the prime minister
[19] The program has a general principle to promote measures to cope with global warming
[20] http://www.env.go.jp/earth/cop3/kanren/suisin2.html

control of CFC substitutes, etc., must be +2%, and the use of the Kyoto mechanisms must be –1.8%, which will be a total of –6%. The policies in the Program are mainly based on command-and-control, and the share of those using the Kyoto mechanisms is 1.8%. These reduction target allocations were not decided based on some rational principle such as minimizing compliance costs. There is no convincing explanation of why the share of the Kyoto mechanism is 1.8%. While the EU and the USA seem to have been acting strategically for their national interests, Japan lacks an international and long-term strategy. The Japanese measures are very passive, and only aim to achieve a 6% reduction from the 1990 emissions level. Japan, which has a relative advantage in energy-saving technologies and so on, should take more strategic actions and measures in order to make the best use of its relative strength. At the same time, Japan should balance emissions reductions with other important policy goals, such as energy market liberalization, energy security, and the preservation of international industrial competitiveness, which are not always compatible with abatement measures.

The Japanese negotiating position changed dramatically in response to the USA's objection to ratification. The requirements for the rules of the Protocol to come into force are (i) that it must be ratified by more than 55 countries, and (ii) that it must be ratified by countries with an emissions cap whose aggregate 1990's emissions is at least 55% of the total CO_2 emissions in 1990 of capped parties. The USA's share in 1990 was 36.1%, and that of Japan was 8.5%. Therefore, the total share, excluding these two countries, will be 55.4%. Since the shares of Canada and Australia are 3.3% and 2.1%, respectively, the Protocol will not be able to come into effect if either of these do not ratify it, following Japan. Canada is wavering between ratifying the Protocol without the USA or waiting for ratification by the USA, while Australia is indicating that it might wait for the USA. This means that Japan has become the pivotal player in deciding the future of the Kyoto Protocol.

In COP-6, held in The Hague in November 2001, no agreement was reached because the proposal of President Pronk was refused by Germany and others. At that conference, one of the main interests of the Japanese government was the treatment of sinks from LULUCF[21] activities. Although it was estimated that there could be a 3.7% reduction below the 1990 level through sinks from forest management in the Program, only 0.5% was to be approved in the Pronk proposal. Although President Pronk offered a compromising proposal of 3% to Japan in June 2001, Japan refused to accept it and demanded more. In addition, Japan did not clarify its attitude toward the ratification of the Protocol, saying the first thing to do was to urge the USA to come back to the Protocol regime. Although it was reasonable for Japan to maintain that attitude in order to proceed with negotiations which would be favorable to itself, the press and NGOs

[21] LULUCF stands for land use and land-use change and forestry, including forest management

were offended. In the resumed session of COP-6 in Bonn in July 2001, the EU made substantial concessions to the umbrella group, and Japan secured a 3.9% upper reduction limit through sinks. It is said that Japan had almost all of its demands granted, despite some remaining uncertainties.

The Japanese government argued for a lax compliance system as well as no limits on the Kyoto mechanisms. If it intends to use the Kyoto mechanisms, including emissions trading, it is essential to design a robust compliance system as well as to set no limits on the use of the mechanisms. In this regard, the argument of the government lacks consistency. In the Marrakech Accords,[22] the following statement is included: "Each party . . . shall maintain, in its national registry, a commitment period reserve which should not drop below 90% of the party's assigned amount calculated pursuant to Article 3, paragraphs 7 and 8, of the Protocol, or 100 per cent, of five times its most recently reviewed inventory, whichever is lowest." This reserve system has some problems, although it seems to be well thought out for preventing the overselling of so-called hot air.[23] Reserves mean supply limits, which makes the price of emissions permits, or the cost of compliance to buyer countries such as Japan, higher than it would be in a case without reserves. The Japanese government tolerated the reserve system, and deferred the issue of a legally binding penalty system.[24]

What, then, is the USA's strategy? The US Senate decided several points by consensus in the summer of 1997. First, developing countries need also to have emissions reduction commitments. Second, the Protocol should not have a severely negative impact on the USA's economy. Third, the ratification of the Protocol cannot be approved unless both of the above requirements are fulfilled. Although in Kyoto the then Vice President Gore accepted a 7% reduction below the 1990 level, the USA's emissions increased by 12.5% above the 1990 level in 1998, reflecting its economic boom in the 1990s. This means that it had to make a 19.5% reduction from the 1998 level. A comparison of before (1972) and after (1982) the two oil shocks shows that CO_2 emissions in the USA have been somewhat reduced. However, they increased by 23.6% over the 1982 level in 1998. It is clear that the USA's Protocol target will be difficult achieve unless something equivalent to continuous oil shocks, inter alia, a high level of domestic carbon tax or a command-and-control type of regulation, is introduced. The USA, which does not favor taxes and regulations, tried to widen the accounting of domestic forest sinks, and aimed at designing a system that would not limit the use of mechanisms that would utilize overseas reductions. However, these attempts were cut

---

[22] COP-7 was held in Marrakech, Morocco, in October to November 2001, and reached the so-called Marrakech Accords, which generally followed the Bonn Agreement

[23] http://www.unfccc.de/cop7/documents/accords_draft.pdf

[24] It has been a central issue to restrict the use of the Kyoto mechanism, which is called supplementarity. See also K. Kaino, T. Saijo, and T. Yamato, Economic consequences of the EU's proposal of quantity restraints on the Kyoto mechanism in Japanese. Energy and Resources, 21(2), 2000, p. 38–42

off by the EU strategy.[25] The abandonment of the Protocol by the Bush administration means that the USA has returned to the position of the 1997 Senate decisions. However, the Bonn Agreement and the Marrakech Accords should facilitate the USA's return to the framework of the Protocol owing to substantial concessions by the EU.

The situations of the EU, the USA, and Japan are often compared to a wet towel. The EU is a wet towel holding a substantial amount of water, which can be wrung out. The problem is whether the towel can be wrung out after the year 2013 when decarbonization, etc., is to be completed. Japan is a towel which is almost dry, with very little water to be wrung out. It is even uncertain whether it can achieve the Protocol target unless there are significant changes in technology or in people's lifestyles. The USA is a wet towel holding abundant water, but it is less willing to wring it out. It was the American way of life and national interests as its excuse. It is also one of the problems of the Protocol that the EU, the USA, and Japan have very similar reduction commitments, although their national conditions are quite different.

After Kyoto

We now consider the problems of the Protocol. The first problem is that it took too much time to reach an agreement. The practical starting point of the Protocol was COP-1 in 1995 in Berlin, followed by COP-3, where the numerical targets were set, and finally the resumed session of COP-6 and COP-7 in 2001, at which operational rules were agreed upon. In the course of negotiations, many Cabinet members were replaced in Japan, and similar changes took place in other countries. If the negotiations are extended, they will be more vulnerable to administrative policy changes in each country, and thus the negotiations could become more complicated and extended. The economic environment and other conditions which set the political positions of each country will also change. For example, the USA has been hit by energy crises such as the electricity crisis in California and economic recession. In addition, the first commitment period will be from 2008 to 2012, when the numerical targets will be applied. It is difficult to predict the economic conditions of each country during the first commitment period. From the political aspect, it is unlikely that an administration that ratifies the Protocol will remain after the end of the first commitment period, when the compliance of the country will become clear. There is the possibility that a country might accept a difficult reduction target as an impermanent political

[25] For the US Climate Change Review—Initial Report, see http://www.whitehouse.gov/news/releases/2001/06/climatechange.pdf. It is impossible to design an institution or a treaty where every party ratifies it when the issue is an international public good such as global warming. As a theoretical framework, see T. Saijo and T. Yamato, A voluntary participation game with a non-excludable public good. Journal of Economic Theory, 84, 1999, p. 227–242

stand of the current administration. At the same time, we should pay attention to the fact that global warming prevention measures need a certain amount of long-term vision. For example, 10 years is not enough if we consider the lead-time that is necessary for the construction of a power plant. In the case of Japan, a power plant construction plan beyond the year 2010 is already set, and it will be difficult to change that plan. In order to lead such investment activities, which need a long-term vision, in a desirable direction, the long-term policy for 10–20 years should be known by the private sector. Therefore, current negotiations should decide the long-term goal within a short time period, and clearly distinguish the difference between showing long-term vision and taking time to make political compromises. The reason why it took such a long time to decide the operational rules of the Protocol is that the negotiations failed to make decisions in the right order; reduction targets had been set before details of the operational rules had been decided. Thus, the order of future decisions should be changed: operational rules should be set first, the economic burden for each party should then be clarified, and in the final stage the reduction targets should be fixed. Otherwise, keen negotiation skills will decide the economic burdens of each party, and an inequality in economic burdens will occur again, as in the Kyoto process.

The second problem is the base year for reductions. The important objective of UNFCCC was to stabilize GHG levels at 1990 levels by 2000. Although this objective was not achieved in most of the major countries, 1990 has tacitly been set as the base year in negotiations over the Protocol. The numerical targets seem to have been set with each country's specific economic situation and with equitability with other countries being taken into consideration, but they still give the impression that the target establishment process was rough-and-ready. Ambassador Raúl A. Estrada-Oyuela of Argentina, who managed to settle the negotiations over the Protocol, cannot show an objective rationale of why the Japanese target is a 6% reduction, while the Australian target is an 8% increase. It is ironic that Japanese companies are trying to earn emissions credits through afforestation in Australia, which won the 8% increase. The numerical reduction targets of the Kyoto Protocol are just political compromises that have no scientific validity. In addition, the targets were set hastily, when the parties did not know their own economic burdens, and at a stage without clear operational rules. As a result, the negotiations after COP-3 became extremely complicated, and eventually the USA seceded from the Kyoto regime. Moreover, we should not overlook the fact that the Protocol places a heavier economic burden on countries that had been incorporating energy-saving measures before 1990, such as Japan, while placing a lighter burden on energy-wasting countries which have made no efforts so far. This harsh reality may send an undesirable message to developing countries that have been observing the whole process. They may learn that early reduction does not pay, and that they should not take early measures before they have accepted future numerical reduction targets.

The third problem is that country-specific figures for sinks were virtually decided in the political negotiations at the resumed session of COP-6 in an artificial way, after the numerical reduction targets had been decided at COP-3.

Forest absorption figures, as well as other types, should be decided based on scientific evidence instead of by negotiation. The problem is that these baseless numerical targets had been decided before detailed rules were set.

The fourth problem is that developing countries do not participate in the Protocol, as was pointed out by the USA. We should not underrate the effect of carbon leakage. Japan, which has to make a drastic emissions reduction while its major trading partners do not have GHG reduction commitments, may face serious problems. Remedies for global warming must be part of a global system that can really reduce GHG emissions globally. Even if one party cleans up its own domain, i.e., complies with the emissions target even if neighboring countries increase their emissions, if other uncommitted countries increase GHG emissions by carbon leakage, such as the transfer of energy-intensive industry from committed countries to uncommitted countries, the aggregate total of global GHG emissions does not decrease, and emissions abatement efforts in developed countries do not contribute to the prevention of global warming. A country that takes emissions reduction measures may encounter political difficulties in obtaining a domestic consensus, because employment and the standard of living may be reduced unnecessarily. There are some options for developing countries which are not likely to restrain their economic growth, such as per unit energy-use targets, and so forth. The distribution of burdens and benefits in the Kyoto Protocol, in which developing country parties have virtually no obligations and one-sidedly receive funds and technologies from developed country parties, make it difficult to conduct negotiations and operations about the Protocol on an equal footing. The following paragraphs are intending to offer some practical proposals to resolve these problems.

Given that the Protocol must be effective, we have to start developing a new protocol for the period after 2013 by 2005. We also have to avoid wasting time in choosing an suitable base year. For that purpose, a country-specific, scientific and transparent GHG emissions path for 2013–2017 must be established. The objective and reasonable establishment of a GHG emissions path can prevent the notorious hot air situation, as well as allow parties to set internationally equal emissions reduction targets. The selection of a path may depend on the reduction efforts that have already been made by the country concerned, its economic performance, its weather and climate, its energy consumption pattern, its forest sinks, its emissions per GDP, and so on. This path must not be decided in a plenary session of the COP, but by a team consisting of experts from three countries, and representing developed countries, semideveloped countries, and developing countries. When an emissions path for Japan is to be decided, the team should consist of experts from three such countries but excluding Japan itself. In other words, the emissions path for any country should not be decided by the country concerned, but by other countries.[26] In this case, Japan would of course be involved

[26] For this type of method in the theory of institutional design, see T. Saijo, Strategy space reduction in Maskin's theorem: sufficient conditions for Nash implementation. Econometrica 56(3), 1998, p 693–700

in the decision by providing information to the three decision-making countries, but the decision-making itself would be done by the three evaluating countries. Through such a process, there would be a much smaller possibility of unfair favorable allowances being made than in processes such as bilateral evaluation. This would also give objectivity to each evaluation. Moreover, it would facilitate decision-making as well as save time compared with a plenary session, as the evaluation would be made by three countries only. This process must be implemented in the countries ratifying the UNFCCC. In the case of a country without sufficient human resources, organizations such as the International Energy Agency would give support.

After country-specific emissions paths have been decided, reduction schedules should be decided at an annual conference of the parties (COP). The emissions paths for developing countries may be approved unconditionally, while developed countries may be required to make greater reductions. It is important that the reduction schedules are set favorably for countries that have been making early voluntary reductions, although in the existing Kyoto Protocol, countries that made efforts before 1990 have been disadvantaged. For example, equating abatement margins from the approved emissions path may ensure international equity. In addition, some encouragment for a country that has achieved higher energy-saving standards, which can be judged based on present energy intensity and so on, would be desirable. This is because a country with a higher energy-saving level may face heavier marginal abatement costs and economic burdens if abatement targets are set at the same level as those for energy-wasting countries. For example, emissions reduction targets for normal developed countries can be set at a 15% reduction from their own emission paths, and a country that achieved higher energy-saving can receive some discounts according to its current per-unit emissions level, while a country that has made less effort has to compensate and make greater reductions.

Global warming is a problem brought about by cumulative GHG emissions, as GHGs remain in the atmosphere for about 50–200 years. If we follow the claims of developing countries that to date global warming has been caused by cumulative GHG emissions mainly from developed countries, we should differentiate emissions reduction targets according to cumulative per capita emissions. For example, in terms of current per capita emissions flow, there is not much difference between Japan, the UK, and Germany. However, if we pay attention to the past 50 years, during which GHG emissions increased explosively, and during which cumulative GHG emissions account for about 80% of the total GHG emissions since the industrial revolution, and assume that per capita cumulative responsibility can be obtained by dividing the past 50 years' cumulative emissions by the current population, it will be found that emissions from Germany and the UK are roughly twice as big as those from Japan (Table 2). Taking into account the fact that considering cumulative emissions means giving advantageous treatment to any country that has taken early voluntary action, this type of adjustment seems to be desirable.

On the other hand, the USA and others deny their cumulative responsibility, insisting that past emissions happened before we recognized the risk of global warming, and that the reason why developed countries should tackle emissions abatement on their own initiative is that developed countries are rich and can afford to take such measures. In the world of criminal law, a widely accepted legal principle is that if an action was not predetermined as a crime, we cannot punish the action retroactively. Although in this context the claim of the USA seems to be appropriate, few people object to providing some incentives for early voluntary emissions reductions.

If we really want to prevent global warming, it is extremely important to establish some mechanisms that give incentives for early voluntary reduction efforts by developing countries. We should establish such tangible mechanisms, and clearly prove that early action for emissions reductions is beneficial for the developing countries. The existing Kyoto scheme discourages early voluntary actions by developing countries because the greater the level of voluntary emissions reductions a country undertakes, the higher the marginal abatement costs will be when developing countries accept emissions reduction targets in the future. The establishment of an objective emissions path is also valid in this context.

Each country would use emissions trading and joint implementation (JI) as flexible mechanisms while implementing domestic reductions. As all countries participate in this system, CDM should be integrated into JI, and CDM-specific problems, including the measurement of emissions reductions generated from the CDM, could be avoided.

However, even in this system there remains the possibility that total global emissions would increase owing to the participation of developing countries. To avoid this, the global total target—which would be set lower than the total of the national targets—would be set separately from country-specific targets, and the administrative organization of emissions trading would retain emission permits supplied from developing countries and could abandon some of them. If the equivalent cost of the value of these abandoned emissions permits is borne by developed countries in proportion to their accumulated emissions amounts in the past, this system can also result in differentiated responsibilities based on cumulative emissions.

Concluding Remarks

We had an opportunity to talk with Mr. Seluka from Tuvalu, which is an island country with coral atolls in the South Pacific, at the resumed session of COP-6 in Bonn. The population of this country is 12 000. He said that their immediate problem was the increase in seawater flowing into wells. He said sorrowfully, "We have to ask for overseas support because we have no technology to remove salts." This would be an unnecessary technology for such a country if there was no

TABLE 2. National cumulative CO_2 emissions of the past 50 years: 1949–1998, from fossil-fuel burning, cement production, and gas flaring (compiled by the authors) (units, million tons of carbon)

Rank	Top 50 countries	Cumulative CO_2 emissions	Share (%)	Population in 1998 (millions)	Per capita cumulative emissions (tons)
	World total	210802	100.0	5838.82	36.1
1	United States of America	54488	25.8	269.09	202.5
	Eu 15	37194	17.6	374.37	99.4
2	Former Soviet Union	32519	15.4	291.68	111.5
3	China[a]	17459	8.3	1238.60	14.1
4	Germany	12239	5.8	82.02	149.2
5	Japan	9163	4.3	126.49	72.4
6	United Kingdom	7859	3.7	59.24	132.7
7	France[b]	4868	2.3	58.85	82.7
8	India	4830	2.3	979.67	4.9
9	Canada	4421	2.1	30.30	145.9
10	Poland	4135	2.0	38.67	106.9
11	Italy[c]	3599	1.7	56.98	63.2
12	South Africa	2549	1.2	41.40	61.6
13	Mexico	2333	1.1	95.68	24.4
14	Australia	2290	1.1	18.75	122.1
15	Spain	1832	0.9	39.37	46.5
16	Brazil	1724	0.8	165.87	10.4
17	Republic of Korea	1576	0.7	46.43	33.9
18	Romania	1567	0.7	22.50	69.6
19	The Netherlands	1561	0.7	15.70	99.4
20	Islamic Republic of Iran	1549	0.7	61.95	25.0
21	Democratic People's Republic of Korea	1479	0.7	23.17	63.8
22	Belgium	1435	0.7	10.20	140.7
23	Saudi Arabia	1293	0.6	20.74	62.3
24	Venezuela	1155	0.5	23.24	49.7
25	Argentina	1132	0.5	36.13	31.3
26	Indonesia	1115	0.5	203.68	5.5
27	Turkey	989	0.5	64.79	15.3
28	Taiwan	848	0.4	21.87	38.8

29	Hungary	818	0.4	10.11	80.9
30	Sweden	796	0.4	8.85	90.0
31	Bulgaria	715	0.3	8.26	86.5
32	Denmark	653	0.3	5.30	123.2
33	Thailand	628	0.3	61.20	10.3
34	Austria	624	0.3	8.08	77.2
35	Nigeria	576	0.3	120.82	4.8
36	Egypt	569	0.3	61.40	9.3
37	Algeria	530	0.3	29.92	17.7
38	Greece	514	0.2	10.51	48.9
39	Finland	489	0.2	5.15	95.0
40	Colombia	473	0.2	40.80	11.6
41	Switzerland	440	0.2	7.11	61.9
42	Iraq	436	0.2	22.33	19.5
43	Philippines	381	0.2	75.17	5.1
44	Chile	326	0.2	14.82	22.0
45	Norway	310	0.1	4.42	70.0
46	Portugal	302	0.1	9.98	30.3
47	Cuba	290	0.1	11.10	26.1
48	Kuwait	288	0.1	1.87	154.1
49	Ireland	284	0.1	3.71	76.4
50	Israel	282	0.1	5.96	47.3

Data sources

G. Marland and T. Boden, Carbon Dioxide Information Analysis Center, Oak Ridge National Laboratory, Oak Ridge, Tennessee, USA. R. J. Andres, University of North Dakota, Grand Forks, North Dakota, USA. National CO_2 Emissions from fossil-fuel burnings, cement manufacture and gas flaring: 1751–1998, July 25, 2001. IEA, key world energy statistics from the IEA 2000 edition, Nov 2000

[a] Excluding Macao

[b] Including Monaco

[c] Including San marino

problem of rising sea levels. This made us think about their future, in which the coral atolls will disappear under water, together with their wonderful culture, and in which the people will be forced to migrate overseas only to experience discrimination and oppression as well as the loss of their identity. We also heard a statement by Prof. Dr. A. Atiq Rahman from the Bangladesh Centre for Advanced Studies that, "we should not choose a path where we would pursue economic development with the help of ODAs, polluting the environment as well as emitting GHGs, only to end up being given GHG reduction technology," at a NGO meeting on technological transfer. The developing countries themselves are also facing problems associated with a free ride. The position of a developing country, whose primary goal is economic development and poverty alleviation, is difficult partly because the prevention of global warming is a typical prisoners' dilemma. Although the payoff from global cooperation is clearly the highest over all, a developing country may obtain higher benefits through taking the opposite measures, or through not taking any action at all, when the majority of other countries take action to prevent global warming. For example, creating an energy-intensive and wasteful economy may be a short cut to economic development and poverty reduction when energy prices go down because of the actions of other countries to prevent global warming.

From the view point of economic efficiency, there are even more difficult problems with measures to prevent global warming. We can assume that the future abatement costs will be lower than they are now if we predict continuous technological development. In this case, early action is relatively expensive and imposes high economic burdens. In addition, we cannot completely deny the idea that adaptation to global warming is more efficient than prevention when we cannot be certain that the negative effects of global warming are irreversible. In terms of the cost benefit of measures to prevent global warming, it is obvious that the biggest beneficiaries of the present measures are future generations in developing countries. In addition, if we calculate the net present value of future benefits from current measures, the net present value of measures to prevent global warming is negligibly small because the benefits will materialize several decades or centuries in the future. As mentioned above, from the view point of pure cost–benefit analysis, there are few convincing economic reasons for immediate action by the developed countries. In addition, uncertainty because of the lack of scientific knowledge makes decisions more difficult.

The Bonn Agreement and the Marrakech Accords also stipulate that the per capita emissions gap between developed and developing countries should be narrowed. Although this is not shown in Fig. 1, if we consider the changes over time (1971–1998), they show that each country has been moving to the left, across the board. In other words, although the efficiency per GDP has improved, CO_2 emissions per capita have not shown any remarkable change. Per capita differences have not narrowed in the past 30 years either within developed countries or between developed and developing countries. The environmental Kuznets curve shows that, in the long run, emissions per capita increase as incomes go up, but they then decrease when incomes reaches a certain level.[27] This effect may not be seen over a period of only 30 years, but we cannot wait for 50 or 100 years to see the effect of the environmental Kuznets curve. It is necessary to design a mechanism that will lead each country to move toward the lower left-hand corner in Fig. 1. The Kyoto Protocol is an agreement among countries that would serve as a first step toward this objective, and it does not directly affect the domestic policies of each country. Although the Kyoto Protocol has many good points, such as the principle of differentiated responsibility with developed countries acting on their own initiative, it also has many problems, such as the absence of any incentives for developing country parties to take early voluntary measures. If we consider the negotiations over the Protocol as a type of game, the lessons learned from the Kyoto process may be that the smart way is to accept an emissions cap after expanding your emissions allowances by inefficient facilities, as can be seen in transition economies, while the less intelligent way is to accept a severe emissions cap that may hamper economic growth after achieving significantly high energy-saving standards, as can be seen in Japan. We can say that transition economies such as Russia have acquired enormous additional fund transfers through hot air. Regarding the fund contributions to assist developing countries, a large voluntary early donor, Japan, which has given large amounts of financial aid to developing countries, as a Kyoto initiative, since 1998, is now in a disadvantageous position compared with other developed countries, including the EU, that have not contributed so much. The existing framework of the Kyoto Protocol never rewards a country that has taken early voluntary action with goodwill. The framework, including the setting of reduction targets, needs to be fundamentally revised in order to favor, and provide incentives for, early voluntary action.

For Japan, it is also true that the Protocol objectives cannot be achieved simply by consumers with an "awareness" that they should contribute to GHG reductions. Although it is, of course, very important that there are public relations activities to increase consumers' awareness, it is more important to design a system that includes a mechanism that can control consumption by increasing the price of goods and services that generate GHGs. In this context, neither a tightly controlled system that tries to watch and control every action of every entity, nor a loose system that cannot prevent free-riders, is desirable.

Aspects of global warming prevention that relate to the North–South problem make this type of argument more difficult. Poverty is worsening the population explosion, and is closely related to wasteful resource usage such as slash-and-

[27] See, for example, J. T. Roberts and P. E. Grimes, Carbon intensity and economic development 1962–91: a brief exploration of the environmental Kuznets curve. World Development, 25(2), 1997, p 191–198, and Noriyuki Goto, Empirical examination of the relationship between carbon emissions and economic development. Proceedings of the Department of Advanced Social and International Studies, Graduate School of Arts and Sciences, University of Tokyo, March 2001, p 111–148

burn farming. When we consider poverty alleviation, which is commonly the first priority of an international donor community, we cannot simply criticize the USA for importing goods that are made in developing countries under poor working conditions with extremely low wages. In developing countries where business conditions are poor, the infrastructure is inefficient, and currency exchange systems are risky, jobs exist *because* wages are low, compensating for the negative aspects of the business environment. If wages are not low enough, people lose their jobs and are likely to suffer more severe poverty under even worse working conditions. The lack of due governance in developing countries also makes finding solutions for this type of problem very complicated. Without appropriate government control, financial and technological aid is never used efficiently or in ways originally intended by donor countries, as donors have learned from past experience.

As mentioned above, many difficult problems exist. However, it is a fact that many experts have their doubts about the sustainability of the American resource-wasting lifestyle, symbolized by mass production, mass consumption, and mass disposal. Thus, it is obvious that we should aim at establishing a sustainable, resource-circulating society. Although it is difficult to create such a society with existing technology, it is also a fact that we are coming to a stage where we must take one further step in that direction. Each country needs to take responsible measures to help prevent global warming. These should be on a global scale, and be sustainable and effective for the foreseeable future.

Acknowledgments. We would like to thank Peter Bohm, Mitsutoshi Hayakawa, Carsten Helm, Yoshie Ishida, Naoki Kojima, Mikako Kokitsu, Hiroki Kudo, Takako Nakajima, Hidetaka Nakanishi, Hidenori Niizawa, Naoko Nishimura, Ichiro Sadamori, Kanako Tanaka, Akinobu Yasumoto, and Takehiko Yamato for their helpful comments and discussions. This research was partially supported by the Nissan Foundation and the Sumitomo Foundation.

Emissions Trading Experiments: Investment Uncertainty Reduces Market Efficiency

TAKAO KUSAKAWA* and TATSUYOSHI SAIJO[†,‡]

Summary. This chapter reviews two emissions trading experiments conducted by Hizen and Saijo (2001, 2002) and Hizen et al. (2001). In Hizen and Saijo, where the abatement investment was reversible and no time-lag effect of investment was introduced, the efficiencies of emissions trading were quite high. In Hizen et al., however, two types of price dynamics were observed following the introduction of irreversible abatement investment and its time-lag effect.

Key words. Emissions trading, Investment irreversibility, Investment time-lag, Bilateral trading, Double auction, Point equilibrium

Introduction

At the Third Conference of Parties (COP-3) to the United Nations Framework Convention on Climate Change, held in Kyoto in December 1997, the Kyoto Protocol[1] was adopted. The protocol establishes national emission targets for greenhouse gases (GHGs) for developed countries and economies in transition. In effect, the protocol calls for an overall emissions reduction of 5.2% from 1990 levels, with Japan, for example, requised to reach 94% of 1990 emissions, the USA 93%, the EU 92%, and Russia 100% during the period 2008–2012. In order to achieve this goal the use of Kyoto mechanisms, which include international emissions trading, joint implementation, and the clean development mechanism, was authorized. At COP-7, held in Marrakesh in November 2001, the parties reviewed the details of the Kyoto Protocol in preparation for its ratification, and adopted

* Graduate School of Economics, Osaka University, Toyonaka, Osaka 560-0043, Japan
† Institute of Social and Economic Research, Osaka University, Ibaraki, Osaka 567-0047, Japan
‡ Research Institute of Economy, Trade and Industry, 1-3-1 Kasumigaseki, Chiyoda-ku, Tokyo 100-8901, Japan
[1] See http://www.unfccc.de/index.html

the Marrakesh Accords.[2] Following these developments, the detailed design of the mechanisms from an economics viewpoint is important in order to achieve further progress in this area.

The focus in this chapter is on emissions trading. Bohm (1997), an initiator of GHG emissions trading experiments, reported a bilateral trading experiment among four Nordic countries using experienced public officials or experts appointed by the energy ministries. The resulting prices were very close to the competitive equilibrium price with an efficiency of allocation of 97%, which is extremely high. Muller and Mestelman (1998) and Godby et al. (1999) published many new findings on general emissions trading experiments. Among other results, they found that (i) allowing the banking of permits over time smoothes contract prices across time periods, and (ii) a trader with market power outside the emissions trading market can influence the emissions trading market, and hence the introduction the emissions market reduces the efficiency of the economy as a whole.[3]

Following these experiments, Hizen and Saijo (2001, 2002) designed an experiment with three controls: (i) trading method, i.e., bilateral trading or double auction; (ii) disclosure of contract information, i.e., price, quantity, identity of buyer and seller; (iii) disclosure of marginal abatement cost curves. By changing these controls, they explored what types of institution can efficiently attain the targets of the Kyoto Protocol in relation to GHG emissions reduction. The main result in Hizen and Saijo is that the efficiency of both bilateral trading and double auction is quite high, regardless of the control of contract information.[4]

Using almost the same experimental design, Hizen et al. (2001) focused on three features that were not analyzed by Hizen and Saijo: the effects of a noncompliance penalty, abatement irreversibility, and the time-lag effect of investment. Hizen et al. obtained four main results. First, the trading sessions can be put into two groups according to the price dynamics of *point equilibrium*, an equilibrium concept first introduced in their paper. If abatement investment is irreversible, the normative equilibrium price at each point in time depends on the previous abatement investment decisions, since abatement investment changes the shape of the marginal abatement cost (MAC) curve, and hence the supply and demand curve. For example, assuming that at a certain point in time a party conducts emissions reductions which are larger than its own competitive equilibrium reduction after other parties have already conducted their own competitive equilibrium reductions exactly, the total amount of reductions would then exceed the total amount of the competitive equilibrium reductions, and hence

[2] In March 2001, President Bush announced that the USA would not ratify the Kyoto Protocol

[3] The experiments by Bohm and Carlén (1999) show that the market power problem is not as serious as Muller and Mestelman (1998) and Godby et al. (1998) suggest, since participants in an emissions trading market can buy *and* sell the permits

[4] The efficiencies of emissions trading were high, i.e., between 91.9% and 99.9% for bilateral trading, and between 99.4% and 99.7% for double auction

the competitive equilibrium price *at this point in time* should be less than the competitive equilibrium price *before* the session began. There are therefore two sequences of price dynamics. One is the actual contract price data, and the other is the "should be" price data derived from the actual abatement investment data. This "should be" price at each point in time is called the point equilibrium price. In sessions that belonged to one of the two groups, the point equilibrium price dropped at an early stage of the transactions[5]: relatively high contract prices and/or fear of noncompliance caused some subjects to conduct excessive domestic reductions at an early stage, which produced an excess supply of emissions permits, and hence the point equilibrium price went down. The efficiencies of these sessions were relatively low. In sessions which belonged to the other group, the point equilibrium price did not drop at an early stage of the transactions: relatively low contract prices at an early stage caused insufficient domestic reduction and the point equilibrium prices in the first half of the period were very close to the competitive equilibrium price. The efficiencies of these sessions were relatively high, but substantially lower than those of Hizen and Saijo.

Second, although no difference in efficiencies between double auction and bilateral trading was observed when sessions were compared as a whole, in each group of sessions the efficiencies of double auction were higher than those of bilateral trading. That is, while an analysis of the type of transaction method alone did not give a clear-cut distinction in terms of efficiency, analysis within each of the two groups did. On the other hand, no clear distinction on efficiency was observed between bilateral trading and double auction by Hizen and Saijo.

Third, compared with bilateral trading, double auction was more likely to result in the low-efficiency group and less likely to result in the high-efficiency group. That is, although double auction was likely to attain high efficiency in each group, it often resulted in the low-efficiency group.

Fourth, although the overall efficiencies of sessions were not very high, on average emissions trading reduces the total costs of achieving the Kyoto targets compared with carrying out domestic reductions only.

This chapter compares the two experiments conducted by Hizen and Saijo (2001, 2002), and Hizen et al. (2001), and shows the effects that investment has on the market. Because the first three results described above were not observed by Hizen and Saijo, it is thought that the introduction of investment irreversibility, investment time-lag, and/or a noncompliance penalty might account for them.

This chapter is organized as follows. The experimental design and procedures are described first, and then point equilibrium is explained. We then show two types of efficiency, before comparing the results using point equilibrium and giving some concluding remarks.

---

[5] Baron (2001) observed the same effect, although since he conducted just one session, it is hard to generalize from the result. However, the price pattern of the sulfur dioxide market in the USA showed the same price dynamics in the first few years

Experimental Design and Procedures

To explore what types of institution can efficiently attain the targets of the Kyoto Protocol in relation to GHG emissions reduction, Hizen and Saijo (2001, 2002) designed an experiment with three controls: (i) trading method, i.e., bilateral trading or double auction; (ii) disclosure of contract information, i.e., price, quantity, and buyer and seller identity; (iii) disclosure of MAC curves.

Based on the results of Hizen and Saijo, Hizen et al. (2001) employed two controls: (i) trading method; (ii) disclosure of contract information. Sessions to disclose MAC curves were not conducted because almost no effect from the disclosure of MAC curves was observed in the Hizen and Saijo experiment.

The main features of the experimental design in Hizen and Saijo are reversible investment, no time-lag investment, and the impossibility of noncompliance. The opposite features were used in the Hizen et al. experiment: irreversible investment, time-lag investment, and the possibility of noncompliance.

Reversible investment is explained in Fig. 1, where the horizontal axis represents the amount of emissions, and the vertical axis represents MAC. The downward sloping curve is a MAC curve. Point *a* corresponds to the initial position of the subject. If the subject decides to invest in emission reductions and moves from *a* to *b* on the MAC curve, the area below the curve between the dotted lines is the abatement cost of moving from *a* to *b*, and the subjest's position moves from its initial position to the new position shown in Fig. 1b. The MAC curve is considered to represent the marginal *cost* when a subject reduces one unit of emissions, but it can be regarded as representing a marginal *benefit* when it increases one unit of emissions. If abatement investment is *reversible*, the benefit of one additional unit of emissions after the abatement investment is the height of point *b* above the horizontal axis. That is, the abatement investment is fully recoverable by increasing the emissions. This rule was applied by Hizen and Saijo. On the other hand, if the investment is totally unrecoverable, the benefit of emitting

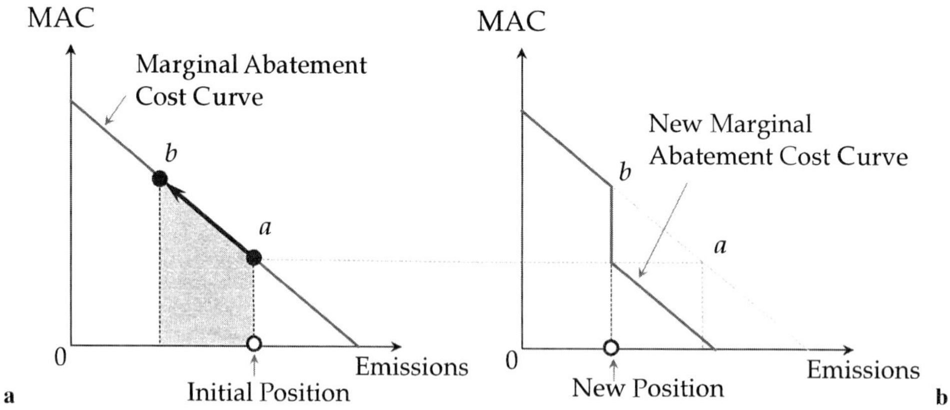

FIG. 1. Abatement irreversibility

one additional unit would be zero. The Hizen et al. experiment took the middle point between these two extremes: after any amount of emission reduction investment, the benefit of emitting one additional unit is always the height of the initial position. Any further benefit of emissions decreases on the right-hand side of point *a* on the MAC curve in Fig. 1a. Therefore, after moving to the new position, the new MAC curve is the curve depicted in Fig. 1b. That is, with irreversible investment, the shape of the MAC curve changes after abatement investment. In the case of abatement reversibility, it does not change.

In the Hizen and Saijo experiment, there was no time-lag between investment and emissions reduction: subjects could reduce their emissions immediately after deciding to do so, even just before the end of the commitment period. In the Hizen et al. experiment, on the other hand, subjects could not reduce their emissions if they decided to do so just before the end of the period, since actual abatement took a considerable time owing to the investment time-lag.

An important issue in the negotiation process of the design of the Kyoto Protocol has been what type of penalty would be appropriate when a party cannot comply with the Kyoto target (noncompliance).[6] In focusing on the types of trading institution, Hizen and Saijo designed their experiment so that noncompliance and over-compliance would not occur in order to exclude these effects: each subject started the experiment having achieved the goal required by the Kyoto Protocol, and when they sold (bought) permits, they were required to reduce (increase) emissions by the same amount. Thus, in Fig. 1a the subjects could reduce their emissions from point *a* to point *b* only when they could sell the same amount of permits to another subject. In the Hizen et al. experiment, noncompliance and over-compliance could occur: each subject started the experiment without achieving their goal, and could sell or buy their permits even if they did not reduce or increase their emissions. Two rules concerning noncompliance and over-compliance were adopted: (i) when a party ended with noncompliance, a penalty approximately 2.5 times higher than the competitive equilibrium price per unit was imposed, and (ii) when a party ended with over-compliance, any remaining permits had no value.

In both experiments, a minimum of six students were recruited for each session by campus-wide advertisements at Osaka University. The students were told that there would be an opportunity to earn money in a research experiment (in the recruitment, no term was used which was peculiar to emissions trading). The sessions were conducted two or three days after recruitment. In each session, the subjects were seated at desks in a relatively large room and listened to a tape-recorded voice giving instructions. During this part, each subject received a sample MAC curve.

Figure 2a is a sample graph used by Hizen and Saijo. The upper half is a sample MAC curve. The horizontal axis represents the amount of an abstract commod-

---

[6] Japan insists that the individual targets are only objectives, and hence no penalty should be imposed for noncompliance. However, most parties, including the USA, the EU, and developing countries, advocate strong penalties, including a monetary punishment

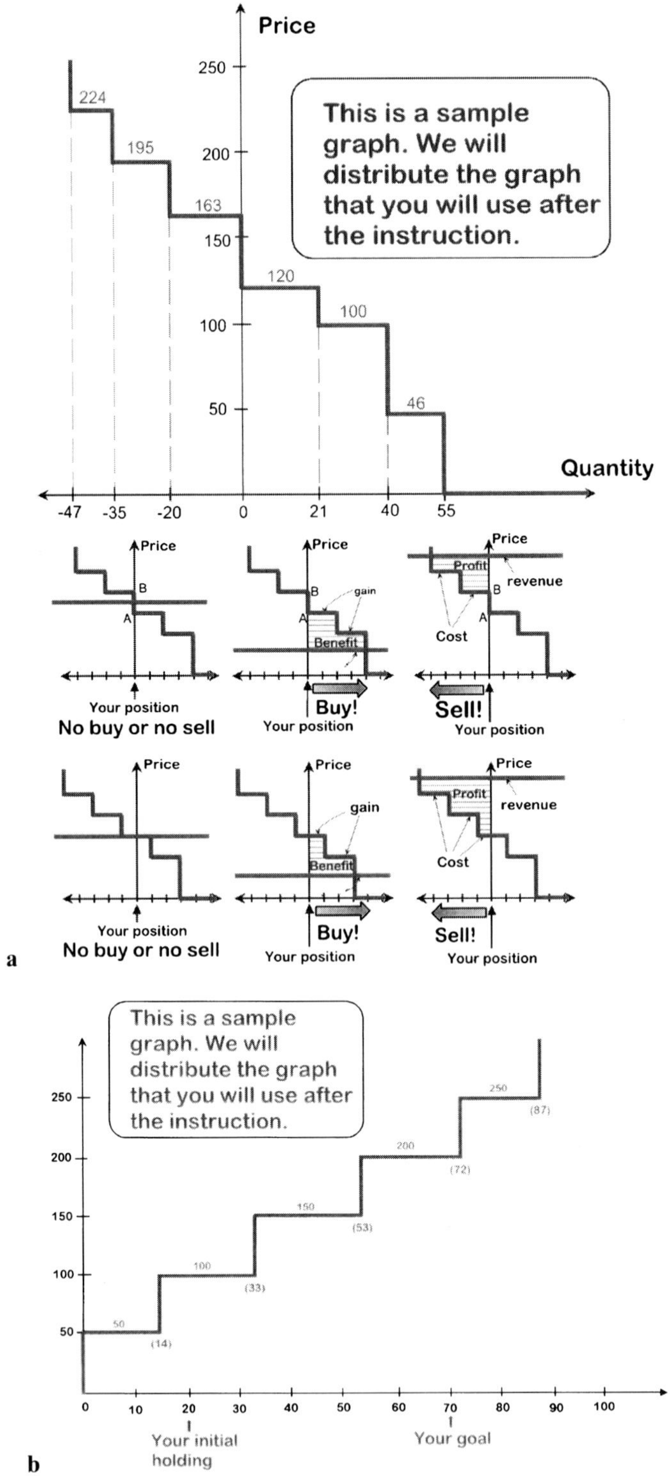

FIG. 2. A sample graph

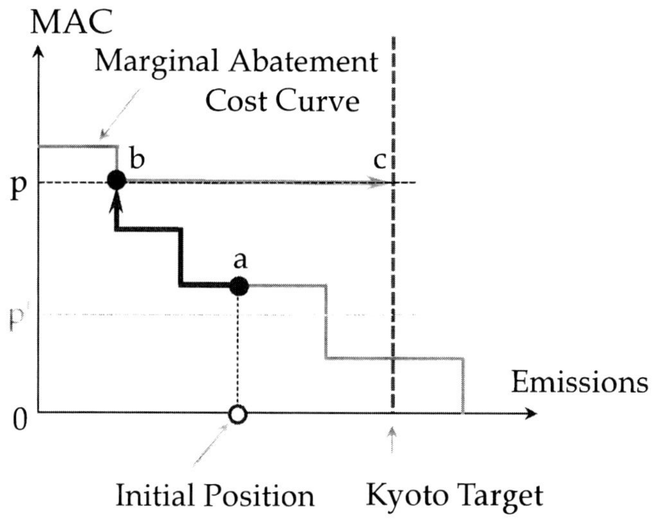

FIG. 3. A possible strategy

ity, and the vertical axis represents the marginal cost. Each subject was told that the initial position is at 0.[7] When they bought (sold) the commodity, they moved to the right (left) on the curve and earned benefits (obtained a profit), which were equivalent to buying (selling) permits and conducting additional emissions (emissions reductions). All possible situations in the Hizen and Saijo experiment are depicted in the lower half of Fig. 2a.

Figure 2b is a sample graph used by Hizen et al. Because the direction of the horizontal axis in this figure is opposite to that in Fig. 2a, the MAC curve is upward sloping. Each subject was told that they would start from "your initial holding" (20 units in the sample figure), which was also called "your initial position," and that they should finish with a number of units more than or equal to the goal (70 units in the sample figure). The initial position represents initial emissions, and "your goal" is the target of the Kyoto Protocol. When subjects move one unit to the right (left) along the curve from the initial position, they obtain (lose) one unit of goods in exchange for paying (receiving) 100 units of money, which is equivalent to a reduction (increase) of one unit of emission from their initial emissions. In the experiment, one unit of emissions reduction was termed "producing the unit," and one unit of increment of emissions was termed "returning the unit to the experimenter." Figure 3 shows some possible situations in the Hizen et al. experiment. In this figure, the Kyoto target is on the right-hand

[7] Hizen and Saijo implicitly presume that position 0 is the position where each country attains the goal required by the Kyoto Protocol. The experimental setting asks what type of trading method should be adopted in order to achieve that goal, and does not address the noncompliance issue

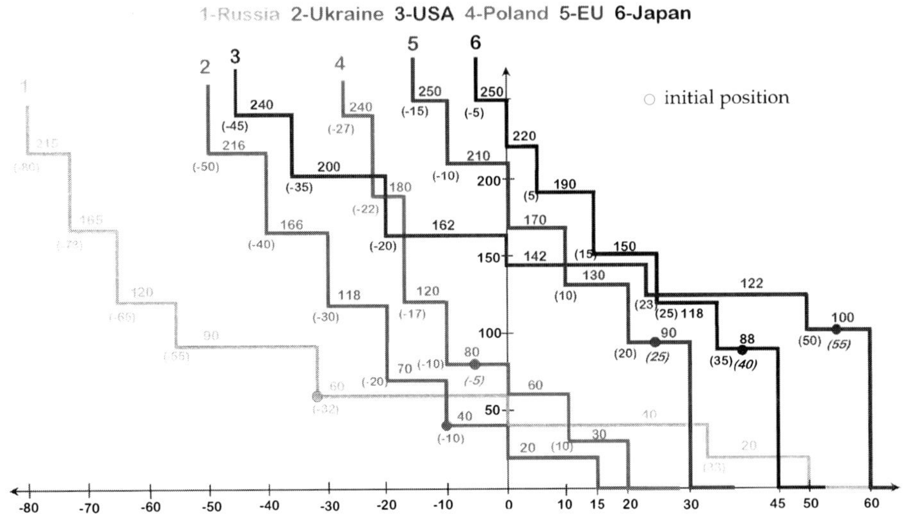

FIG. 4. All marginal abatement cost curves. No country names were given to the subjects in the experiment

side of the initial position. Assuming that the price level of permits is at p, for maximum gain the subject should reduce their emissions until the MAC is equalized to the price, i.e., from a to b, and should then sell the difference between the Kyoto target and the new position, i.e., from b to c. Additional cases where the price level was at p' and where the target was on the left-hand side of the initial position were also shown to the subjects.

After instruction, all subjects took a test to check their understanding of the instructions. The top six subjects continued the session, and the rest of the subjects were asked to leave the room and were given \$13.16 (\$1 = 114 yen). The six remaining subjects were assigned a number between 1 and 6 (equivalent to the roles of Russia, Ukraine, the USA, Poland, the EU, and Japan) and received their own MAC curves (in the MAC curves disclosure sessions in the Hizen and Saijo experiment, they also received the MAC curves of the other subjects). Figure 4 shows the MAC curves of six subjects. In this figure, their MAC curves are shifted so that their Kyoto targets are at the origin, or 0. An MAC curve can be regarded as an excess demand curve for emissions permits if we consider the Kyoto target as the new origin: on its left-hand side, the MAC curve becomes the supply curve of the permits, and on its right-hand side, the MAC curve becomes the demand curve. The initial position for each subject in the Hizen and Saijo experiment is the origin, and the initial position in the Hizen et al. experiment is the solid circle on each curve. The subjects were given 15 min to consider their strategy before the 60-min trading period started.[8]

[8] Each session had just one period

In bilateral trading, each subject could move around the room freely to find a subject with whom to transact. To avoid information leakage, subjects were not allowed to talk during negotiations: numbers (price and quantity) and "yes" and "no" symbols were exchanged on their negotiation sheets. Once a pair had reached agreement, they reported the price, quantity, and their subject numbers to an experimenter. In bilateral trading with open contract information sessions, the experimenter announced this information and wrote it on a blackboard.

In a double auction, an auctioneer called on the subject who raised their hand earliest. The subject then provided the subject number, sell or buy decision, quantity, and price per unit (e.g., "Subject five wants to sell ten units at $100 per unit"). The auctioneer projected the proposal onto a screen using an overhead projector, and the subject who raised the hand earliest could either accept the proposal or make another one. The accepted quantity had to be smaller than or equal to the proposed quantity. The "improvement rule" was imposed on proposals, i.e., asks (bids) had to be successively lower (higher).[9]

In the Hizen et al. experiment, in order to reduce or increase emissions, a subject informed the experimenter of the amount of emissions reduction or additional emissions; this information was not known to the other subjects. These reductions or increases could be carried out only in the first half of the 60-min trading period. After half an hour, the point searched on the MAC curve is fixed for the rest of the period. If a subject ended the experiment with noncompliance, they paid a penalty of 300 per over-emissions unit; if a subject ended with over-compliance, any remaining permits had no value.

In the Hizen and Saijo experiment, each session lasted approximately 160 min, with a mean payoff per subject of $31.52 (maximum payoff $66.67, minimum payoff $17.54). In the Hizen et al. experiment, each session lasted approximately 3 h, with a mean payoff per subject of $34.61 (maximum payoff $66.34, minimum payoff $17.54).

Point Equilibrium

In the Hizen et al. experiment, owing to changes in the shapes of the MAC curves caused by abatement investment irreversibility, as described above, the normative competitive equilibrium price also changes, as illustrated in Fig. 5. This figure is similar to Fig. 1 but the position of the Kyoto target is shown explicitly, so that the downward-sloping curve is regarded as a demand curve for emissions permits, and the supply curve (the upward-sloping curve in the figure) is derived from the MAC curves of other parties. Before any abatement investment, the competitive equilibrium is at e and the competitive equilibrium price is at p^*. However, if a demander conducts excessive abatement investment and moves from point a to point b, the normative competitive equilibrium *after* this investment would no longer be at e but at e', and the new competitive equilibrium price is at p'. At

[9] See Davis and Holt (1993, p 41)

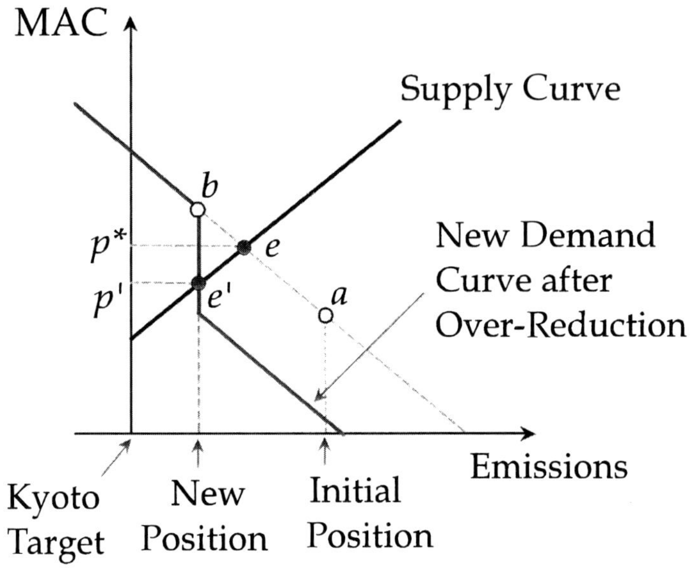

FIG. 5. Point equilibrium

each point in time, therefore, the normative competitive equilibrium can change as abatement investments proceed. The normative competitive equilibrium, given previous actions of subjects at a certain time, is called the *point equilibrium* at that time.

In the Hizen and Saijo experiment, the competitive equilibrium price and the point equilibrium price always coincide throughout the period because abatement investment is reversible and the shapes of the MAC curves cannot be changed. In the Hizen et al. experiment, on the other hand, although these prices coincide at the starting point, they do not always coincide throughout the period because the shapes of the MAC curves can change as time proceeds. There are therefore three price sequences in a session: the first is the sequence of real contract price data; the second is the normative competitive equilibrium price sequence which is constant throughout the period; the third is the point equilibrium price sequence. While the first two sequences are usually compared in experimental economics literature, it is shown later that Hizen et al. enrich the analysis of price dynamics by the introduction of the point equilibrium price sequence.

Two Definitions of Efficiency

In published work, efficiency is defined as

$$\frac{\text{The sum of surplus extracted in the experiment}}{\text{The sum of surplus extracted at competitive equilibrium}}$$

This is a standard measure of efficiency, which is usually measured as the percentage of the realized sum of surplus to the maximum possible sum of surplus which is attained at competitive equilibrium. In the Hizen and Saijo experiment, where noncompliance and over-compliance are not possible, this value is measured uniquely. In the Hizen et al. experiment, on the other hand, two types of efficiency measures are employed, depending on how "the sum of surplus extracted in the experiment" is measured when noncompliance and/or over-compliance occur. The first method is to measure "the sum" as the sum of the actual payoffs obtained during the experiment. For example, when a subject ended with over-compliance, the cost of the unnecessary abatement was subtracted from the payoff, and hence from "the sum." On the other hand, when a subject ended with noncompliance, a penalty was subtracted from the payoff and hence from "the sum."

There are two problems with this measure of efficiency. First, the permits left over under over-compliance will have some value, since the Kyoto Protocol allows the banking of permits, and over-compliance represents a reduction of GHGs beyond the targets of the Kyoto Protocol, which means it would be natural to assign a value to the leftover permits. Second, penalties from noncompliance would be handed to an international body and would be distributed to other parties, so that the total sum of the penalties would not be a source of loss of efficiency.

Given these two problems, efficiency in the Hizen et al. experiment is modified in the following manner. When some subjects finish with over-compliance and nobody finishes with noncompliance, one unit of over-compliance is reevaluated using the value of the subject who has the highest marginal emissions benefit. Hypothetically, therefore, the subject obtains benefit by using one unit of over-compliance. The same reevaluation is then carried out for the next unit of over-compliance using the value of the subject who has the second highest marginal emissions benefit. The process is then applied to subsequent units until the last unit of over-compliance is reevaluated. Since the shape of the marginal abatement cost curve (or the marginal emission benefit curve) of each subject will have been changed at the end of a session owing to abatement irreversibility, reevaluation is carried out using this reshaped marginal emissions benefit curve.

When some subjects finish with noncompliance and nobody finishes with over-compliance, one penalty unit of noncompliance is replaced with the marginal abatement cost of the subject who has the lowest marginal abatement cost. Hypothetically, the subject reduces one unit of emission instead of another subject paying one unit of penalty. The same replacement is then carried out as a penalty for the next unit of noncompliance using the second lowest marginal abatement cost, and the same process is applied to subsequent units until the penalty for the last unit of noncompliance is replaced.[10]

Figure 6 shows an example where at the end of a session subject A has over-compliance and subject B complies with the target exactly. Both marginal abatement cost curves have kinks at the end positions of the session owing to

[10] Hizen et al. did not have a session where both a subject with over-compliance and a subject with noncompliance coexisted

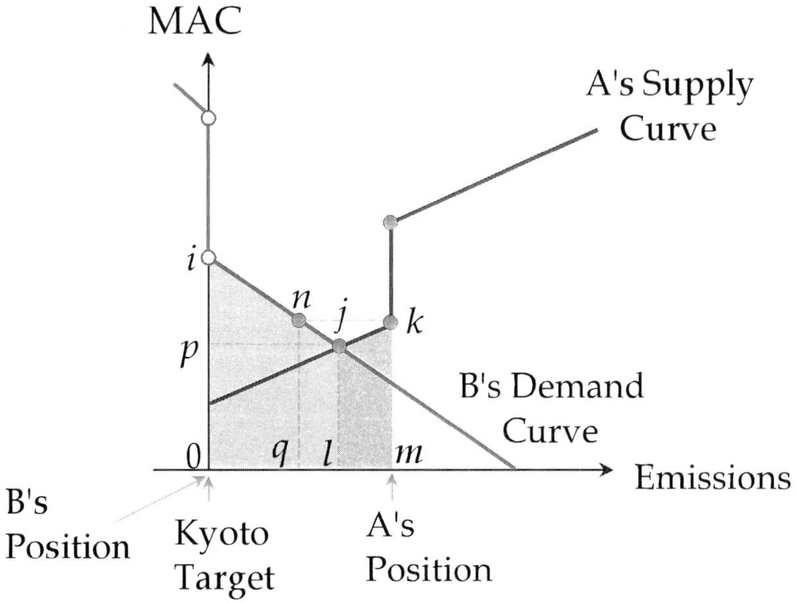

FIG. 6. Re-evaluation of permit surplus

abatement irreversibility. The amount of over-compliance for subject A is 0-*m*, and the end position of subject B is exactly at the Kyoto target. The subject who has the highest marginal emissions benefit is subject B. That is, the height of *i* is greater than the height of *k*. Hence, the benefit from emitting from *i* to *n* is the area 0–*q*–*n*–*i*. At *n*, the marginal benefit of subject B becomes exactly the same as that of subject A at *k*. From this point, the benefit can be obtained from both subjects A and B, and is the area *q*–*m*–*k*–*j*–*n*. In total, the over-compliance of 0-*m* generates the benefit depicted by the area 0–*m*–*k*–*j*–*i*.

The reevaluation of over-compliance in Fig. 6 can be interpreted as a hypothetical trade after the end of the session. Immediately after the session, subject A emits the amount *m*-*l*, and hence obtains the benefit *l*–*m*–*k*–*j*. Subject A then sells the amount 0-*l* to subject B at the point equilibrium price *p*, and obtains the benefit 0–*l*–*j*–*p*. On the other hand, subject B pays subject A 0–*l*–*j*–*p* and obtains the benefit 0–*l*–*j*–*i* by emitting the amount 0-*l*, for a net benefit of *p*–*j*–*i*. In total, the sum of the surplus is equivalent to the area 0–*m*–*k*–*j*–*i*.

Results

Analyses with Efficiency Table

In this section we analyze the experimental results using efficiency Tables 1–3, which provide a summary of the sessions.

TABLE 1. Efficiency of bilateral trading in the Hizen and Saijo experiment

Subject No.	OO1	OO2	OX1	OX2	XO1	XO2	XX1	XX2
1 (2555)	1420	1870	960	1710	1510	1100	1460	1600
(Russia)	0.556	0.732	0.376	0.669	0.591	0.431	0.571	0.626
2 (1290)	1140	914	360	1665	1320	940	1536	2370
(Ukraine)	0.884	0.709	0.279	1.291	1.023	0.729	1.191	1.837
3 (610)	685	683	2060	372	1846	615	583	550
(USA)	1.123	1.120	3.377	0.610	3.026	1.008	0.956	0.902
4 (390)	520	570	850	530	500	555	910	500
(Poland)	1.333	1.462	2.179	1.359	1.282	1.423	2.333	1.282
5 (620)	800	1105	1300	755	−150	1080	81	150
(EU)	1.290	1.782	2.097	1.218	−0.242	1.742	0.131	0.242
6 (1525)	2425	1800	1450	1844	1400	2700	2390	1800
(Japan)	1.590	1.180	0.951	1.209	0.918	1.770	1.567	1.180
Sum (6990)	6990	6942	6980	6876	6426	6990	6960	6970
	1	0.993	0.999	0.984	0.919	1	0.996	0.997

The bilateral trading experiment in Hizen and Saijo has two controls: (i) disclosure or closure of contracted prices, and (ii) disclosure or closure of marginal abatement cost curves. Therefore, there are four treatments. Repeating the same treatment twice yields eight sessions. In what follows, "O" represents disclosure and "X" represents closure. For example, OX2 indicates session 2 in the price disclosure, marginal abatement cost curve closure treatment.

In this experiment, the competitive equilibrium price ranges from 118 to 120, so that 119, the midpoint between 118 and 120, is regarded as the competitive equilibrium price. At this price, the total amount of benefit and profit (i.e., the maximum amount that these six subjects can enjoy) is 6990. In Table 1, the top row indicates the name of the session, the left-hand column shows the ID numbers of the subjects, and the numbers in parentheses are their benefits or profits at the competitive equilibrium price. In each cell, the upper figure is the actual benefit or profit that the subject earned, and the lower figure is the efficiency of this subject. For example, the 0.732 figure for subject 1 in session OO2 is the ratio 1870:2555, which is termed the individual efficiency. As Table 1 shows, with the exception of XO1, the efficiency of each session is quite high. The reason for the low efficiency level in session XO1 is that subject 5 traded with other subjects despite suffering a loss.

Result 1: *When investment is reversible and does not have a time-lag, the efficiency of bilateral trading is statistically larger than zero, and is almost one.*

In a double auction, all proposals, including contract prices and quantities, are disclosed. Therefore, the double auction experiment in Hizen and Saijo has only one control: disclosure or closure of marginal abatement cost curves. There were five sessions in total, three disclosure sessions and two closure sessions. In Table 2, these sessions are denoted by O3, X2, and so on. As Table 2 shows, the efficiency of each session is quite high.

TABLE 2. Efficiency of a double auction in the Hizen and Saijo experiment

Subject No.	O1	O2	O3	X1	X2
1 (2555)	2410	2410	1981	2260	2865
(Russia)	0.943	0.943	0.775	0.885	1.121
2 (1290)	1320	1320	520	1770	1120
(Ukraine)	1.023	1.023	0.403	1.372	0.868
3 (610)	850	865	1144	681	1270
(USA)	1.393	1.418	1.875	1.116	2.082
4 (390)	200	350	230	209	355
(Poland)	0.513	0.897	0.590	0.536	0.910
5 (620)	750	500	1380	700	0
(EU)	1.210	0.806	2.226	1.129	0.000
6 (1525)	1430	1515	1695	1350	1360
(Japan)	0.938	0.993	1.111	0.885	0.892
Sum (6990)	6960	6960	6950	6970	6970
	0.996	0.996	0.994	0.997	0.997

Result 2: *When investment is reversible and does not have a time-lag, the efficiency of a double auction is statistically larger than zero, and is almost one.*

In the Hizen et al. experiments, there are two experimental controls: the trading method and the disclosure of contract information. There are therefore three types of session: (i) a bilateral trading session with contract information closed ("Bc" session), (ii) a bilateral trading session with contract information open ("Bo" session), and (iii) a double auction session with contract information open ("D" session). Since each type was conducted four times, there are 12 sessions in total.

Table 3 gives results from the 12 sessions. In addition to the two figures in each cell as given in Tables 1 and 2, Table 3 gives additional figures at the bottom of each cell. The number in the left-hand parentheses at the bottom of each cell in the first column corresponds to the initial position in Fig. 4. The first number in the right-hand parentheses is the position in Fig. 4 to which it should reduce its emissions at a price of 119, and the second number is the amount of noncompliance. If the subject follows the transaction at competitive equilibrium, this should be 0. For example, Russia, whose subject number is 1, can make the maximum surplus of 2555 at the price of 119 when it reduces its emissions from its initial position of –32 to the point –55, and sells her permits until the amount of noncompliance becomes 0. In the second and subsequent columns, the first number at the bottom of each cell shows the final position in Fig. 4, and the second number shows the amount of noncompliance at the end of the session. In the case of over-compliance, this number is negative. In session Bc3, for example, Russia achieved a surplus of 620; the ratio of this surplus to the surplus it extracts at competitive equilibrium is 620 : 2555 = 0.243. It reduced emissions to the point –65, i.e., 10 units of over-reduction, and the amount of noncompliance was –22, i.e., 22 units of over-compliance. A bold square around the cell indicates over-

TABLE 3. Efficiency in the Hizen et al. experiment

Subject No.	Bilateral trading								Double auction			
	Bc1	Bc2	Bc3	Bc4	Bo1	Bo2	Bo3	Bo4	D1	D2	D3	D4
1 (2555)	1535	1600	620	656	1415	384	1825	1465	1425	2435	1360	2060
(Russia)	0.601	0.626	0.243	0.257	0.554	0.150	0.714	0.573	0.558	0.953	0.532	0.806
(−32) (−55,0)	−40, 0	−32, 0	−65, −22	−42, 0	−55, −3	−52, 0	−33, 0	−52, 0	−55, 0	−65, 0	−44, 0	−60, 0
2 (1290)	766	1175	1820	700	−565	2625	1285	2200	1195	−30	850	−1925
(Ukraine)	0.594	0.911	1.411	0.543	−0.438	2.035	0.996	1.705	0.926	−0.023	0.659	−1.492
(−10) (−30,0)	−28, 0	−20, 0	−30, 0	−20, 0	−20, −15	−20, 0	−25, 0	−20, 0	−30, 0	−30, −27	−20, 0	−30, −37
3 (610)	1046	220	556	1416	−4130	−4094	481	316	890	641	769	−404
(USA)	1.715	0.361	0.911	2.321	−6.770	−6.711	0.789	0.518	1.459	1.051	1.261	−0.662
(55) (50,0)	23, 0	30, 3	23, 0	23, 0	−20, −30	50, 23	23, 0	23, −2	40, 0	23, 0	23, 0	23, 0
4 (390)	240	100	20	94	77	500	300	450	165	275	375	763
(Poland)	0.615	0.256	0.051	0.241	0.197	1.282	0.769	1.154	0.423	0.705	0.962	1.956
(−5) (−10,0)	−5, 0	−10, 0	−17, 0	−10, 0	−10, 0	−10, 0	−13, 0	−10, 0	−10, 0	−10, 0	−11, 0	−17, 0
5 (620)	−650	375	850	850	1002	975	630	965	760	−900	770	682
(EU)	−1.048	0.605	1.371	1.371	1.616	1.573	1.016	1.556	1.226	−1.452	1.242	1.100
(25) (20,0)	5, −5	10, 0	25, 0	20, 0	20, 0	20, 0	20, 0	20, −2	20, 0	10, −10	20, 0	20, 0
6 (1525)	2175	2130	−3100	1710	1931	2040	1625	340	2515	25	1822	1200
(Japan)	1.426	1.397	−2.033	1.121	1.266	1.338	1.066	0.223	1.649	0.016	1.195	0.787
(40) (25,0)	35, 0	25, 0	15, −25	35, 6	35, 0	25, 0	25, 0	30, −5	35, 0	25, −10	25, −5	25, 0
Sum (6990)	5112	5600	766	5426	−270	2430	6146	5736	6950	2446	5946	2376
Efficiency	0.731	0.801	0.110	0.776	−0.039	0.348	0.879	0.821	0.994	0.350	0.851	0.340
Sum (6990)	5612	6230	4136	6686	3140	6680	6146	6596	6950	5856	6426	5186
Modified	0.803	0.891	0.592	0.957	0.449	0.956	0.879	0.944	0.994	0.838	0.919	0.742

compliance, and a bold square with gray shading in the cell indicates non-compliance. The bottom four rows of the table show the sum of the surplus each subject extracted and the efficiency of the session both before and after the modification of efficiency described above. In session Bc1, for example, the sum of the surplus is 5112 and the efficiency is 5112:6990 = 0.731. The surplus after modification is 5612 and the modified efficiency is 5612:6990 = 0.803.

Results given in Table 3 show that when investment is irreversible and has a time-lag, efficiency drops considerably. However, on average, the market achieves both positive efficiency and positive modified efficiency, indicating that emissions trading reduces the total costs of the Kyoto target at the market level compared with the case where only domestic reductions occur.

Result 3: *When investment is irreversible and has a time-lag, the efficiency and modified efficiency are statistically larger than zero.*

In the Hizen and Saijo experiment, no differences were observed between the efficiency of bilateral trading sessions and that of double auction sessions, between the efficiency of contract information disclosure sessions and that of contract information closure sessions, and between the efficiency of MAC information disclosure sessions and that of MAC information closure sessions.

In the Hizen et al. experiment also, no difference was observed between the modified efficiencies of the two trading institutions when comparing data from all sessions. However, when the sessions are classified into two groups according to their dynamic processes (such as the path of the point equilibrium price), a difference between the modified efficiency of the double auction and that of bilateral trading was observed.[11]

Classification of the Sessions by their Dynamic Processes

Figure 7 (session Bc1) illustrates the dynamic processes of transactions, emissions reductions, and additional emissions as time passes. In this figure, the horizontal axis represents minutes, and the vertical axis represents MAC and price. The squares in the figure indicate transactions. The left-hand (right-hand)-side letter near a square represents the initial letter of the seller (buyer), and the number under the square is the quantity traded. Diamonds (triangles) show the emissions reduction (additional emissions). The letter near a diamond (triangle) represents the initial letter of the subject who conducted the emissions reduction (additional emissions), and the number under the diamond (triangle) indicates the amount of emissions reduction (additional emissions). The gray horizontal line shows the competitive equilibrium price, which ranges from 118 to 120. The black line represents the point equilibrium price path up to 30 min. It has some thickness until 14 min, and then becomes 120 until just before 30 min, when it drops to 85. The

[11] No significant effects for price disclosure were observed even after the classification

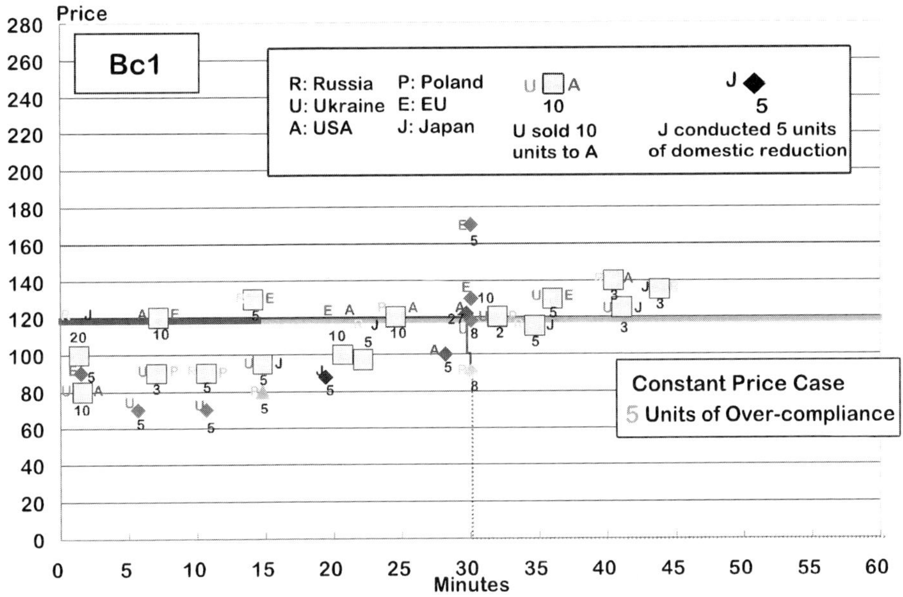

FIG. 7. Constant price case

dotted line shows the point equilibrium price path after 30 min, which is zero if over-compliance occurs, 300 if noncompliance occurs, and 0–300 if compliance is exact.

By comparing these figures for each session, it can be seen that the pattern of the point equilibrium path up to 30 min varies considerably from session to session: in some sessions it drops early, and in others it is almost the same as the competitive equilibrium price. That is, sessions are characterized by the pattern of their point equilibrium path. As a measure of these path patterns, Hizen et al. introduced the concept of *discrepancy area*, which is the area of the region enclosed by the midpoint of the competitive equilibrium price up to 30 min, i.e., 119, and the sequence of the midpoints of the point equilibrium prices up to 30 min. This area becomes larger depending on how early the discrepancy between the competitive equilibrium price and point equilibrium prices occurs and/or how large this discrepancy is. An example is given in Fig. 8, where the discrepancy area is shaded. In this session, the point equilibrium price dropped early and the degree of drop was large, so that the discrepancy area is also large. In session Bc1 (see Fig. 7), however, the discrepancy between the competitive equilibrium price and the point equilibrium prices occurred just before 30 min, so that the discrepancy area is almost zero.

After normalizing the discrepancy area and the modified efficiency, Hizen et al. used cluster analysis to classify the sessions. The 12 sessions are first divided into two groups, i.e., session Bo1 and the other 11 sessions, and then these

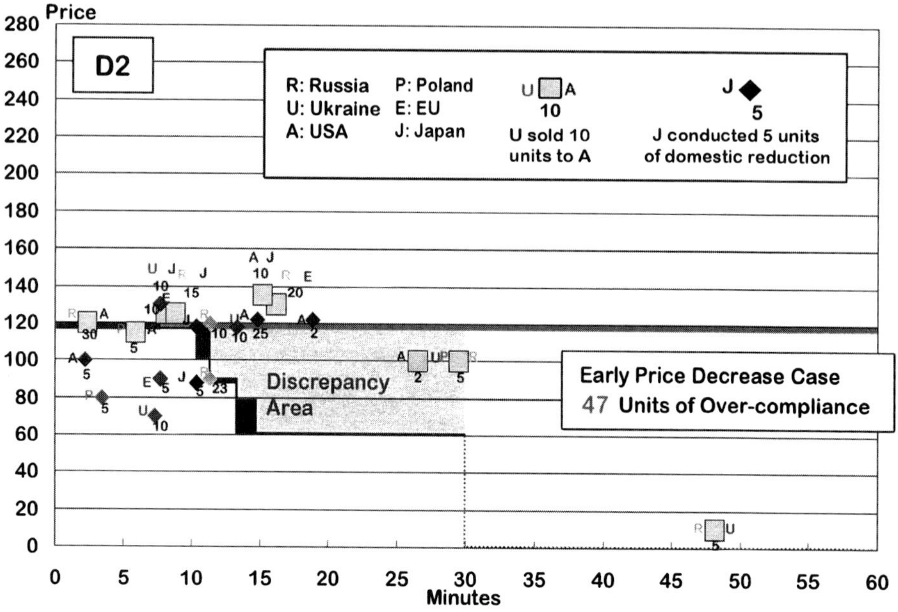

FIG. 8. Early price decrease case

11 sessions are further divided into two groups, i.e., sessions Bc3, D2, D3, and D4, and sessions Bc1, Bc2, Bc4, Bo2, Bo3, Bo4, and D1.

Figure 7 (session Bc1) is an example from the largest group, i.e., sessions Bc1, Bc2, Bc4, Bo2, Bo3, Bo4, and D1, and is termed the constant point equilibrium price case (constant price case). In this session, low contract prices at the early stage caused insufficient emissions reduction for suppliers such as Russia, Ukraine, and Poland. Therefore, just before 30 min the USA and the EU, who could not buy enough permits, reduced their emissions to levels where their MACs exceeded the point equilibrium price 120. These excessive reductions caused the point equilibrium price to drop to 90. This type of dynamic process also applies to the other sessions that belong to this group. That is, relatively low contract prices at the early stage caused insufficient emissions reduction by suppliers, and as a result, in many cases, demanders conducted excessive reductions just before 30 min in order to avoid a noncompliance penalty. Although this caused efficiency losses, the losses were minor.

Figure 8 (session D2) is an example from the second largest group, i.e., sessions Bc3, D2, D3, and D4, which is termed the early point equilibrium price decrease case (early price decrease case). In this session, because of the relatively high contract prices around the first 10 min, Japan and Russia conducted excessive reductions at that time. Although this caused a drop in the point equilibrium price, contract prices did not drop immediately owing to inertia of contract prices. Accordingly, Ukraine and the USA continued to reduce their emissions at around 15 min based on the former contract prices, so that the point equilibrium price dropped even further. This type of dynamic process also applies to the other sessions that belong to this group. That is, at the early stage high contract prices and/or the expectation of high contract prices in the future caused some subjects to reduce their emissions to levels where their MACs exceeded the competitive equilibrium price. Accordingly, the point equilibrium price immediately decreased, but owing to inertia, contract prices did not drop immediately. Therefore, some subjects continued reducing their emissions to levels where their MACs exceeded the new point equilibrium price, not knowing that the point equilibrium price had decreased. These excessive reductions caused the point equilibrium price to decrease even further. This cycle caused great loss of efficiency.

In the third and smallest group, session Bo1, the USA reduced her emissions at extremely high MACs at the early stage, and as a result the point equilibrium price dropped heavily. However, the contract prices remained at about the same level as the contract prices at the early stage because of inertia, and as a result subjects continued to conduct excessive reductions. Because these dynamic processes are the same as those of the second largest group, these two groups are combined into one.

The total sessions are thus divided into two by cluster analysis using two variables, i.e., modified efficiency and discrepancy area, and by reclassification according to the dynamic processes of transactions and emissions reduction.

Result 4: *When investment is irreversible and has a time-lag, two patterns of price dynamics are observed.*

Figure 9 shows a scatter diagram of the 12 sessions, where the horizontal axis represents modified efficiency, and the vertical axis represents the discrepancy area. The number above the session name is the modified efficiency, and the number below is the efficiency. Where the two efficiencies are the same, only the modified efficiency is included. The figure shows that sessions belonging to the constant price case are densely located around the southeast corner, and sessions belonging to the early price decrease case are located further away from that corner. This visual impression confirms the cluster analysis described above.

Using this classification, Hizen et al. obtained the following two results.

Result 5: *Bilateral trading is statistically more likely to result in the constant price case and less likely to result in the early price decrease case than double auction.*

Result 6: *In both groups, the modified efficiency of double auction is statistically higher than that of bilateral trading.*

In comparing all sessions, there was no significant difference between the modified efficiencies. This can be explained from Results 5 and 6, which show that although a double auction attains higher modified efficiency than bilateral trading, the early price decrease case (which is likely to result in low modified efficiency) occurs more frequently in double auctions, so that the two effects offset each other.

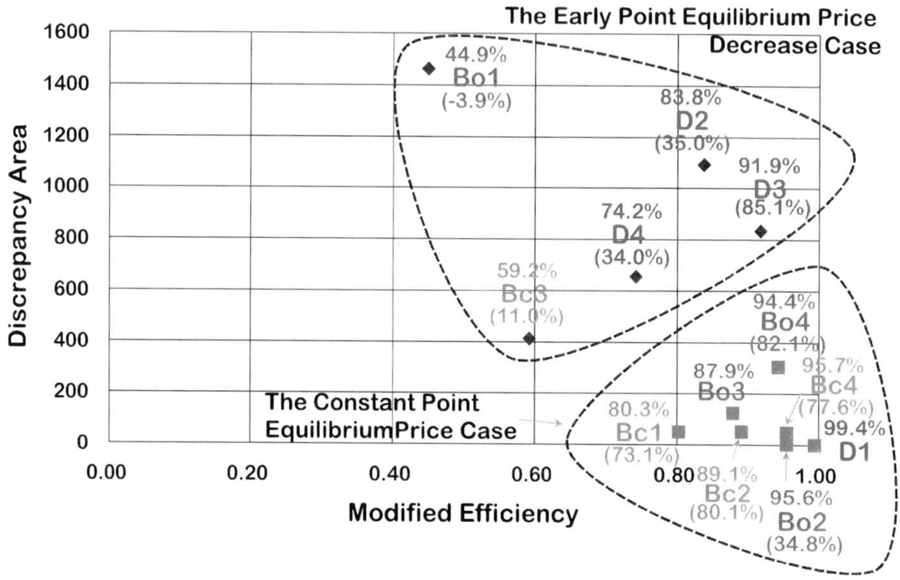

FIG. 9. Two groups

Concluding Remarks

The two experiments by Hizen and Saijo (2001, 2002) and Hizen et al. (2001) show that uncertainty in investment, such as irreversibility and time-lag, plays an important role in terms of efficiency of emissions trading. Although the efficiency of the experiments in Hizen and Saijo is high, two types of price dynamic were observed in Hizen et al. In the constant point equilibrium price case, relatively low prices at an early stage caused insufficient emissions reduction by suppliers, and demanders therefore conducted excessive reductions immediately before the end of the investment period. Efficiency was relatively high in this case. On the other hand, in the early point equilibrium price decrease case, owing to fear of noncompliance, some subjects conducted excessive domestic reductions at an early stage of the transactions, and hence the point equilibrium price dropped, although contract prices did not drop immediately because of price inertia. The efficiency of this pattern was relatively low. In each pattern, the modified efficiency of double auction sessions was higher than that of bilateral trading, but this does not necessarily imply that the double auction is more effective than bilateral trading, since only one double auction session belongs to the constant point equilibrium price case where modified efficiency is higher than that of the early point equilibrium price decrease case.

It is not easy to determine policy implications from the results of these experiments, but it seems that a strategy based on "carrying out reduction investment by responding immediately to the market price" was not successful. Rather, it seems that a party should *gradually* purchase emissions permits if the market price is cheaper than the marginal abatement cost, and it should *gradually* carry out abatement investment otherwise. In order to verify this statement, however, further experiments are needed.

Experimental economists have found that the double auction is one of the best institutions for trading. However, in an environment incorporating investment decisions explicitly, it seems that the double auction is not always the best institution. For example, in bilateral trading, the market price does not move in one direction since information such as price and quantity is not centralized, and it takes a considerable amount of time for it to be disseminated to participants. Further study is also needed on this point.

In order to provide a solid basis for the design of emissions trading institutions, further experiments must also be conducted on areas such as (1) oral double auction, (2) banking, (3) liability and (4) market power.

References

Baron R (2001) International emission trading: from concept to reality. OECD

Bohm P (1997) A joint implementation as emission quota trade: an experiment among four Nordic countries. Nord 1997:4 by the Nordic Council of Ministers, June

Bohm P, Carlén B (1999) Emission quota trade among the few: laboratory evidence of joint implementation among committed countries. Resource Energy Econ 21:1

Davis DD, Holt CA (1993) Experimental economics. Princeton University Press, Princeton

Godby RW, Mestelman S, Muller RA (1999) Experimental tests of market power in emission trading markets. In: Petrakis E, Sartzetakis E, Xepapadeas A (eds) Environmental regulation and market structure. Edward Elgar, Cheltenham, pp. 67–94

Hizen Y, Saijo T (2001) Designing GHG emissions trading institutions in the Kyoto Protocol: an experimental approach. Environ Model Software 16:533–543

Hizen Y, Saijo T (2002) Price disclosure, marginal abatement cost information and market power in a bilateral GHG emissions trading experiment. In: Zwick R, Rapoport A (eds) Experimental business research. Kluwer, Oordsecht, pp. 231–251

Hizen Y, Kusakawa T, Niizawa H, Saijo T (2001) Two patterns of price dynamics were observed in greenhouse gases emissions trading experiments: an application of point equilibrium. May, mimeo

Muller RA, Mestelman S (1998) What have we learned from emissions trading experiments? Manag Decision Econ 19:225–238

A Simple Model of CDM Low-Hanging Fruit

Jiro Akita

Summary. Some of the prospective host countries of the clean development mechanism (CDM) scheme under the Kyoto Protocol are concerned that if they lose their "low-hanging fruit" (LHF) i.e., relatively inexpensive greenhouse gas (GHG) emissions abatement projects, by CDM today, they may end up being left with only more costly projects tomorrow when they may assume their own abatement obligations. This chapter attempts to explore the background of this CDM–LHF argument using a simple analytical model that demonstrates conditions under which the argument may be valid or invalid.

Key words. Climate change, Kyoto Protocol, Clean development mechanism (CDM)

Introduction

The Kyoto Protocol and CDM

Article 12 of the Kyoto Protocol (KP) introduced a clean development mechanism (CDM) in order to assist non-Annex I parties to the protocol to achieve sustainable economic development and to contribute to the ultimate objective of the United Nations Framework Convention on Climate Change (UNFCCC 1997). It will also assist Annex I parties to achieve compliance with their quantified emissions limitations and reduction commitments under KP Article 3. In other words, CDM is a scheme of project-based activities where Annex I parties, which are mostly developed countries, invest in CDM projects taking place in non-Annex I parties, which are mostly less developed countries (LDCs). On the one hand, Annex I investor countries introduce more advanced and efficient greenhouse gas (GHG) emissions abatement technologies into non-Annex I host countries, and receive in return some part of the certified emissions reductions

Graduate School of Economics and Management, Tohoku University, Kawauchi, Aoba-ku, Sendai 980-8576, Japan

(CERs) generated by such project activities. On the other hand, non-Annex I host countries benefit from such activities because they are helped to reconcile global climate objectives and their long-run economic development objectives.

The introduction of the CDM scheme was promoted in recognition of the projection that the ultimate global climate objective of the UNFCCC is not really achievable solely through GHG emissions control efforts on the part of Annex I parties, and that it is imperative to assist non-Annex I parties in achieving their "sustainable economic development" by introducing more global-climate-friendly technology options.

The "Low-Hanging Fruit" Problem

Apprehension About LHF

While the objectives and ideals of the CDM were well received, some non-Annex I parties have remained apprehensive about the idea of participating in the CDM prematurely for various reasons, one of which has been referred to as the "low-hanging fruit" (LHF) problem. Suppose a non-Annex I LDC chooses to join the CDM today. Then the argument is that the CDM activities may effectively exhaust most of the relatively inexpensive GHG emissions abatement options, leaving only more expensive abatement options by the time that LDC chooses to join Annex I and assume its own emissions reduction obligations in the future. In the usual metaphor, it is feared that the CDM may reap all "low-hanging fruit," leaving only "high-hanging fruit" which are more difficult to reap when the LDC joins Annex I and needs these fruit in the future.

The State of the Literature

The LHF problem has been widely and repeatedly addressed in the UNFCCC–KP-related discussions, and the term "low-hanging fruit" seems already to have secured a firm position in the UNFCCC–KP vocabulary. However, the notion of LHF has been subject to far less analytical scrutiny. In particular, the question of why the LHF problem is a problem at all has never been properly addressed to the best of my knowledge. Certainly LDCs may be left only with fruit that are "high hanging" if CDM projects selectively remove LHFs, but will this always make the LDCs worse off? And if so, worse off than what? We believe that these questions have remained unanswered. Thus, the primary purpose of this chapter is to remedy this situation.[1]

---

[1] The literature is quite limited when it comes to the formal analysis of the LHF problem. One notable exception is the work of Rose et al. (1999). They constructed a continuous time model of a CDM-joint implementation (JI) in LDCs, where the instantaneous GHG emissions abatement cost at each point in time is assumed to depend not only on the flow-rate of abatement, but also on the cumulative abatement until that time. They then worked out the optimal dynamic abatement programming, and examined how "the cumulative abatement effect" manifests itself in the Euler equation that governs the evolution of the

Organization of this Chapter

The rest of this chapter is arranged as follows. In the next section, we examine the background premises of the LHF argument. In the third section, we propose a simple analytical model of the CDM–LHF situation capturing the premises discussed in the second section. We set up our model in such a way that a CDM host country can choose between the LHF and the "high-hanging fruit" (HHF) to be subject to the CDM. Using the model, we demonstrate that the LHF argument does not always hold. That is, a LDC may voluntarily, with no external coercion, choose to subject LHF to the CDM, which contradicts the LHF argument. By the same token, we also explore the conditions under which the LHF argument is indeed justifiable. The final section sets out our conclusions.

Premises of the LHF Argument

The CDM–LHF problem refers to the apprehension that participating in the CDM today may effectively wipe out all inexpensive GHG emissions abatement options (i.e., "low hanging fruit," LHF), leaving only higher-cost options (i.e., "high-hanging fruit," HHF) in the future when LDCs assume their own obligations, and thus must reduce emissions on their own. While this LHF argument, among others, has contributed significantly to the general reluctance on the part of LDCs to join the CDM, the argument seems to involve various loose ends. Thus, we found it useful first to spell out some essential logical premises underlying the whole LHF argument.

Future Possibilities that LDCs May Join Annex I and the Accessibility of Technology via JI Thereafter

Future Possibility that a LDC Joins Annex I

First, the LHF argument would not make much sense unless there is some possibility at all the LDCs currently belonging to the non-Annex I group may at some future time choose to join the Annex I group, thus assuming their own emissions abatement obligations. That is, if a non-Annex I LDC is determined to remain with that status indefinitely, there will be no LHF problem to begin with.

Accessibility of Technology After a LDC Joins Annex I

This poses the question of what might happen after a LDC has joined the Annex I parties, in particular, with regard to accessibility to technology. As KP

---

optimal trajectory. However, they did not address the question of whether or not the LDC is worse off by letting the CDM focused on LHF. Their cost-structure assumptions preclude us from conducting such a thought experiment

Article 12 states, one of the central virtues of the CDM is that it enables the non-Annex I LDCs to access more advanced and efficient GHG emissions abatement technologies from the Annex I developed countries. Will the future shift of LDCs from non-Annex I to Annex I status upset this accessibility to technology?

We take the view that the answer is "no". It is clear that such LDCs can no longer rely on the CDM for that purpose, since the Kyoto Protocol stipulates that the CDM should take place between Annex I parties and non-Annex I parties. However, KP Article 6 defines another mechanism known as joint implementation (JI), which is analogous to the CDM except that the former takes place among Annex I parties that have assumed Article 3 emissions abatement obligations. Thus, after the transition, JI instead of the CDM will provide LDCs with access to the technologies of the developed countries.

Low-Hanging Fruit Versus High-Hanging Fruit

Second, the LHF argument presumes (i) that CDM projects today will concentrate on the LHF, and (ii) that such concentration will hurt the host LDC in the long run. We therefore asked what would happen if the CDM projects concentrated on HHF rather than on LHF.[2] If the latter option is more beneficial to the host LDC than the former, then we may conclude that the LHF argument is indeed valid. However, if it is found that the LDC gains less by making the CDM today gather the HHF rather than the LHF, then the LHF argument has no case. We consider this hypothetical question to be central in properly addressing the LHF problem. In other words, to justify the CDM–LHF argument, it is not enough to claim that the LHF will be gone, leaving only HHF, if the CDM takes away LHF, since this is simply a tautology.

Technological Progress on the Part of LDCs

Third, as the LHF argument sees it as a problem that CDM activities today will remove all LHF from a LDC, there is an implication that tomorrow the LDC may regret that the LHF are already gone. Clearly this presumes that the LHF, if not reachable by the LDC today, will be within its reach tomorrow. That is, the LHF argument presupposes a progress from today to tomorrow in the LDC's abatement technology. If the LDC's technology should remain unchanged, then the fruit that are beyond the reach of the LDC today will still be out of reach tomorrow, and there would be nothing to regret. Thus, LHF that might possibly

[2] Note that this thought experiment about gathering HHF before LHF is a nonstandard one. In the theoretical literature on optimal emissions reduction, it is more or less taken for granted that we gradually climb up along a fixed marginal abatement cost (MAC) curve from lower MAC to higher MAC. That we need to consider this rather unconventional thought experiment characterizes the peculiarity of the CDM–LHF question

pose a problem are only the fruit that are beyond the reach of the LDC today but become within its reach tomorrow.

LDCs' Future Emissions Abatement Obligations

Fourth, the LHF argument presumes that the LHF that are gathered via the CDM today will be "lost" to the host LDC in one sense or another. However, we ought to question carefully exactly in what sense they would be deemed to be "lost."

LHF "Lost" from LDCs in What Sense?

It is clear that a project which has been used cannot be used again. In that sense, it is "lost," but this is not limited to low-cost projects. HHF can be lost is the same way as LHF. Therefore, the issue is not about the disappearance of fruit per se. Perhaps it concerns the disappearance of LHF only, not just any fruit. Again it is clear that a low-cost project, once used, cannot be used again, but this is not special to LDCs. We cannot use a low-cost project in a LDC twice, but the same applies to abatement projects in developed countries. Thus, the issue cannot be about the disappearance of LHF per se either.

Therefore, the issue must be about the involvement of developed countries in harvesting LHF that belong to LDCs. Granted that CDM projects and the associated advanced technologies do not come free, the host LDC needs to give up a portion of CDM-generated CERs to the investing countries. However, the question of whether or not this constitutes a genuine loss on the part of the LDC cannot be discussed independently of what emissions reduction obligations the LDC may assume tomorrow.

The Question of LHF "Loss" and LDCs' Future Obligations

Under a Scenario that Avoids Double-Counting of CERs. Suppose, for example, that a LDC agrees to join the Annex I parties at the start of tomorrow by assuming some GHG emissions reduction obligations. Furthermore, suppose that the LDC's commitment is such that it will have to reduce its emissions by the end of tomorrow down to $k\%$ of some predetermined level, say that of the start of today, which is independent of its reductions during today. Of course, some of the reductions that take place in the LDC today may be CDM-based. Now, if all such CDM-based reductions (without taking account of the portion of CERs accruing to the investing Annex I parties) are considered to contribute to the emissions reductions in the LDC by the end of tomorrow, then such CERs will in effect be counted "twice," first for the purpose of the Annex I parties' compliance today, and second for the purpose of the LDC's compliance tomorrow. Should we wish to avoid such "double counting," we must rule that such CERs as accrue to Annex I parties may not be counted again as a component of the emissions reductions in the LDC, and used for the purpose of its own compliance at the end of tomorrow. By the same token, the LDC should be allowed to use the remainder of the CERs generated by CDM projects today after it joined Annex I to fulfill her obligation tomorrow. Under this scenario, the double-counting of CERs is avoided, and the LHF that are gathered today may indeed rightfully be deemed as "lost" from the viewpoint of the LDC.

Under a Scenario that Permits Double-Counting of CERs. In reality, however, there has been little serious discussion about how the emissions reduction obligation is to be determined if a currently non-Annex I party decides to join the Annex I group tomorrow.[3] In fact, unlike the previous scenario, it is even conceivable that a form of $k\%$ rule may be applied not to the emissions level as at the start of today, but to that of the end of today.[4] Note, however, that the LHF argument becomes much harder to justify under this alternative scenario. In terms of the low-hanging fruit metaphor, the situation is analogous to the case where the fruits are nonstorable, and only tomorrow's harvest is valued. Then fruit gathered today will be wasted, regardless of whether they hang low or high. For that matter, fruit gathered today by LDCs on their own will also be wasted, implying that whatever the consequences, they have nothing to do with the so-called LHF problem.

CDM-Induced Technology Improvements

The Spirit of KP Article 12: "Sustainable Economic Development"

Fifth, and finally, we take note of the importance of CDM-induced technology transfer. While KP Article 12 on the CDM is not explicit on this point, its spirit may well encompass the possibility that advanced GHG emissions abatement technologies should not only be borrowed, but should also help to improve the domestic technologies of the CDM host LDCs, in order to assist them to adjust the course of their economic development in a more "sustainable," i.e., a more global climate-friendly, manner. Taking this possibility into account adds an important dimension to the whole discussion of the LHF problem, since technologies capable of gathering higher-hanging fruit would presumably be more difficult to replicate and to transplant into LDCs.

Why Not Wait for JI Tomorrow?

In fact, without considering these CDM-induced technology improvements properly, it is somewhat puzzling that a LDC should ever wish to join the CDM today. This is despite the fact that it is under no abatement obligation today, and will have access to even better foreign advanced technologies tomorrow via JI. CDM-

---

[3] For that matter, even for Annex I parties, little has been agreed when it comes to the second commitment period abatement obligations

[4] It is perhaps more sensible to say that a $\sqrt{k}\%$ rule applies to the emissions level at the end of today, corresponding to the $k\%$ rule that applies to the level at the start of today, provied that today and tomorrow are the same length. However, we use the term "$k\%$ rule" rather loosely here

induced improvements in domestic technologies are not only in accordance with the spirit of KP Article 12, but are also highly relevant to incentives which induce LDCs to take part in the CDM today.

A Simple Model of LHF

In this section, we introduce a simple model of the LHF problem, including the background premises examined in the previous section.

The Model

Two Countries (a LDC and a Developed Country) and Two Periods (Before and After the LDC Joins Annex I)

There are two periods, $t = 1, 2$, and two countries, $i = 1, 2$. Country 1 is a LDC, while country 2 is a developed country. Country 1 is a non-Annex I party in period 1, but becomes an Annex I party at the beginning of period 2. That is, country 1 is subject to no GHG emissions abatement obligation whatsoever in period 1, but must achieve a cumulative abatement of 2 units by the end of period 2. On the other hand, country 2 is an Annex I party throughout the two periods.

Condition 1: *Country* 1, *which is a non-Annex I country during period* 1, *is committed to join Annex I at the beginning of period* 2, *and must achieve* 2 *units of GHG emissions abatement by the end of period* 2.

Two Projects in a LDC: HHF and LHF

Country 1 has two GHG emissions abatement projects, j = L, H, each of which is capable of generating 1 unit of abatement. During period 1, country 1 can either carry out projects on its own, or do so jointly with country 2 via the CDM. Likewise, during period 2, country 1 can abate emissions by itself, or jointly with country 2 via JI. In either case, CDM/JI allows country 1 to rely on the more advanced and efficient abatement technologies of country 2. In each period, should it choose to rely on CDM/JI to carry out a project, a fraction $\delta \in (0,1)$ out of 1 unit of abatement will accrue to country 2 in the form of CDM/JI "certified emissions reductions" (CERs), and country 1 must purchase this amount of emissions reduction from the international emissions trade (ET) market to fulfill its obligation at the end of period 2. For simplicity, we assume no uncertainty and zero real interest rate.

Condition 2: *Country* 1 *has two GHG emissions abatement projects*, H, L, *each of which generates* 1 *unit of abatement. It is free to carry out either project in either period, either through the CDM (in period* 1) *or JI (in period* 2) *or on its own. However, if it chooses to rely on CDM/JI, δ units of emissions reduction per project* *are subtracted from its own reduction, and that amount must be replenished through the purchase of an emissions permit at the international emissions trade market at price* p. *There is no uncertainty, and the rate of interest is zero.*

Abatement Cost Specifications

Period 1. In period 1, if country 1 carries out projects L or H by itself, the abatement costs are a_1l, a_1h, respectively, where $h > l$ (> 0).

Condition 3: *If country* 1 *chooses to carry out projects* L *or* H *by itself in period* 1, *the abatement costs are* a_1l, a_1h, *respectively, where* $h > l > 0$.

However, if it does so via the CDM in period 1, the costs are b_1l,b_1h, respectively. We assume that $a_1 > b_1$, meaning that country 2 technologies are more cost-effective than their country 1 counterparts. However, a CDM project will leave country 1 with only $1 - \delta$ units of CERs, adding an extra cost of $p\delta$ per project to compensate for the loss of CERs, where p is the competitive unit price of emissions reductions on the ET market at the end of period 2. Thus, the total costs of a period-1 CDM applied to projects L, H are $b_1l + p\delta$, $b_1h + p\delta$, respectively.

Condition 4: *If country* 1 *chooses to carry out projects* L *or* H *via the CDM in period* 1, *the abatement costs, including the cost of replenishing the CERs paid to country* 2, *are* $b_1l + p\delta$, $b_1h + p\delta$, *respectively, where* $a_1 > b_1$.

Period 2. The Effect of a Period-1 CDM Experience on a LDC's Period 2 Technologies. In period 2, projects that we carried out in period 1 are no longer available. If country 1 (now an Annex I country) chooses to depend on JI (with country 2), the total costs associated with projects L, H (if they are still available) are $b_2l + p\delta$, $b_2h + p\delta$, respectively. We assume that $b_1 > b_2$, which represents the progress of country 2 technologies from period 1 to period 2.

Condition 5: *If country* 1 *chooses to carry out projects* L *or* H *(if they are still available) via JI in period* 2, *the abatement costs, including the cost of replenishing the CERs paid to country* 2, *are* $b_2l + p\delta$, $b_2h + p\delta$, *respectively, where* $b_1 > b_2$.

The abatement costs when country 1 chooses to use its own technologies in period 2 depend on its history during period 1.

On the one hand, if it had chosen not to depend on the CDM, then the costs in period 2 would be a_2l, a_2h for projects L, H, respectively, if they are still available. We assume $a_1 > a_2$ to represent autonomous progress in country 1 technologies.

On the other hand, had it chosen to depend on the CDM in period 1 for project L, then the cost of the remaining project H in period 2 would be reduced to ηa_2h, where $\eta \in (0, 1)$. Likewise, had it used the CDM for project H, the cost of project L would have been reduced to ηa_2l in period 2.

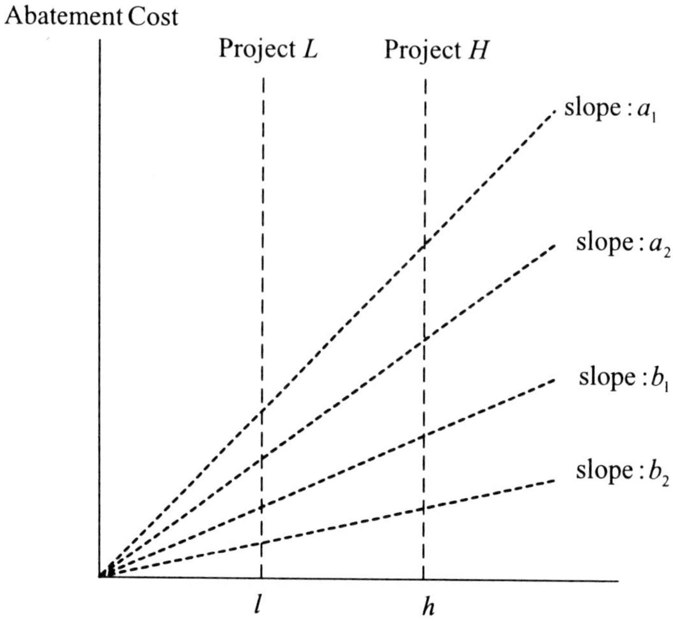

FIG. 1. Abatement cost structure

We assume that $a_2 > b_2$. That is, country 2 technologies in period 2 would dominate country 1 technologies were it not for any cost reductions due to the period 1 CDM experience.

Condition 6: *If country* 1 *chooses to carry out projects* L *or* H *(if they are still available) on its own in period* 2, *the abatement costs are (i)* a_2l, a_2h, *respectively, if it has chosen to carry out projects* H *or* L *in period* 1 *on its own, or (ii)* $\eta a_2 l$, $\eta a_2 h$, *respectively, if it had chosen to carry out projects* H *or* L *in period* 1 *via the CDM, where* $a_1 > a_2 > b_2$, $0 < \eta < 1$.

Remark 1: *These assumptions about the abatement costs are depicted in Fig. 1, except that we do not assume* $a_2 > b_1$ *a priori.*

Solving the Model

By assumption, country 1 is free to carry out either project in either period, and either through CDM/JI or on its own. Thus, if it chooses to carry out both projects on its own, there are four possible ways of doing it.[5] Likewise, if it chooses to carry out both projects via the CDM and/or JI, there are also four possible

[5] Project H may be placed in either period 1 or period 2. Project L may be placed in either period 1 or period 2 independently of the placement of project H

cases,[6] and if it chooses to carry out one project on its own and the other via the CDM/JI, there are eight possible cases.[7]

Thus, in total there exist 16 possible alternative courses of action from which country 1 can choose in order to minimize its costs. However, it turns out that it needs to consider only the following five options out of the total of 16 possible options, for the other options are consistently dominated by at least one of those five options.

Lemma 1: *The optimal choice by country* 1 *is one of the following five candidate cases.*[8]

(1) Carrying out both projects on its own in period 2, *in which case the cost is*

$$c_{\mathrm{SB}} \equiv a_2 l + a_2 h$$

(2) Carrying out both projects via JI in period 2, *in which case the cost is*

$$c_{\mathrm{JB}} \equiv (b_2 l + p\delta) + (b_2 h + p\delta)$$

(3) Carrying out project L *in period* 1 *on its own, and carry out project* H *in period* 2 *via JI, in which case the cost is*

$$c_{\mathrm{JH}} \equiv a_1 l + (b_2 h + p\delta)$$

(4) Carrying out project L *in period* 1 *via the CDM, and carrying out project* H *in period* 2 *on its own, in which case the cost is*

$$c_{\mathrm{CL}} \equiv (b_1 l + p\delta) + \eta a_2 h$$

(5) Carrying out project H *in period* 1 *via the CDM, and carry out project* L *in period* 2 *on its own, in which case the cost is*

$$c_{\mathrm{CH}} \equiv (b_1 h + p\delta) + \eta a_2 l$$

In particular, it is never optimal to carry out project H *on its own in period* 1, *and project* L *via JI in period* 2.

For the proof see Appendix A.

Comparing c_{SB}, c_{JB}, c_{JH}, c_{CL}, and c_{CH}

We now compare the abatement costs c_{SB}, c_{JB}, c_{JH}, c_{CL}, and c_{CH} associated with the above five candidate policies. For this purpose, we must make a total of ten pair-wise comparisons.

---

[6] Project H may be placed in either period 1 (CDM) or period 2 (JI). Project L may be placed in either either period 1 (CDM) or period 2 (JI), independently of the placement of project H

[7] Project H may be placed in either period 1 or period 2. Project L may be placed in either period 1 or period 2, independently of the placement of project H. So there are four possible ways to place H, L, for each of which there are two cases depending on which one is conducted via the CDM/JI

[8] The indices attached to each cost, SB, JB, JH, CL, CH, stand for self–both, JI–both, JI–HHF, CDM–LHF, CDM–HHF, respectively

Lemma 2: *Comparing c_{SB}, c_{JB}, c_{JH}, c_{CL}, and c_{CH}, we find that*

$$c_{CL} \gtrless c_{CH} \Leftrightarrow \eta \gtrless \frac{b_1}{a_2} \equiv \eta_{CL,CH}$$

$$c_{CL} \gtrless c_{JB} \Leftrightarrow \eta \gtrless \frac{b_2}{a_2} + \frac{p\delta - (b_1 - b_2)l}{a_2 h} \equiv \eta_{CL,JB}$$

$$c_{CH} \gtrless c_{JB} \Leftrightarrow \eta \gtrless \frac{b_2}{a_2} + \frac{p\delta - (b_1 - b_2)h}{a_2 l} \equiv \eta_{CH,JB}$$

$$c_{JB} \gtrless c_{SB} \Leftrightarrow p\delta \gtrless (a_2 - b_2)\frac{h+l}{2}(>0) \equiv p\delta_{JB,SB}$$

$$c_{CL} \gtrless c_{SB} \Leftrightarrow p\delta \gtrless (a_2 - b_1)l + (1-\eta)a_2 h \equiv p\delta_{CL,SB}$$

$$c_{CH} \gtrless c_{SB} \Leftrightarrow p\delta \gtrless (a_2 - b_1)h + (1-\eta)a_2 l \equiv p\delta_{CH,SB}$$

$$c_{JH} \gtrless c_{SB} \Leftrightarrow p\delta \gtrless (a_2 - b_2)h \equiv p\delta_{JH,SB}$$

$$c_{JH} \gtrless c_{JB} \Leftrightarrow p\delta_{JH,JB} \equiv (a_2 - b_2)l \gtrless p\delta$$

$$c_{JH} \gtrless c_{CL} \Leftrightarrow \eta_{JH,CL} \equiv \frac{1}{a_2}\left[(a_2 - b_1)\frac{l}{h} + b_2\right] \gtrless \eta$$

$$c_{JH} \gtrless c_{CH} \Leftrightarrow \eta_{JH,CH} \equiv \frac{1}{a_2}\left[a_2 + (b_2 - b_1)\frac{h}{l}\right] \gtrless \eta$$

For the proof see Appendix B.

We deal with the above ten inequalities in steps.

Comparing $\eta_{CL,CH}$, $\eta_{CL,JB}$, and $\eta_{CH,JB}$

First, we shall deal with the first set of inequalities

$$c_{CL} \gtrless c_{CH} \Leftrightarrow \eta \gtrless \frac{b_1}{a_2} \equiv \eta_{CL,CH}$$

$$c_{CL} \gtrless c_{JB} \Leftrightarrow \eta \gtrless \frac{b_2}{a_2} + \frac{p\delta - (b_1 - b_2)l}{a_2 h} \equiv \eta_{CL,JB}$$

$$c_{CH} \gtrless c_{JB} \Leftrightarrow \eta \gtrless \frac{b_2}{a_2} + \frac{p\delta - (b_1 - b_2)h}{a_2 l} \equiv \eta_{CH,JB}$$

Comparing the three threshold levels of η, namely $\eta_{CL,CH}$, $\eta_{CL,JB}$, and $\eta_{CH,JB}$, we make the following observation.

Lemma 3: *In the relationship among the three threshold levels, $\eta_{CL,CH}$, $\eta_{CL,JB}$, and $\eta_{CH,JB}$, there exist the following two possible cases depending on the level of $p\delta$.*

$$(i)\ p\delta > (b_1 - b_2)(h+l) \equiv p\delta_{CL,CH,JB}$$

$$\Leftrightarrow \eta_{CL,CH} < \eta_{CL,JB} < \eta_{CH,JB}$$

$$(ii)\ p\delta < (b_1 - b_2)(h+l) \equiv p\delta_{CL,CH,JB}$$

$$\Leftrightarrow \eta_{CH,JB} < \eta_{CL,JB} < \eta_{CL,CH}$$

Furthermore, we find that

$$c_{\mathrm{CL}} = c_{\mathrm{CH}} = c_{\mathrm{JB}} \Leftrightarrow \begin{cases} p\delta = (b_1 - b_2)(h+l) \equiv p\delta_{\mathrm{CL,CH,JB}} \\ \eta = \dfrac{b_1}{a_2} \equiv \eta_{\mathrm{CL,CH}} \end{cases}$$

For the proof see Appendix C.

Remark 2: *The relationship among* $\eta_{\mathrm{CL,CH}}$, $\eta_{\mathrm{CL,JB}}$, *and* $\eta_{\mathrm{CH,JB}}$ *in* $(\eta, p\delta)$ *space is illustrated in Fig. 2. The entire* $(\eta, p\delta)$ *space is divided into three regions in which min* $(c_{\mathrm{CL}}, c_{\mathrm{CH}}, c_{\mathrm{JB}})$ *is* c_{CL}, c_{CH}, *and* c_{JB}, *respectively. The point of intersection* $(\eta_{\mathrm{CL,CH}}, p\delta_{\mathrm{CL,CH,JB}})$ *is where we find* $c_{\mathrm{CL}} = c_{\mathrm{CH}} = c_{\mathrm{JB}}$.

Comparing c_{CL}, c_{CH}, c_{JB}, and c_{SB}

Second, we consider c_{SB} and deal with the second set of inequalities

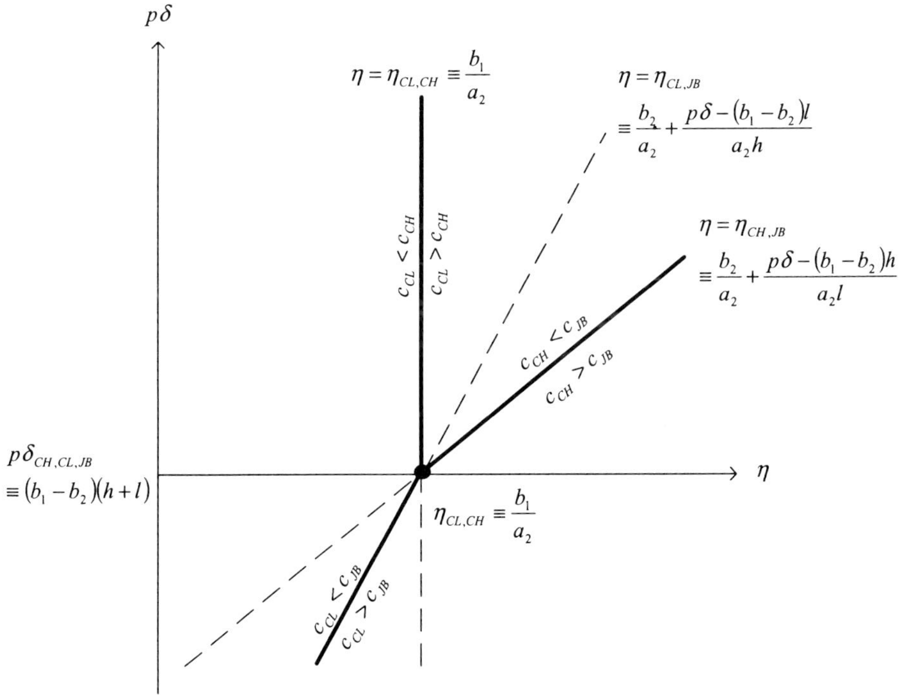

FIG. 2. C_{JB}, C_{CH}, and C_{CL}

$$c_{JB} \gtreqless c_{SB} \Leftrightarrow p\delta \gtreqless (a_2 - b_2)\frac{h+l}{2}(>0) \equiv p\delta_{JB,SB}$$
$$c_{CL} \gtreqless c_{SB} \Leftrightarrow p\delta \gtreqless (a_2 - b_1)l + (1-\eta)a_2 h \equiv p\delta_{CL,SB}$$
$$c_{CH} \gtreqless c_{SB} \Leftrightarrow p\delta \gtreqless (a_2 - b_1)h + (1-\eta)a_2 l \equiv p\delta_{CH,SB}$$

In the $(\eta, p\delta)$ space, it is convenient to identify three reference points.

Definition 1: *The following three reference points in the $(\eta, p\delta)$ space, namely $(\eta_{CL,CH}, p\delta_{SB,CH,CL})$, $(\eta_{CH,JB,SB}, p\delta_{SB,JB})$, and $(\eta_{CL,JB,SB}, p\delta_{SB,JB})$, are defined by the conditions $c_{SB} = c_{CL} = c_{CH}$, $c_{SB} = c_{JB} = c_{CH}$, and $c_{SB} = c_{JB} = c_{CL}$, respectively.*

$$c_{SB} = c_{CL} = c_{CH} \Leftrightarrow \begin{cases} \eta = \dfrac{b_1}{b_2} = \eta_{CL,CH} \\ p\delta = (a_2 - b_1)(l+h) \equiv p\delta_{SB,CH,CL} \end{cases}$$

$$c_{SB} = c_{JB} = c_{CH} \Leftrightarrow \begin{cases} \eta = \dfrac{1}{2a_2 l}[(a_2+b_2)l + (a_2+b_2-2b_1)h] \equiv \eta_{CH,JB,SB} \\ p\delta = (a_2 - b_2)\dfrac{h+l}{2} \equiv p\delta_{SB,JB} \end{cases}$$

$$c_{SB} = c_{JB} = c_{CL} \Leftrightarrow \begin{cases} \eta = \dfrac{1}{2a_2 h}[(a_2+b_2)h + (a_2+b_2-2b_1)l] \equiv \eta_{CL,JB,SB} \\ p\delta = (a_2 - b_2)\dfrac{h+l}{2} \equiv p\delta_{SB,JB} \end{cases}$$

Comparing these three reference points with the point $(\eta_{CL,CH}, p\delta_{CL,CH,JB})$, we make the following observation.

Lemma 4: *Comparing each of the three reference points $(\eta_{CL,CH}, p\delta_{SB,CH,CL})$, $(\eta_{CH,JB,SB}, p\delta_{SB,J\ B})$, and $(\eta_{CL,JB,SB}, p\delta_{SB,JB})$ with the point $(\eta_{CL,CH}, p\delta_{CL,CH,JB})$, we find that two possible cases exist.*

(i) If $a_2 + b_2 - 2b_1 > 0$, then

$$\eta_{CL,CH} < \eta_{CL,JB,SB} < \eta_{CH,JB,SB}$$
$$p\delta_{CL,CH,JB} < p\delta_{SB,JB} < p\delta_{SB,CH,CL}$$

(ii) If $a_2 + b_2 - 2b_1 < 0$, then

$$\eta_{CL,CH} > h_{CL,JB,SB} > h_{CH,JB,SB}$$
$$p\delta_{CL,CH,JB} > p\delta_{SB,JB} > p\delta_{SB,CH,CL}$$

Furthermore, we find that

$$\begin{aligned} & p\delta_{\mathrm{SB,JB}} - p\delta_{\mathrm{CL,CH,JB}} \\ & = p\delta_{\mathrm{SB,CH,CL}} - p\delta_{\mathrm{SB,JB}} \\ & = \frac{1}{2}(a_2 + b_2 - 2b_1)(l + h) \end{aligned}$$

For the proof see Appendix D.

Remark 3: *The two cases above are depicted in Fig. 3 (when $a_2 + b_2 - 2b_1 > 0$) and Fig. 4 (when $a_2 + b_2 - 2b_1 < 0$). When $a_2 + b_2 - 2b_1 > 0$, the $(\eta, p\delta)$ space is divided into four regions in which* min *($c_{\mathrm{CL}}, c_{\mathrm{CH}}, c_{\mathrm{JB}}$, and c_{SB}) are $c_{\mathrm{CL}}, c_{\mathrm{CH}}, c_{\mathrm{JB}}$, and c_{SB}, respectively. It is only in the central triangular region where c_{CH} is a minimum. Conversely, when $a_2 + b_2 - 2b_1 < 0$, the $(\eta, p\delta)$ space is divided into three regions in which* min *($c_{\mathrm{CL}}, c_{\mathrm{CH}}, c_{\mathrm{JB}}$, and c_{SB}) are $c_{\mathrm{CL}}, c_{\mathrm{JB}}$, and c_{SB}, respectively. In this case, c_{CH} is never a minimum.*

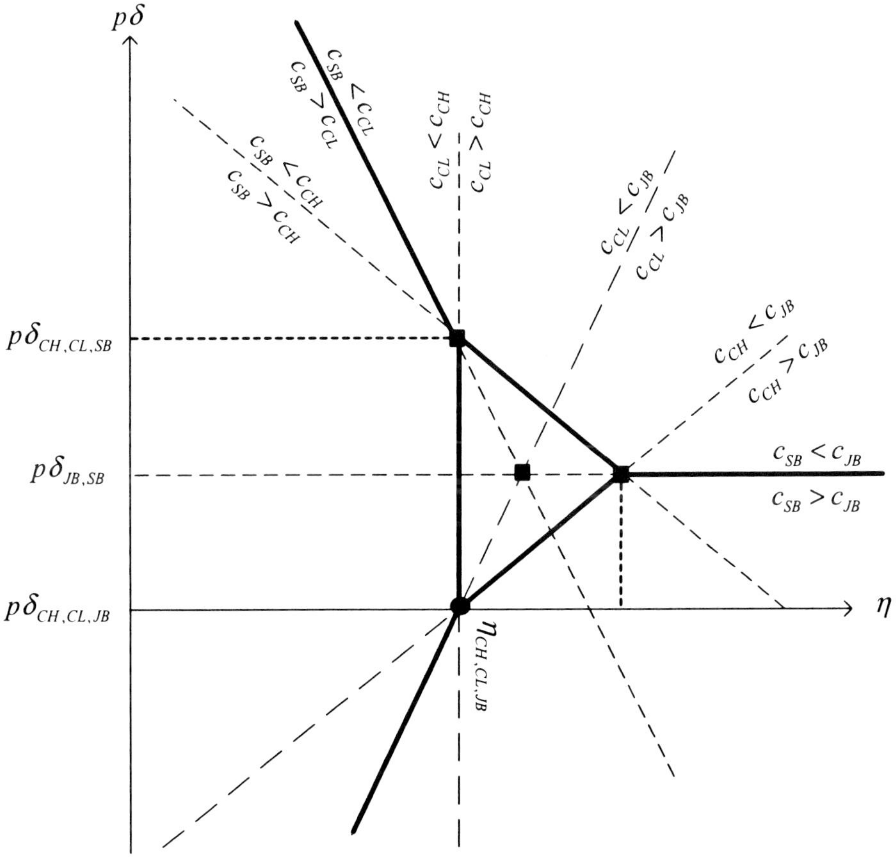

FIG. 3. C_{SB}, C_{JB}, C_{CH}, and C_{CL} when $\dfrac{a_2 + b_2}{2} > b_1$

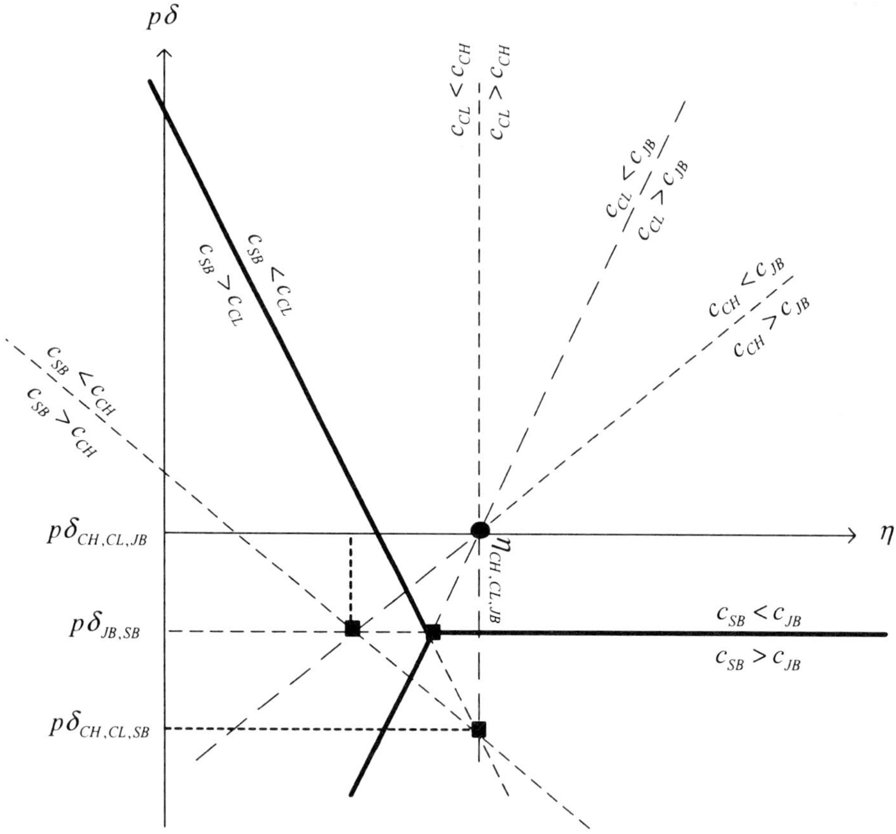

FIG. 4. C_{SB}, C_{JB}, C_{CH}, and C_{CL} when $\frac{a_2+b_2}{2} < b_1$

Comparing c_{CL}, c_{CH}, c_{JB}, c_{SB}, and c_{JH}

Third, we consider c_{JH} and deal with the last set of inequalities

$$c_{\mathrm{JH}} \gtrless c_{\mathrm{SB}} \Leftrightarrow p\delta \gtrless (a_2 - b_2)h \equiv p\delta_{\mathrm{JH,SB}}$$
$$c_{\mathrm{JH}} \gtrless c_{\mathrm{JB}} \Leftrightarrow p\delta_{\mathrm{JH,JB}} \equiv (a_2 - b_2)l \gtrless p\delta$$
$$c_{\mathrm{JH}} \gtrless c_{\mathrm{CL}} \Leftrightarrow \eta_{\mathrm{JH,CL}} \equiv \frac{1}{a_2}\left[(a_2 - b_1)\frac{l}{h} + b_2\right] \gtrless \eta$$
$$c_{\mathrm{JH}} \gtrless c_{\mathrm{CH}} \Leftrightarrow \eta_{\mathrm{JH,CH}} \equiv \frac{1}{a_2}\left[a_2 + (b_2 - b_1)\frac{h}{l}\right] \gtrless \eta$$

The two threshold levels $\eta_{\mathrm{JH,CL}}$, $\eta_{\mathrm{JH,CH}}$ that appear in the last two inequalities are first compared with $\eta_{\mathrm{CL,CH}}$.

Lemma 5: *By comparing the two threshold levels $\eta_{\mathrm{JH,CL}}$, $\eta_{\mathrm{JH,CH}}$ with $\eta_{\mathrm{JH,CL}}$, we find that two possible cases exist.*

(1) If $(b_2 - b_1)\,h + (a_2 - b_1)\,l > 0$, *then*

$$\eta_{CL,CH} < \eta_{JH,CL} < \eta_{JH,CH}$$

(2) If $(b_2 - b_1)\,h + (a_2 - b_1)\,l < 0$, *then*

$$\eta_{JH,CH} < \eta_{JH,CL} < \eta_{CL,CH}$$

Note that if $\frac{b_1}{a_2} > 1$, *then we necessarily have* $(b_2 - b_1)\,h + (a_2 - b_1)\,l < 0$.

For the proof see Appendix E.

Next, we compare $\eta_{JH,CL}$ with $\eta_{CH,JB,SB}$ and $\eta_{CL,JB,SB}$.

Lemma 6: *Comparing* $\eta_{JH,CL}$ *with* $\eta_{CH,JB,SB}$, *we find that*

$$\eta_{JH,CL} - \eta_{CH,JB,SB} = \frac{1}{2}\left[\frac{(2b_1 - b_2 - a_2)h}{a_2 l} + 2\frac{(a_2 - b_1)l}{a_2 h} + \frac{b_2 - a_2}{a_2}\right]$$

where the sign of the expression is generally ambiguous since the signs of $2b_1 - b_2 - a_2$, $a_2 - b_1$ *are ambiguous, while* $b_2 - a_2 < 0$. *However, note that*

$$\frac{a_2 + b_2}{2} > b_1, \frac{b_1}{a_2} > 1 \Rightarrow \eta_{JH,CL} < \eta_{CH,JB,SB}$$

Comparing $\eta_{JH,CL}$ *with* $\eta_{CL,JB,SB}$, *we find that*

$$\eta_{JH,CL} - \eta_{CL,JB,SB} = \frac{1}{2a_2 h}(b_2 - a_2)(h - l) < 0$$

unambiguously since $b_2 - a_2 < 0$.

For the proof see Appendix F.

Likewise, we compare $\eta_{JH,CH}$ with $\eta_{CH,JB,SB}$ and $\eta_{CL,JB,SB}$.

Lemma 7: *Co mparing* $\eta_{JH,CH}$ *with* $\eta_{CH,JB,SB}$, *we find that*

$$\eta_{JH,CH} - \eta_{CH,JB,SB} = \frac{1}{2a_2 l}(b_2 - a_2)(h - l) < 0$$

unambiguously since $b_2 - a_2 < 0$.

Comparing $\eta_{JH,CH}$ *with* $\eta_{CL,JB,SB}$, *we find that*

$$\eta_{JH,CH} - \eta_{CL,JB,SB} = \frac{1}{2}\left[\frac{(2b_1 - a_2 - b_2)l}{a_2 h} + 2\frac{(b_2 - b_1)h}{a_2 l} + \frac{a_2 - b_2}{a_2}\right]$$

where the sign of the expression is generally ambiguous since the sign of $2a_1 - a_2 - b_2$ *is ambiguous, and* $a_2 - b_2 > 0$, $b_2 - b_1 < 0$.

For the proof see Appendix G.

The Optimal Choice by Country 1

Putting together the above results, we put forward the following proposition regarding the optimal choice by country 1. Depending on the parameter configurations, we find six distinct cases, depicted by Figs. 5–10, where the $(\eta, p\delta)$ space is divided into regions indicating which of the five previously identified candidate policies is the optimal one.

Proposition 1: *The optimal choice by country* 1 *depends on the configuration of* $(\eta, p\delta)$ *as described by Figs. 5–10, which correspond to six distinct possible cases that can take place depending on the parameter of the model. In each figure, the* $(\eta, p\delta)$ *space is divided into regions indicating which of* c_{SB}, c_{JB}, c_{HJ}, c_{CL}, *or* c_{CH} *yields the minimum cost.*

(i) When $(b_2 - b_1)\,h + (a_2 - b_1)\,l > 0$ *(thus when* $\frac{b_1}{a_2} < 1$*), we have* $\eta_{CL,CH} < \eta_{JH,CL} < \eta_{JH,CH}$*, and the following possible cases.*

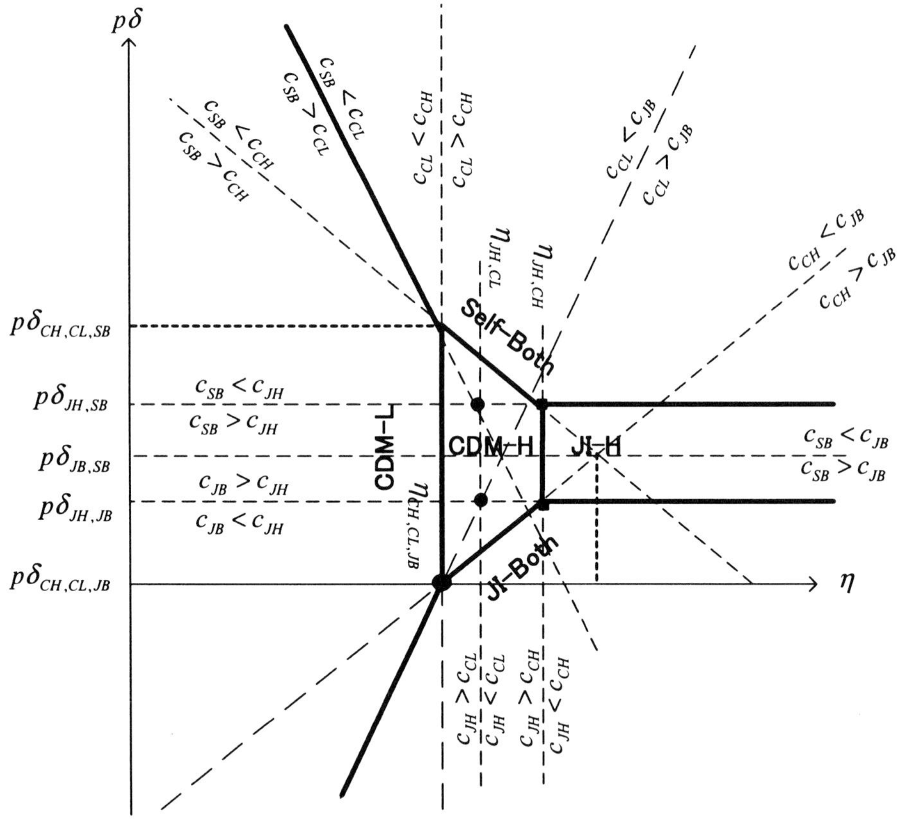

FIG. 5. C_{SB}, C_{JB}, C_{CH}, C_{CL}, and C_{JH} when $\frac{a_2 + b_2}{2} > b_1$. *CDM*, clean development mechanism; *JI*, joint implementation

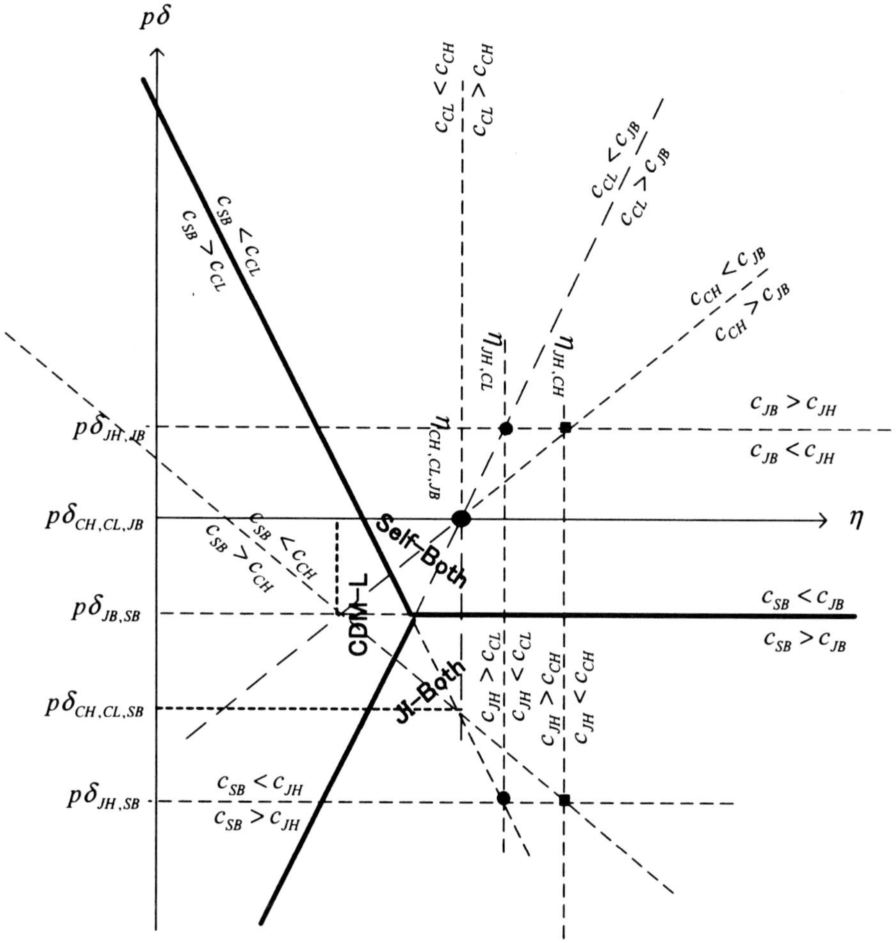

FIG. 6. C_{SB}, C_{JB}, C_{CH}, C_{CL}, and C_{JH} when $\frac{a_2+b_2}{2} < b_1$

	$\frac{a_2+b_2}{2} > 0$	$\frac{a_2+b_2}{2} < 0$
$\eta_{CL,CH} < \eta_{JH,CL} < \eta_{JH,CH} < \eta_{CH,JB,SB}$	*Fig.* 5	*Fig.* 6

(ii) When $(b_2 - b_1)\,h + (a_2 - b_1)\,l > 0$, *we have* $\eta_{JH,CH} < \eta_{JH,CL} < \eta_{CL,CH}$, *and the following possible cases.*

	$\frac{a_2+b_2}{2} > 0$	$\frac{a_2+b_2}{2} < 0$
$\eta_{JH,CL} < \eta_{CL,JB,SB} < \eta_{CL,CH}$	*Fig.* 7	*Fig.* 8
$\eta_{CL,JB,SB} < \eta_{JH,CL} < \eta_{CL,CH}$	*Fig.* 9	*Fig.* 10

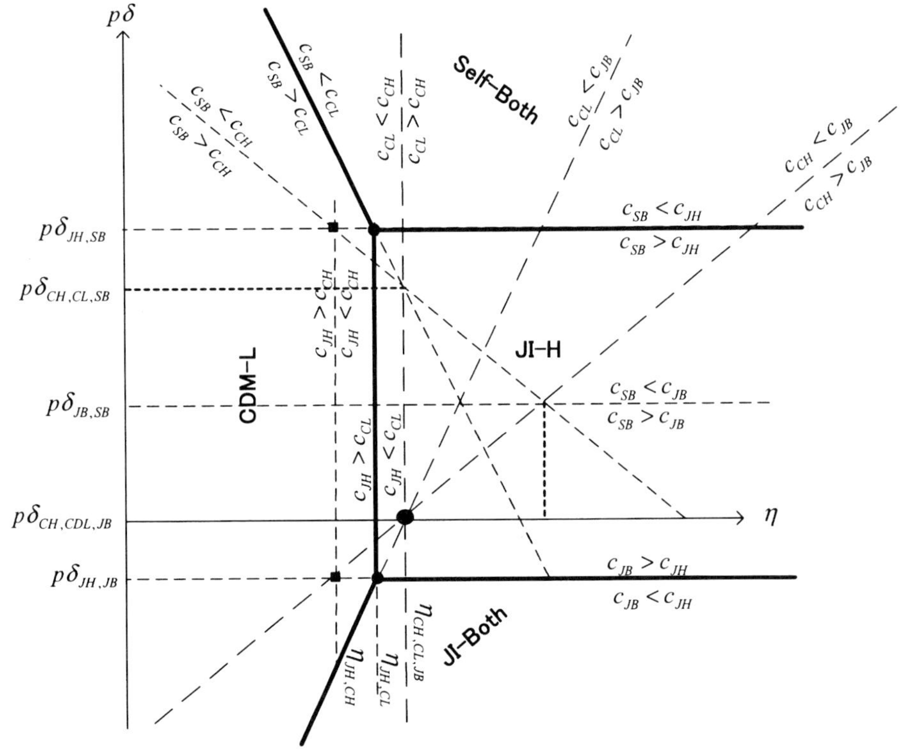

FIG. 7. C_{SB}, C_{JB}, C_{CH}, C_{CL}, and C_{JH} when $\frac{a_2 + b_2}{2} > b_1$

Remark 4: *In all figures, note that the* $c_{JB} = c_{JH}$ *line must pass through (i) the intersection of the* $c_{JH} = c_{CH}$ *line and the* $c_{CH} = c_{JB}$ *line, and (ii) the intersection of the* $c_{JH} = c_{CL}$ *line and the* $c_{CL} = c_{JB}$ *line. Likewise, the* $c_{SB} = c_{JH}$ *line must pass through (i) the intersection of the* $c_{JH} = c_{CH}$ *line and the* $c_{CH} = c_{SB}$ *line, and (ii) the intersection of the* $c_{JH} = c_{CL}$ *line and the* $c_{CL} = c_{SB}$ *line.*

The Case for the CDM–LHF Argument

For the low-hanging fruit argument to make sense, it is necessary and sufficient that

$$\min(c_{CL}, c_{CH}, c_{JB}, c_{JH}, c_{SB}) = c_{CH}$$

Otherwise, country 1 is free either to stay way from the CDM and/or JI, or willing to subject project L to the CDM. Hence, there is no case for the low-hanging fruit argument.

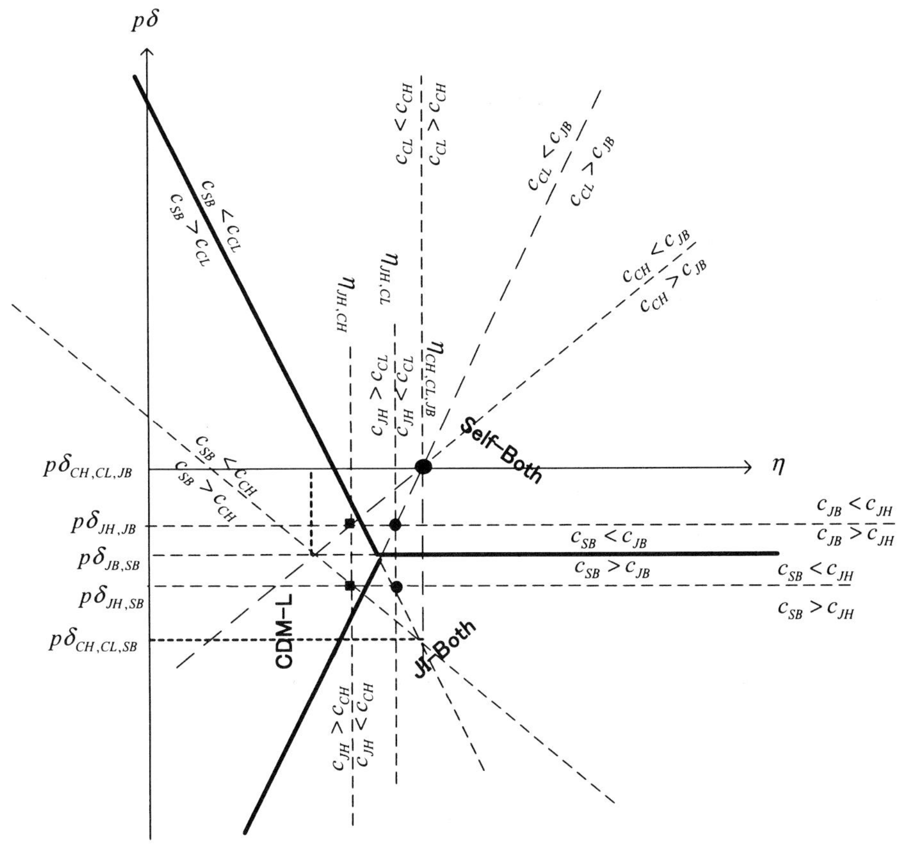

FIG. 8. C_{SB}, C_{JB}, C_{CH}, C_{CL}, and C_{JH} when $\frac{a_2 + b_2}{2} < b_1$

Necessary Conditions for the LHF Argument

From Figs. 5–10, we note that is only in Figs. 5 and 7 that we have regions where c_{CH} is the minimal cost. We first identify a few necessary conditions for these cases to be possible.

Proposition 2: *For* min $(c_{CL}, c_{CH}, c_{JB}, c_{JH}, c_{SB}) = c_{CH}$ to *hold, each of the following conditions is necessary.*

$$\frac{a_2 + b_2}{2} > b_1 \tag{1}$$

That is, the abatement costs of country 2 *(investor) in period* 1 *must be higher than the world average cost in period* 2*. Otherwise, it would be better for country* 1 *to carry out both projects on its own in period* 2*.*

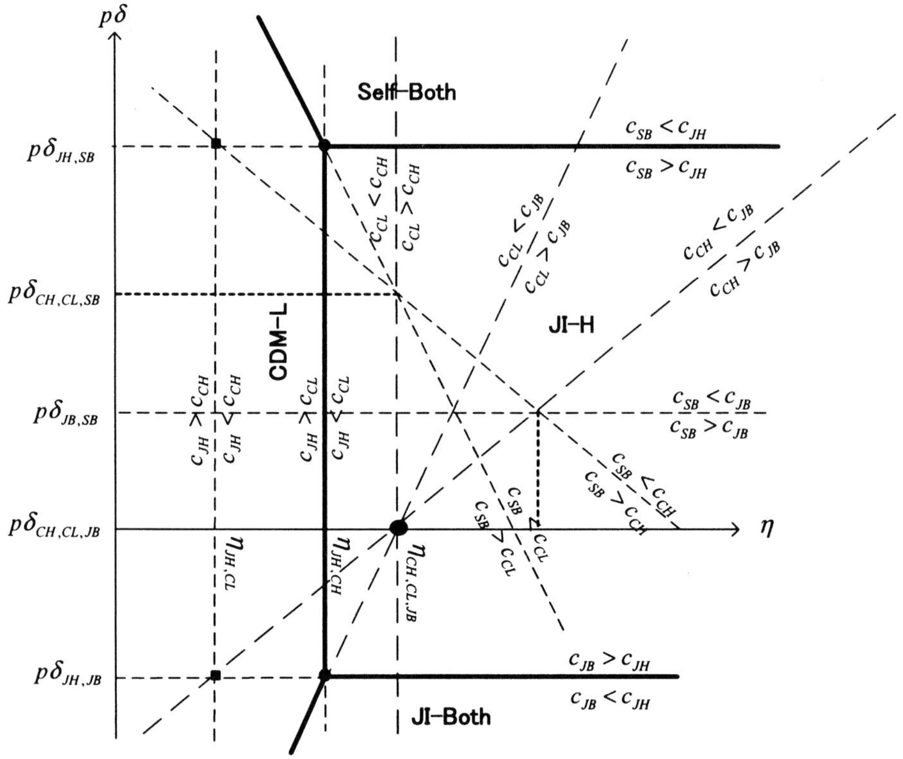

FIG. 9. C_{SB}, C_{JB}, C_{CH}, C_{CL}, and C_{JH} when $\frac{a_2 + b_2}{2} > b_1$

$$\frac{b_1}{a_2} = \eta_{CL,CH} < 1 \tag{2}$$

That is, the abatement technology of country 2 *(investor) in period* 1 *must be better than that of country* 1 *(host) in period* 2*. Otherwise, the CDM_*H *region becomes irrelevant.*

$$(b_2 - b_1)h + (a_2 - b_1)l > 0, \qquad \text{i.e.,} \frac{a_2 - b_1}{b_1 - b_2} > \frac{h}{l} > 0 \tag{3}$$

where $b_1 - b_2 > 0$*. For this to be possible, it is necessary that the above condition* $b_1/a_2 \equiv \eta_{CL,CH} < 1$ *holds. Given that condition, the requirement is that the cost gap between project* H *and project* L *should not be too large. Otherwise, country* 1 *is better off subjecting project* H *to* JI *in period* 2.

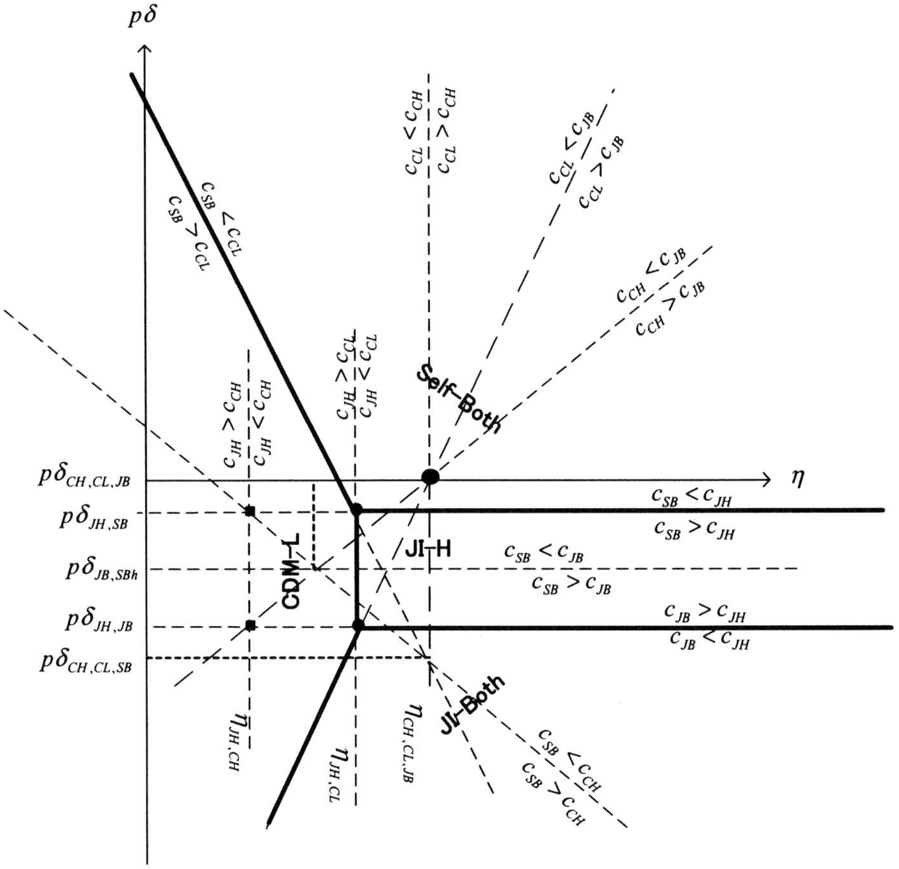

FIG. 10. C_{SB}, C_{JB}, C_{CH}, C_{CL}, and C_{JH} when $\dfrac{a_2 + b_2}{2} < b_1$

Necessary and Sufficient Conditions for the LHF Argument

Given that the necessary conditions described above are satisfied, it is necessary and sufficient for c_{CH} to be the minimum cost that $(\eta, p\delta)$ is located within the triangular region in Fig. 5, or in the trapezoidal region in Fig. 7.

Theorem 1: *Assume that*

$$\frac{b_1}{a_2} \equiv \eta_{\mathrm{CL,CH}} < 1, \qquad \frac{a_2 - b_1}{b_1 - b_2} > \frac{h}{l} > 0, \qquad \frac{a_2 + b_2}{2} > b_1$$

Then, for $\min(c_{\mathrm{CL}}, c_{\mathrm{CH}}, c_{\mathrm{JB}}, c_{\mathrm{JH}}, c_{\mathrm{SB}}) = c_{\mathrm{CH}}$ *to hold, it is necessary and sufficient that* $(\eta, p\delta)$ *satisfies*

$$\eta_{\mathrm{CL,CH}} \le \eta \le \min(\eta_{\mathrm{CH,JB}}, \eta_{\mathrm{CH,JH}})$$

$$p\delta_{\mathrm{CL,CH,JB}} \le p\delta \le (a_2 - b_1)h + (1-\eta)a_2 l$$

where

$$\eta_{\mathrm{CL,CH}} \equiv \frac{b_1}{a_2}$$

$$\eta_{\mathrm{CH,JB}} \equiv \frac{b_2}{a_2} + \frac{p\delta - (b_1 - b_2)h}{a_2 l}$$

$$\eta_{\mathrm{CH,JH}} \equiv \frac{1}{a_2}\left[a_2 + (b_2 - b_1)\frac{h}{l}\right]$$

$$p\delta_{\mathrm{CL,CH,JB}} \equiv (b_1 - b_2)(h + l)$$

Intuitions

The above theorem claims that for the LHF argument, i.e.,

$$\min(c_{\mathrm{CL}}, c_{\mathrm{CH}}, c_{\mathrm{JB}}, c_{\mathrm{JH}}, c_{\mathrm{SB}}) = c_{\mathrm{CH}}$$

to hold, both η and $p\delta$ must assume somewhat intermediate values. We now attempt to provide an intuitive account of this.

First, let us recall what η and $p\delta$ stand for. For $\eta \in (0, 1)$, note that a higher η implies smaller benefits from the period 1 CDM experience of the domestic technology of country 1 in period 2. For $p\delta$, note that $p\delta$ is the cost of buying back δ units of emissions that drains to country 2 in the form of CERs when country 1 resorts to CDM/JI.

Second, also recall that in the CDM_H scenario, country 1 uses the CDM for project H in period 1, but carries out project L on its own in period 2.

When $p\delta$ is too High. When $p\delta$ is too high, then c_{CH} becomes larger than c_{SB}. Since $p\delta$ is the cost of using the CDM in period 1 (aside from the benefit of the CDM experience), this is readily understandable.

When $p\delta$ is too Low. When $p\delta$ is too low, then c_{CH} becomes larger than c_{JB}. In the JI scenario, country 1 carries out both projects via JI in period 2. Conversely, the CDM_H scenario leaves only project H for the CDM. Hence, the former pays the cost $p\delta$ of using the CDM/JI twice, while the latter pays it only once. Thus, a smaller $p\delta$, ceteris paribus, tends to make the former relatively less expensive.

When η is too High. When η is too high, c_{CH} becomes larger than c_{JB}, c_{SB}, or c_{JH}. The only reason why it may ever be optimal for country 1 to join the CDM in period 1 is the beneficial effect of the CDM experience on the domestic technology in period 2. Thus, too high an η, i.e., too small a benefit, will make the CDM_H scenario suboptimal.

When η is too Low. When η is too low, c_{CH} becomes larger than c_{CL}. In choosing between the CDM_H scenario and the CDM_L scenario, the key consideration of country 1 is which project to use for to the CDM in period 1, and which project to carry out on its own in period 2. If η is not sufficiently low, then the CDM experience will not improve the domestic technology in period 2 by very much, and thus it is better to use the foreign (more cost-effective) technology for project H, which is a more costly project. Conversely, if η is small enough, then the improved domestic technology ηa_2 in period 2 should be used for the more costly project H in period 2, so that it is better for country 1 to leave the less costly project L to the CDM in period 1.

Conclusion

We have examined the so-called CDM low-hanging fruit (LHF) argument, which has frequently been used against the CDM under the Kyoto Protocol. The LHF argument suggests that it is suboptimal for a LDC to give its LHF to the CDM today, and thus that it should preserve them until tomorrow when it assumes its own GHG emissions abatement obligations. In order to check the validity of this argument, we needed to model the situation in which a LDC has complete freedom to choose which fruit to harvest in which way and when.

The simple analytical model constructed in this chapter demonstrates that there is indeed a case for this LHF argument, but not without qualifications. In particular, we found that both the cost of employing foreign technology and the extent of home technology improvements induced by the CDM experience must be at intermediate levels for the LHF argument to hold. Otherwise, it is in the interests of the host country (i) to stay away from the CDM/JI completely, (ii) to subject both LHF and HHF to JI, or (iii) to subject only LHF to JI.

Finally, we take note of a few of the limitations of the present analysis. First, this model presumes a specific form of GHG emissions abatement obligations to be imposed on a LDC when it shifts to Annex I status in the second period. While it can be argued that this form of obligation, or something similar, would be necessary for the LHF argument to make sense at all, it still remains clear that the presumed setting is only hypothetical at this stage. The present analysis is contingent on the current state of affairs, and may require significant modifications in accordance with the future development of negotiations on the Kyoto Protocol. Second, in order to keep the model as simple as possible, this model treats some economic parameters as exogenously given from the viewpoint of a LDC–CDM host country, while in fact they may well be deemed endogenous. In particular, the CDM–CERs share parameter δ is presumably determined via negotiations and/or bargaining between the host country and the investing country. Our analysis shows how this share parameter may affect the incentives for LDCs to join the CDM today, while ruling out any potential strategic interactions involving the terms of the CDM. Overall, this analysis is only an initial attempt to state the LHF question in an analytically comprehensible manner. Adding more realistic features is left for future research.

Acknowledgments. The author thanks Prof. Hauro Imai of the Institute of Economic Research, Kyoto University, and Prof. Hidenori Niizawa of the Institute of Economic Research, Kobe University of Commerce, for their valuable comments and suggestions. However, the author is solely responsible for any residual erros.

Appendix A. Proof of Lemma 1

Proof: There are three possibilities. First, country 1 may choose not to use the CDM (in period 1) or JI (in period 2) at all, and carry out both projects on its own. Second, country 1 may choose not to carry out either project on its own, and depend entirely on either the CDM or JI. Third, country 1 may choose to partially depend on the CDM and/or JI.

In the first case where country 1 is to carry out both projects on its own, it should carry out both projects in period 2, on its own, with a cost of $c_{SB} \equiv a_2h + a_2l$. Carrying out either project in period 1 will be more costly, since $a_1 > a_2$.

In the second case where country 1 is to depend entirely on the CDM and/or JI, it should carry out both projects in period 2 via JI with a cost of $c_{JB} \equiv (b_2l + p\delta) + (b_2h + p\delta)$. Since it is not to use domestic technology, the benefit of the CDM experience is irrelevant. Then carrying out either project in period 1 is more costly because $b_1 > b_2$.

In the third case where country 1 partly depends on the CDM/JI, there are eight possible policies.

TABLE A1. The eight possible policies in the third case

	Period 1	Period 2	
Policy 1	Project H for the CDM	Project L on its own	*
Policy 2	Project L for the CDM	Project H on its own	*
Policy 3	Project H on its own	Project L for JI	8
Policy 4	Project L on its own	Project H for JI	7
Policy 5	Project H for the CDM Project L on its own	None	1
Policy 6	Project L for the CDM Project H on its own	None	2
Policy 7	None	Project H for JI Project L on its own	*
Policy 8	None	Project L for JI Project H on its own	7

However, policy 3 is clearly dominated by policy 8, and policy 4 by policy 9, since $a_1 > a_2$. Likewise, policy 5 is dominated by policy 1, and policy 6 by policy 2 because $a_1 > a_2$ and $\eta \le 1$. Finally, policy 8 is dominated by policy 7 because the cost of the former is $a_2h + (b_2l + p\delta)$, which is unambiguously larger than the cost of the latter $c_{JH} \equiv a_2l + (b_2h + p\delta)$ because $a_2 > b_2$. ■

Appendix B. Proof of Lemma 2

Proof: Comparing c_{CL} and c_{CH}, we find that

$$\begin{aligned} & c_{CH} - c_{CL} \\ & \quad = \{(b_1h + p\delta) + \eta a_2 l\} - \{(b_1 l + p\delta) + \eta a_2 h\} \\ & \quad = (b_1 - \eta a_2)(h - l) \end{aligned}$$

Since $h - l > 0$ by assumption, we obtain

$$c_{CH} > c_{CL} \Leftrightarrow \eta < \frac{b_1}{b_2} \equiv \eta_{CL,CH}$$

Comparing c_{CL} and c_{JB}, we find that

$$\begin{aligned} & c_{CL} - c_{JB} \\ & \quad = \{(b_1 l + p\delta) + \eta a_2 h\} - \{(b_2 l + p\delta) + (b_2 h + p\delta)\} \\ & \quad = \{b_1 l + \eta a_2 h\} - \{b_2 l + (b_2 h + p\delta)\} \\ & \quad = (b_1 - b_2)l + (\eta a_2 - b_2)h - p\delta \end{aligned}$$

Thus we obtain

$$c_{CL} < c_{JB} \Leftrightarrow \eta < \frac{b_2}{a_2} + \frac{p\delta - (b_1 - b_2)l}{a_2 h} \equiv \eta_{CL,JB}$$

Comparing c_{CH} and c_{JB}, we find that

$$\begin{aligned} & c_{CH} - c_{JB} \\ & \quad = \{(b_1 h + p\delta) + \eta a_2 l\} - \{(b_2 l + p\delta) + (b_2 h + p\delta)\} \\ & \quad = b_1 h + \eta a_2 l - \{b_2 l + b_2 h + p\delta\} \\ & \quad = (b_1 - b_2)h + (\eta a_2 - b_2)l - p\delta \end{aligned}$$

Thus we obtain

$$c_{CH} < c_{JB} \Leftrightarrow \eta < \frac{b_2}{a_2} + \frac{p\delta - (b_1 - b_2)h}{a_2 l} \equiv \eta_{CH,JB}$$

Comparing c_{SB} and c_{JB}, we find that

$$\begin{aligned} & c_{JB} - c_{SB} \\ & \quad = \{(b_2 l + p\delta) + (b_2 h + p\delta)\} - \{a_2 l + a_2 h\} \\ & \quad = (b_2 - a_2)(l + h) + 2p\delta \end{aligned}$$

Since $h + l > 0$, $b_2 - a_2 < 0$ by assumption, we obtain

$$c_{JB} > c_{SB} \Leftrightarrow p\delta > (a_2 - b_2)\frac{h+l}{2}(> 0)$$

Comparing c_{SB} and c_{CL}, we find that

$$\begin{aligned} & c_{CL} - c_{SB} \\ & = \{(b_1 l + p\delta) + \eta a_2 h\} - \{a_2 l + a_2 h\} \\ & = (b_1 - a_2)l + (\eta - 1)a_2 h + p\delta \end{aligned}$$

We then obtain

$$c_{CL} > c_{SB} \Leftrightarrow p\delta > (a_2 - b_1)l + (1 - \eta)a_2 h$$

Note that while we have $l > 0$, $1 > \eta$, we do not necessarily have $b_2 - a_2 < 0$.

Comparing c_{SB} and c_{CH}, we find that

$$\begin{aligned} & c_{CH} - c_{SB} \\ & = \{(b_1 h + p\delta) + \eta a_2 l\} - \{a_2 l + a_2 h\} \\ & = (b_1 - a_2)h + (\eta - 1)a_2 l + p\delta \end{aligned}$$

We then obtain

$$c_{CH} > c_{SB} \Leftrightarrow p\delta > (a_2 - b_1)h + (1 - \eta)a_2 l$$

Comparing c_{JH} and c_{SB}, we find that

$$\begin{aligned} & c_{JH} - c_{SB} \\ & = \{a_2 l + (b_2 h + p\delta)\} - \{a_2 l + a_2 h\} \\ & = (b_2 - a_2)h + p\delta \end{aligned}$$

We thus obtain

$$c_{JH} > c_{SB} \Leftrightarrow p\delta > (a_2 - b_2)h$$

Comparing c_{JH} and c_{JB}, we find that

$$\begin{aligned} & c_{JH} - c_{JB} \\ & = \{a_2 l + (b_2 h + p\delta)\} - \{(b_2 l + p\delta) + (b_2 h + p\delta)\} \\ & = (a_2 - b_2)l - p\delta \end{aligned}$$

We thus obtain

$$c_{JH} > c_{JB} \Leftrightarrow (a_2 - b_2)l > p\delta$$

Comparing c_{JH} and c_{CL}, we find that

$$\begin{aligned} & c_{JH} - c_{CL} \\ & = \{a_2 l + (b_2 h + p\delta)\} - \{(b_1 l + p\delta) + \eta a_2 h\} \\ & = (a_2 - b_1)l + (b_2 - \eta a_2)h \end{aligned}$$

We thus obtain

$$c_{\mathrm{JH}} > c_{\mathrm{CL}} \Leftrightarrow \frac{1}{a_2}\left[(a_2 - b_1)\frac{l}{h} + b_2\right] > \eta$$

Comparing c_{JH} and c_{CH}, we find that

$$\begin{aligned} & c_{\mathrm{JH}} - c_{\mathrm{CH}} \\ &= \{a_2 l + (b_2 h + p\delta)\} - \{(b_1 h + p\delta) + \eta a_2 l\} \\ &= (a_2 - \eta a_2) l + (b_2 - b_1) h \end{aligned}$$

We thus obtain

$$c_{\mathrm{JH}} > c_{\mathrm{CH}} \Leftrightarrow \frac{1}{a_2}\left[a_2 + (b_2 - b_1)\frac{h}{l}\right] > \eta$$

■

Appendix C. Proof of Lemma 3

Proof: Recalling the definitions

$$\eta_{\mathrm{CL,CH}} \equiv \frac{b_1}{a_2}$$

$$\eta_{\mathrm{CL,JB}} \equiv \frac{b_2}{a_2} + \frac{p\delta - (b_1 - b_2) l}{a_2 h}$$

$$\eta_{\mathrm{CL,JB}} \equiv \frac{b_2}{a_2} + \frac{p\delta - (b_1 - b_2) h}{a_2 l}$$

we find that

$$\begin{aligned} & \eta_{\mathrm{CL,JB}} - \eta_{\mathrm{CL,CH}} \\ &= \left(\frac{b_2}{a_2} + \frac{p\delta - (b_1 - b_2) l}{a_2 h}\right) - \frac{b_1}{a_2} = \frac{p\delta - (b_1 - b_2)(h + l)}{a_2 h} \end{aligned}$$

$$\begin{aligned} & \eta_{\mathrm{CH,JB}} - \eta_{\mathrm{CL,CH}} \\ &= \left(\frac{b_2}{a_2} + \frac{p\delta - (b_1 - b_2) h}{a_2 l}\right) - \frac{b_1}{a_2} = \frac{p\delta - (b_1 - b_2)(h + l)}{a_2 l} \end{aligned}$$

$$\begin{aligned} & \eta_{\mathrm{CL,JB}} - \eta_{\mathrm{CH,JB}} \\ &= \left(\frac{b_2}{a_2} + \frac{p\delta - (b_1 - b_2) l}{a_2 h}\right) - \left(\frac{b_2}{a_2} + \frac{p\delta - (b_1 - b_2) h}{a_2 l}\right) \\ &= -(h - l)\frac{p\delta - (b_1 - b_2)(h + l)}{a_2 h l} \end{aligned}$$

Thus, we have two possible cases. If $p\delta > (b_1 - b_2)\,(h + l)$, we find that

$$\eta_{\mathrm{CL,CH}} < \eta_{\mathrm{CL,JB}} < \eta_{\mathrm{CH,JB}}$$

Conversely, if $p\delta < (b_1 - b_2)(h + l)$, we find that

$$\eta_{CH,JB} < \eta_{CL,JB} < \eta_{CL,CH}$$

■

Appendix D. Proof of Lemma 4

Proof: Comparing the point $(\eta_{CL,CH}, p\delta_{SB,CH,CL})$, where $c_{SB} = c_{CL} = c_{CH}$, with the point $(\eta_{CL,CH}, p\delta_{CL,CH,JB})$, where $c_{CL} = c_{CH} = c_{JB}$, we find that $\eta = \eta_{CL,CH}$ is common, and

$$\begin{aligned} &p\delta_{SB,CH,CL} - p\delta_{CL,CH,JB} \\ &= (a_2 - b_1)(l + h) - (b_1 - b_2)(h + l) = (a_2 + b_2 - 2b_1)(l + h) \end{aligned}$$

Comparing the point $(\eta_{CH,JB,SB}, p\delta_{SB,JB})$, where $c_{SB} = c_{JB} = c_{CH}$, with the point $(\eta_{CL,CH}, p\delta_{CL,CH,JB})$, where $c_{CL} = c_{CH} = c_{JB}$, we find that

$$\begin{aligned} &\eta_{CH,JB,SB} - \eta_{CL,CH} \\ &= \frac{1}{2a_2 l}[(a_2 + b_2)l + (a_2 + b_2 - 2b_1)h] - \frac{b_1}{a_2} \\ &= \frac{1}{2a_2 l}(a_2 + b_2 - 2b_1)(l + h) \end{aligned}$$

$$\begin{aligned} &p\delta_{SB,JB} - p\delta_{CL,CH,JB} \\ &= (a_2 - b_2)\frac{h + l}{2} - (b_1 - b_2)(h + l) = \frac{1}{2}(a_2 + b_2 - 2b_1)(l + h) \end{aligned}$$

Comparing the point $(\eta_{CH,JB,SB}, p\delta_{SB,JB})$, where $c_{SB} = c_{JB} = c_{CL}$, with the point $(\eta_{CL,CH}, p\delta_{CL,CH,JB})$, where $c_{CL} = c_{CH} = c_{JB}$, we find that

$$\begin{aligned} &\eta_{CL,JB,SB} - \eta_{CL,CH} \\ &= \frac{1}{2a_2 h}[(a_2 + b_2)h + (a_2 + b_2 - 2b_1)l] - \frac{b_1}{a_2} \\ &= \frac{1}{2a_2 h}(a_2 + b_2 - 2b_1)(l + h) \end{aligned}$$

■

Appendix E. Proof of Lemma 5

Proof: Recalling

$$\eta_{JH,CL} \equiv \frac{1}{a_2}\left[(a_2 - b_1)\frac{l}{h} + b_2\right]$$

$$\eta_{JH,CH} \equiv \frac{1}{a_2}\left[a_2 + (b_2 - b_1)\frac{h}{l}\right]$$

$$\eta_{\mathrm{CL,\,CH}} \equiv \frac{b_1}{a_2}$$

we find that

$$\eta_{\mathrm{JH,\,CL}} - \eta_{\mathrm{CL,\,CH}}$$
$$= \frac{1}{a_2}\left[(a_2 - b_1)\frac{l}{h} + b_2\right] - \frac{b_1}{a_2} = \frac{1}{a_2 h}[(b_2 - b_1)h + (a_2 - b_1)l]$$

$$\eta_{\mathrm{JH,\,CH}} - \eta_{\mathrm{CL,\,CH}}$$
$$= \frac{1}{a_2}\left[a_2 + (b_2 - b_1)\frac{h}{l}\right] - \frac{b_1}{a_2} = \frac{1}{a_2 l}[(b_2 - b_1)h + (a_2 - b_1)l]$$

■

Appendix F. Proof of Lemma 6

Proof: Recalling the definitions

$$\eta_{\mathrm{JH,\,CL}} \equiv \frac{1}{a_2}\left[(a_2 - b_1)\frac{l}{h} + b_2\right]$$

$$\eta_{\mathrm{CH,\,JB,\,SB}} \equiv \frac{1}{2a_2 l}[(a_2 + b_2)l + (a_2 + b_2 - 2b_1)h]$$

$$\eta_{\mathrm{CL,\,JB,\,SB}} \equiv \frac{1}{2a_2 h}[(a_2 + b_2)h + (a_2 + b_2 - 2b_1)l]$$

and comparing $\eta_{\mathrm{JH,CL}}$ with $\eta_{\mathrm{CH,JB,SB}}$, we find that

$$\eta_{\mathrm{JH,CL}} - \eta_{\mathrm{CH,JB,SB}}$$
$$= \frac{1}{a_2}\left[(a_2 - b_1)\frac{l}{h} + b_2\right] - \frac{1}{2a_2 l}[(a_2 + b_2)l + (a_2 + b_2 - 2b_1)h]$$
$$= \frac{1}{2}\left[\frac{(2b_1 - b_2 - a_2)h}{a_2 l} + 2\frac{(a_2 - b_1)l}{a_2 h} + \frac{b_2 - a_2}{a_2}\right]$$

Comparing $\eta_{\mathrm{JH,CL}}$ with $\eta_{\mathrm{CL,JB,SB}}$, we find that

$$\eta_{\mathrm{JH,CL}} - \eta_{\mathrm{CL,JB,SB}}$$
$$= \frac{1}{a_2}\left[(a_2 - b_1)\frac{l}{h} + b_2\right] - \frac{1}{2a_2 h}[(a_2 + b_2)h + (a_2 + b_2 - 2b_1)l]$$
$$= \frac{1}{2a_2 h}(b_2 - a_2)(h - l) < 0$$

■

Appendix G. Proof of Lemma 7

Proof: Recalling the definitions

$$\eta_{\mathrm{JH,\,CH}} \equiv \frac{1}{a_2}\left[a_2 + (b_2 - b_1)\frac{h}{l}\right]$$

$$\eta_{\mathrm{CH,\,JB,\,SB}} \equiv \frac{1}{2a_2 l}[(a_2 + b_2)l + (a_2 + b_2 - 2b_1)h]$$

$$\eta_{\mathrm{CL,\,JB,\,SB}} \equiv \frac{1}{2a_2 h}[(a_2 + b_2)h + (a_2 + b_2 - 2b_1)l]$$

and comparing $\eta_{\mathrm{JH,CH}}$ with $\eta_{\mathrm{CH,JB,SB}}$, we find that

$$\begin{aligned}
&\eta_{\mathrm{JH,CH}} - \eta_{\mathrm{CH,JB,SB}} \\
&= \frac{1}{a_2}\left[a_2 + (b_2 - b_1)\frac{h}{l}\right] - \frac{1}{2a_2 l}[(a_2 + b_2)l + (a_2 + b_2 - 2b_1)h] \\
&= \frac{1}{2a_2 l}(b_2 - a_2)(h - l) < 0
\end{aligned}$$

Comparing $\eta_{\mathrm{JH,CH}}$ with $\eta_{\mathrm{CL,JB,SB}}$, we find that

$$\begin{aligned}
&\eta_{\mathrm{JH,CH}} - \eta_{\mathrm{CL,JB,SB}} \\
&= \frac{1}{a_2}\left[a_2 + (b_2 - b_1)\frac{h}{l}\right] - \frac{1}{2a_2 h}[(a_2 + b_2)h + (a_2 + b_2 - 2b_1)l] \\
&= \frac{1}{2}\left[\frac{(2b_1 - a_2 - b_2)l}{a_2 h} + 2\frac{(b_2 - b_1)h}{a_2 l} + \frac{a_2 - b_2}{a_2}\right]
\end{aligned}$$

■

References

Climate Change Secretariat, UNFCCC (1997) The Kyoto Protocol to the convention on climate change. December

Rose A, Bulte E, Folmer H (1999) Long-run implications for developing countries of joint implementation of greenhouse gas mitigation. Environ Resource Econ 14:19–31

On the Additionality of GHG Reduction

HIDENORI NIIZAWA

Summary. In the Kyoto Protocol, annex I countries which have made a commitment to reach their emissions targets, can acquire additional emissions reductions from countries which have not made this commitment. This mechanism is called the clean development mechanism (CDM). The additional emissions reductions are the subject of this chapter. We must estimate the amount of emissions which will occur if the CDM project is not implemented. These are the baseline emissions. In this chapter, the history of the negotiations and the agreed rules for baselines and additional emissions reductions are briefly reviewed. Then the baselines will be defined, and additional emissions reductions will be interpreted based on a cost–benefit analysis model of investment. At the same time, some institutional design issues related to the baseline decision are also examined from the viewpoints of incentives, uncertainty, and equity.

Key words. Kyoto Protocol, Clean development mechanism, Additionality, Baseline

Introduction

The clean development mechanism (CDM), in article 12 of the Kyoto Protocol, is a multipurpose mechanism. One of its purposes is "to assist Parties included in annex I in achieving compliance with their quantified emission limitations and reduction commitments under Article 3." That is, Parties who made a commitment to emission targets can acquire certified emission reductions (CERs) from Parties who did not make such a commitment, and can add them to their assigned amounts on condition that the emission reductions are "*additional* to any that would occur in the absence of the certified project activity (Article 12, 5(c))."

Institute of Economic Research, Kobe University of Commerce, 8-2-1 Gakuen-nishimachi, Nishi-ku, Kobe 651-2197, Japan

Parties which acquired CERs can increase their emissions by the same amount. The CDM itself will not increase or decrease the total amount of emissions in the world if it works in an ideal way. The emissions that would occur in the absence of the certified project activity are called the baseline emissions. If the baseline emissions are overestimated, however, part of the reductions that would have occurred otherwise will be verified and transferred as CERs. This means that the CDM would increase the total amount of emissions in the world. The evil of excessive CERs is not only an increase in the total emissions in the world, but also a decrease in the profit of real emissions reductions.

There is an inherent difficulty in assuming counterfactual baseline emissions amounts which are never realized when the CDM project is implemented. The reduction in a project may be larger than the reduction in the nation as a whole because there are repercussive increases in emissions outside the project boundary. The increase in emissions is called leakage.

The additionality issue is transitional because all Parties are expected to commit to emission targets in the future. If all Parties make a commitment, they can trade part of their assigned amounts. There will be no additional issues in this situation. The CDM itself is a compromise (Tietenberg and Victor 1994 p. 6; Pearce 1995).

Most of the rules of the Kyoto Protocol were agreed upon in the Marrakesh Accords at COP-7 in 2001. One of the remaining details to be worked out relating to the CDM is the development of general guidance on methodologies relating to baselines and monitoring.

The purpose of this chapter is to clarify the concept of additional emissions reductions by Parties which have not made a commitment to emission targets. It will be emphasized that investment additionality is an economic behavioral model of environmental additionality, which is explicitly required by the Protocol. Some institutional design issues related to baseline decisions are also examined from the viewpoints of incentives, uncertainty, and equity.

The structure of this chapter is as follows. In next section, an incentive structure to exaggerate emission reductions and the uncertainty of CDM projects are identified. The solution of the Marrakesh Accords on incentives and uncertainty is examined. In the following section, the history of negotiations for baselines and additional emissions reductions is briefly reviewed to promote an understanding of the agreed rules. The rules are explained in the next section, which is followed by a section in which baseline emissions and additional emissions reductions are defined based on a cost–benefit analysis model.

Incentive Structure of the CDM

Two Parties participate in a typical CDM project. The one is the investor Party who supplies money and technology. The other is the host Party, i.e., the country in which the CDM project is implemented. The host Party is a non-annex I country, which has not made a commitment to an emissions target. This type of CDM is called a bilateral CDM. When several Parties jointly invest in a project, it is called a multilateral CDM. In both cases, there is incentive to overestimate the emissions reductions for both the investor Parties and the host Party because the investor Parties want to acquire more CERs and the host Party wants to sell more CERs. A host Party will not have any trouble even if it sells too large an amount of CERs because it does not have a commitment to an emissions target. Therefore, both Parties have incentives to claim false reductions as real reductions.

This incentive problem does not occur in cases based on article 6 and article 17, i.e., joint implementation (JI) and emissions trading (ET). Both JI and ET trade part of an assigned amount. In these cases, the Parties must reduce their emissions by the amount they transfer. If a Party transfers too much, it will find it difficult to reduce as much as it transfers.

Overestimation of emission reductions is caused by both overestimation of baseline emissions and underestimation of the CDM project emissions.

Article 12 requires that emissions reductions are verified and certified by independent operational entities. These operational entities are accreditated and designated by the executive board.

Baseline emissions depend on the discount rate, the enforcement level of environmental regulations and their shadow prices, the maintenance costs of old facilities, energy prices, and energy policies (e.g., subsidy to domestic energy). They are difficult for operational entities to observe (USEPA 1997; Chomitz 1998). The operational entity must verify the baseline by checking the methodology used to produce the baseline, as well as the monitoring data which are submitted by the participants. Participants can manipulate these data. There are two opinions on this issue. Some claim it is fatal; others observe that it is possible to use reliable data.

If the executive board finds that excess CERs are issued, the operational entity must compensate for the excess CERs. If significant deficiencies are identified in the relevant report for which the participating entity is responsible, the operational entity can evade the liability.

Non-annex I countries which have not made a commitment to emissions targets are allowed to participate in a CDM project activity as investors if they are Parties to the Kyoto Protocol. That means that non-annex I countries are allowed to invest in CDM projects in their own territories, and sell CERs to annex I countries. Non-annex I countries are allowed to finance and invest in CDM projects in other non-annex I host countries. The former type of CDM is called a unilateral CDM, and the latter is called a south–south CDM. In both cases, participants have incentives to overestimate emissions reductions.

Uncertainty

Overestimation of baselines is also caused by uncertainty. Baseline methodology must be included in the project design document (PDD) which is required for the validation and registration of projects.

All relevant data necessary for determining the baseline are collected during the crediting period by the project participants. The operational entity will review the monitored results and determine the reductions. This means that reductions by CDM projects are determined after investment and operation.

In the case of a power plant construction project, for example, the emissions from the new power plant depend on the demand for electricity. On the other hand, the baseline of the power plant is defined in terms of product of emission rates, CO_2 ton per kWh, and the demand for electricity. The issue of whether ex-post values or ex-ante values of the electricity demand should be used to calculate the baseline has not been clarified.[1]

The Marrakesh Accords say that a baseline shall be established in a transparent and conservative manner regarding the choice of approaches, assumptions, methodologies, parameters, data sources, key factors, and additionality, and taking uncertainty into account. The operational guidelines for a baseline which the Netherlands requires suggest that a possible range of baselines is presented, and the lower end of that range is used to get a conservative baseline estimate.[2]

The parameters which influence baseline emissions will change, and then the baseline emissions will change. The experience of activities implemented jointly (AIJ) shows that there are several patterns of baselines change even in the same category of projects (Ellis 1999). A project which was not profitable enough to invest in at the beginning may become profitable in the future when the parameters change. Figure 1 shows an example of a change in baseline emissions. The baseline emissions decrease and become equivalent to the CDM project emissions in year *T*. In this case, the shaded area is verified as an additional emissions reduction. The period during which additional emissions reductions are certified is not necessarily equal to the period of the project operation. It is difficult to predict baselines for a long period, and renewal of the baseline is inevitable. The interval of renewal is a compromise between the cost of renewal and the benefit of real reductions.

Equity

The same technologies do not necessarily create the same additional emissions reductions in every country because there are different baselines in each country. However, the methodologies, for establishing baselines will be standardized by guidelines to promote consistency, transparency, and predictability.

---

[1] See the chapter by H. Imai and J. Akita in this volume

[2] Ministry of Housing, Spatial Planning and the Environment of The Netherlands, Operational guidelines for baseline studies, validation, monitoring and verification of clean development mechanism project activities. Vol. 2a, Baseline studies, monitoring and reporting. A guide for project developers, Version 1.0, October 2001

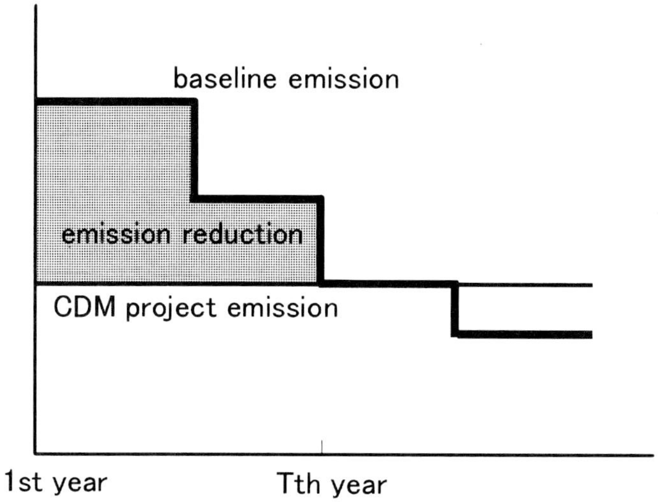

FIG. 1. Change in baseline and additional emissions reductions

History of the Negotiations

The discussion of additionality began when the rules of AIJ were made, and these were adopted at COP-1 in 1995. The following statement appeared in decision 5/CP.1.[3]

The activities implemented jointly should bring about real, measurable and long-term environmental benefits related to the mitigation of climate change that would not have occurred in the absence of such activities. (paragraph 1(d))

Various interpretations of additionality have been proposed since the adoption of the Kyoto Protocol in 1997.

Emissions Additionality (Environmental Additionality)

CDM projects should achieve real, measurable, and long-term GHG emissions reductions.[4]

Investment Additionality

Projects which are commercially viable without emissions credits should not qualify as CDM projects.[4]

---

[3] FCCC/CP/1995/7/Add.1, June 6, 1995. See also United Nations, activities implemented jointly under the pilot phase, FCCC/SBSTA/1977/INF.3, October 13, 1997
[4] Republic of Korea, 1999, Submission by the Republic of Korea on the CDM (Addendum) FCCC/SB/1999/MISC.10/Add.1, September 27, 1999

Investments made below the level of the normal internal rate of return should be considered as CDM investments.[4]

Investment additionality is given if the risk-adjusted internal rate of return of the CDM project activity is below [x] per cent. The executive board determines a country-specific risk adjustment factor and the value of [x].[5]

Financial Additionality

Official development assistance (ODA) projects should not be considered as CDM projects.[4] Funding for CDM project activities shall be additional to ODA, Global Enironmental Facility (GEF), and other financial commitments to the developed country concerned.[6]

Technological Additionality

The technology for CDM projects should be appropriate for non-annex I Parties, and should meet the best available technology standards.[4]

Although emissions additionality is required by the Protocol, it is not operational per se as a criterion because it does not define a real reduction. Investment additionality is an economic behavioral model of environmental additionality. Investment additionality can be assumed to be a feasible criterion of environmental additionality because it defines a real reduction. This will be considered later.

ODA projects are likely to pass the investment additionality test because it can be assumed that the internal rate of return for ODA projects is lower than the normal rate for private projects. Therefore, the financial additionality is independent of investment additionality.

The best available technologies are less likely to be adopted without CERs. This means that investment additionality is satisfied by adopting the best available technologies. Investment additionality can indicate which are the best available technologies. That is, these criteria overlap each other.

In 2000, the United States insisted that because the investment additionality criterion was easy to falsify and was an imperfect proxy for whether a project would have occurred otherwise it was not workable. The United States proposed a requirement that the project activity should achieve a level of performance with respect to emissions reductions or the enhancement of removals that was significantly better than average compared with recently undertaken activities or facilities.[7] This proposal was called the threshold criterion.

---

[5] FCCC/CP/2000/CRP.2/Add.1, November 24, 2000

[6] Group of 77 and China, 1999, Submission on Article 12 of the Kyoto Protocol. The clean development mechanism. FCCC/SB/1999/MISC.10/Add.2, October 30, 1999

[7] United States of America, 2000, Submission by the United States on baselines, monitoring and related provisions for the clean development mechanisms. FCCC/SB/2000/MISC.4/Add.1, September 9, 2000

To alleviate falsification, partial crediting was proposed (Imai et al. 1999; Chomitz 1998; Meyers 1999). However, this proposal was not taken into account in the negotiations. We will see why investment additionality is an imperfect proxy later.

These proposals for additionality and threshold tests assumed that there were two steps in determining additional reductions. First, a proposed project must pass the above-mentioned ex ante additionality test. Second, the baseline emissions amount is estimated by assuming what would have happened in the absence of the CDM project. The difference between the baseline and the CDM project emissions is the additional reduction. The purpose of the additionality test is to screen out "anyway" tones (Lazarus et al. 2000).

The Bonn Agreement and the Marrakesh Accords

Financial Additionality

The Implementation of the Buenos Aires Plan of Action,[8] which was adopted at COP-6bis in 2001, specified that the Conference of the Parties agreed to emphasize that public funding for CDM projects from Parties included in annex I was not to result in the diversion of ODA, and was to be separate from, and not counted toward, the financial obligations of Parties included in annex I. However, implementation of this rule is difficult, because it does not refer to what constitutes *diversion*. Moreover, it should be pointed out that if ODA projects are not allowed to earn CERs, the incentive to reduce greenhouse gases (GHGs) will disappear for ODA projects.

Additionality and Baselines

Participants must submit a PDD to the operational entity for the purpose of project validation. A description of how the anthropogenic emissions of GHGs by sources will be reduced below those that would have occurred in the absence of the registered CDM project activity is required in the PDD. This requirement is no longer an additionality test. An early draft of the text of COP-6bis in July 2001[9] states that the PDD should include an "explanation of how the project activity meets the additionality requirements." However, this sentence had disappeared by the end of the negotiations. The definitions of additionality and baseline are given below.

— A CDM project activity is *additional* if anthropogenic emissions of greenhouse gases by sources are reduced below those that would have occurred in the absence of the registered CDM project activity.

[8] Decision 5/CP.6, FCCC/CP/2001/L.7, July 24, 2001
[9] FCCC/CP/2001/CRP.11, July 27, 2001

— The *baseline* for a CDM project activity is the scenario that reasonably represents the anthropogenic emissions by sources of greenhouse gases that would occur in the absence of the proposed project activity.

Therefore, the difference between CDM project emissions and baseline emissions is the additional reduction.

In choosing a baseline methodology for a project activity, the project participants shall select from among the following approaches the one deemed most appropriate for the project activity, taking into account any guidance from the executive board, and justifying the appropriateness of their choice.

1. Existing actual or historical emissions, as applicable.
2. Emissions from a technology that represents an economically attractive course of action, taking into account barriers to investment.
3. The average emissions of similar project activities undertaken in the previous 5 years, in similar social, economic, environmental, and technological circumstances, and whose performance is among the top 20% of their category.

Among these options, option 2 is special because it explains the grounds or reasons for a baseline. Option 2 corresponds to the investment additionality. Option 3 looks like the threshold proposal by USA.

Participants can propose a new baseline methodology in the PDD. If the new methodology is approved by the executive board, it can be used. The executive board will develop general guidance on methodologies for choosing baselines.

Multiproject Baseline

Two types of baseline have been discussed. One is multiproject and the other is project-specific. In the case of a multiproject baseline, various types of potential project are grouped into a single category with a corresponding single baseline. The multiproject baseline is also called the benchmark. The umbrella group[10] proposed multiproject baselines based on the experience of project-specific baselines in AIJ. They criticized the fact that project-specific baselines entailed a degree of specificity and detail that could be costly, lack transparency, give a false sense of precision, invite falsification, and require a difficult and subjective review process (Lazarus et al. 2000). The Organization for Economic Cooperation and Development (OECD) and the International Energy Agency (IEA) devoted many resources to developing the multiproject baselines (OECD, IEA 2000).

On the other hand, some Parties, for example China, opposed multiproject baselines. China was afraid of multiproject baselines because they might lead China to a commitment. Eventually it was agreed that, except for small-scale CDM projects, baselines should be established on a project-specific basis.

Apart from China's concerns, it is recognized that multiproject baselines have inherent problems. To make multiproject baselines worthwhile, the grouping or

[10] A negotiation group of non-EU developed countries

aggregation should be broad enough to encompass many CDM projects and reduce transaction costs, but not so broad that baseline accuracy is compromised, excessive CERs are awarded, or significant investment opportunities are lost. A single sector-wide baseline may provide no incentive for CDM projects that improve the efficiency of relatively carbon-intensive options (e.g., upgrading the efficiency of a coal plant). Such projects are likely to have carbon intensities above sector-wide baselines, and thus not be credible. Conversely, a sector-wide baseline might overestimate CERs if a project is merely improving upon an already low-carbon activity (e.g., improving the efficiency of a natural gas plant where the sector-wide baseline reflects coal or oil-based power generation). A more stringent multiproject baseline can reduce excess CERs. However, it eliminates the chance that some legitimate projects will be issued CERs. This problem has not yet been resolved (OECD, IEA 2000).

Project-specific baselines could also be standardized to some extent by specifying some required procedures or assumptions (e.g., for oil prices). Guidelines for the standardization of baseline methodologies will be prepared by the executive board.

Definition of Baselines, and an Interpretation of Additional Reduction Based on a Cost–Benefit Analysis Model

Here, the definition of a baseline is related to option 2 of approved baseline methodologies: emissions from a technology that represents an economically attractive course of action, taking into account barriers to investment. The *economically attractive course of action* can be interpreted as follows.

Each country decides the amount of GHG emissions based on the domestic cost and benefit even if that country has not committed to emissions targets in the Protocol. That amount is the baseline. For example, because energy saving, i.e., a reduction of carbon dioxide emissions, is profitable, it must be achieved to some extent. The extent of energy savings depends on energy prices. Environmental regulations on fuel-derived domestic pollutants such as SO_2 and NO_2 often reduce energy consumption. The baseline is defined as the amount of GHG emissions which occur when only domestic costs and benefits are considered. Then further reductions of GHGs, which are justified by taking account of global external benefits, are additional for that country.[11]

Private companies invest to maximize profits under the various regulations, such as environmental regulations. Therefore, the amount of emissions when a company achieves maximum profit without CERs is the baseline. By-product

[11] The same argument was applied to decide the *incremental costs* of the global environmental facility (GEF). Incremental costs, GEF/C.7/Inf.5, February 29, 1996. See also Pearce (1995 Ch. 11) and Bruce et al. (1996 p. 158). The GEF is a fund to finance the incremental costs which are paid by developing countries to conserve the global environment. Developed countries contribute to the fund

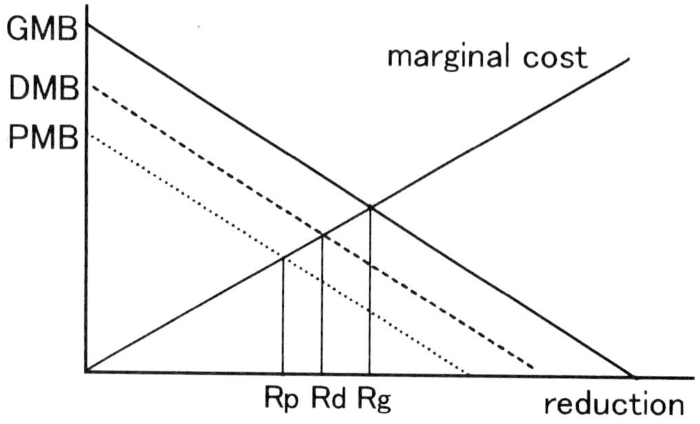

FIG. 2. Baseline and additionality.
PMB, private marginal benefit; *DMB*, domestic marginal benefit; *GMB*, global marginal benefit

reductions of GHGs from a profitable project are not assumed as additional reductions because the project was invested in without CERs.

The same argument can be applied to sink projects. Each country decides the amount of forest based on domestic costs and benefits. That amount is the baseline. For example, a forest is used to produce timber, and is necessary to prevent floods and to maintain the domestic environment. These forests sequester CO_2. This is not additional sequestration, however, because the sequestration would be realized without the CDM.

Figure 2 shows this definition of a baseline. The horizontal axis shows the GHG reductions. The line labeled PMB shows the private marginal benefit from GHG reductions. An example of this benefit is energy cost savings. The line labeled DMB shows the domestic marginal benefit. DMB is the sum of the PMB and the benefit of reducing domestic pollutants. The line labeled GMB shows the global marginal benefit, which is the sum of the DMB and the benefit of reducing GHG emissions. Because each country is assumed to maximize its domestic net benefit, the baseline emission would be R_d, where the marginal cost of reducing GHG emissions is equal to the DMB if there is not a policy failure. Any reduction beyond R_d is *additional*. R_g is the efficient reduction when the global marginal benefit of reducing GHG emissions is taken into account. The difference between R_g and R_d is the *additional* emissions reduction.

If there is an international emissions trading market, assigned amount units (AAU) will have a market price. The CERs produced by CDMs will also be traded at the market price. The difference between the marginal reduction cost and the price, that is area α in Fig. 3, is the profit of the host country. The domestic benefits of local pollution reductions and energy savings are areas β and γ, respectively. It is assumed that each country could maximize its domestic net benefit. However, this assumption may not always be true.

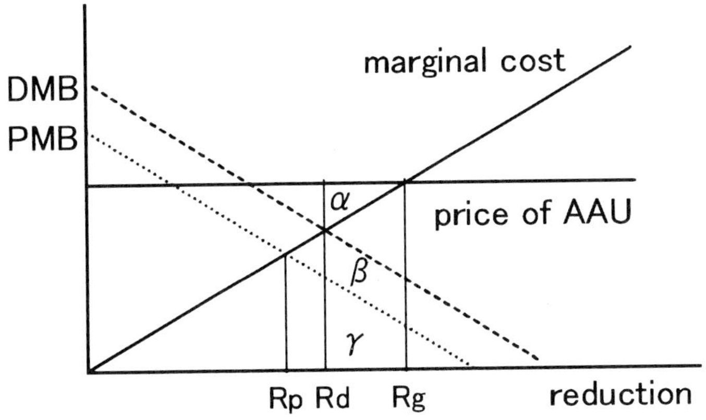

FIG. 3. Profit of a host country in the global emissions trading market. *AAU*, assigned amount units. Other abbreviations as for Fig. 2

It is said that there are many GHG reduction projects which are not implemented although their costs are negative. These are called no-regrets projects (Michaelowa 1998; UNCTAD 1998 p. 51). The reasons why no-regrets projects do not receive investment are lack of technology, regulations, lack of financing, lack of information, organizational inflexibility, etc. These are called barriers.[12] These barriers are the reason why the USA criticized investment additionality as an imperfect proxy for whether a project would otherwise have occurred. The approved baseline methodology (2) refers to "taking into account barriers to investment." This means that no-regrets projects can produce CERs even if the cost is negative. In such a case, the baseline emissions are larger than those of the no-regrets project.

By taking barriers to establishing baselines into account, participants may have incentives to insist on barriers which do not exist to justify larger baseline emissions. Moreover, participants may have a disincentive to overcome the barriers if barriers are taken into account (Carter 1997).

If there were policy failures in a host country, the net benefits to the country would not be maximized. Baselines should not be based on policy failures, but should provide incentives for the host country to correct their national policy failures (Pearce 1995 p.174).

The cost–benefit analysis uses various data. Whether the data from host countries should be used or not is an issue for discussion (Chomitz 1998 p. 21). For example, subsidies for domestic energy sources contribute to low-energy efficiency in some countries. Should we use their emissions as the baseline? It is well

[12] The IEA (1997) proposed a barrier removal method to judge the additionality of such projects. Barriers were taken into account in the AIJ guideline of the US Guidelines for a USIJI project proposal

known that deforestation is partly caused by national policies such as migration, taxation, and property laws. In the case of GEF, if there were distortions in local prices because of subsidies or regulations, the international price would be used instead of the local price. This means that the GEF fund is not used to justify policy failures. The World Bank, based on experiences of the prototype carbon fund (PCF),[13] suggested the need for policy guidelines which would be employed by host countries, and the baseline should be decided based on the policies in the guidelines.[14] However, what constitutes policy failure is a sensitive issue for host countries.

The PCF controlled by the World Bank uses the above-mentioned cost–benefit analysis method to calculate the baselines of some projects.[15] In the case of an energy conversion project from coal to geothermal power for the district heating system in Pyrzyce, Poland, the following data were used (Chomitz 1998):

— investment cost for 2 years is 15.31 million dollars;
— saving in energy costs is 890 thousand dollars per year;
— saving in operating costs is 130 thousand dollars per year;
— value of reducing pollution is 1.97 million dollars;
— reduction of carbon dioxide is 68618 tons.

The net present value of this project can be positive or negative. This depends on the assumption of discount rates and the shadow prices of sulfur dioxides and nitrogen dioxides. If the net present value is negative, unless the value of carbon dioxide is included, the reduction of carbon dioxide by the project will be additional because the project will not have received investment as a non-CDM project.

References

Bruce JP, Lee H, Haites EF (eds) (1996) Contribution of working group III to the second assessment report of the Intergovernmental Panel on Climate Change (IPCC). Climate change 1995. Economic and social dimensions of climate change. Cambridge University Press, Cambridge

Carter L (1997) Modalities for the operationalization of additionality. Prepared for presentation at: UNEP/German Federal Ministry of the Environment, International Workshop on Activities Implemented Jointly, Leipzig, Germany, March 5–6, 1997

Chomitz KM (1998) Baselines for greenhouse gas reductions: problems, precedents, solutions. Draft 1.3 rev. Prepared for the Carbon Offsets Unit, World Bank

[13] Prototype carbon fund. Many companies and countries jointly finance this fund, which is invested in various CDM and JI projects. Earned CERs and emission reduction units (ERUs) are distributed to the contributors

[14] Learning from the implementation of the prototype carbon fund. Prototype carbon fund occasional papers series, No. 1, October 3, 2000

[15] This approach is called investment analysis by the World Bank. Baseline methodologies for PCF projects. PCF implementation note No. 3, Version of April 21, 2000

Ellis J (1999) Experience with emission baselines under the AIJ pilot phase. OECD information paper

Imai H, Akita J, Niizawa H, Sawa T (1999) Some consistency and incentive issues in CDM. Presented at a side event for COP-5: Kyoto Mechanisms from the Asian Point of View

IEA (1997) Activities implemented jointly. Partnerships for climate and development. Organization for Economic Cooperation and Development (OECD) and International Energy Agency

Lazarus M, Kartha S, Bernow S (2000) Key issues in benchmark baselines for the CDM: aggregation, stringency, cohorts, and updating. Prepared for USEPA, Tellus Institute and Stockholm Environment Institute

Meyers S (1999) Additionality of emissions reductions from clean development mechanism projects: issues and options for project-level assessment. Ernest Orlando Lawrence Berkeley National Laboratory, LBNL-43704

Michaelowa A (1998) Joint implementation—the baseline issue, economic and political aspects. Global Environ Change 8:81–92

OECD, IEA (2000) Emission baselines, estimating the unknown. OECD and IEA

Pearce D (1995) Blueprint 4. Capturing global environmental value. Earthscan

Tietenberg T, Victor DG (1994) Possible administrative structures and procedures. Combating global warming: possible rules, regulations and administrative arrangements for a global market in CO_2 emission entitlements. United Nations Conference on Trade and Development (UNCTAD)

UNCTAD (1998) Greenhouse gas emissions trading. United Nations Conference on Trade and Development, draft

US Environmental Protection Agency (USEPA) (1997) Activities implemented jointly: 2nd report to the secretariat of the United Nations Framework Convention on Climate Change

On the Incentive Consequences of Alternative CDM Baseline Schemes

Haruo Imai* and Jiro Akita†

Summary. This chapter addresses incentive-related issues concerning clean development mechanims (CDM) introduced under the Kyoto Protocol. In particular, the consequences of alternative schemes to set CDM baselines, namely the ex-ante baseline and the ex-post baseline, are studied in the context of short-run incentives on the part of CDM investors, as well as longer-run incentives on the part of currently non-Annex I parties to eventually join the Annex I camp, i.e., the context of so-called evolution.

Key words. Clean development mechanism (CDM), Baseline, Risk

Introduction

The clean development mechanism (CDM), one of the so-called Kyoto mechanisms, allows Annex I parties to the protocol to obtain emission reduction credits from an emission-reducing project carried out by non-Annex I parties, and to utilize the credits, combined with their assigned amount of emissions, for the purpose of fulfilling their own greenhouse gas (GHG) emissions reduction obligation under the Kyoto Protocol. In order to determine the due amount of such credits, one needs emissions baselines indicating the hypothetical emission levels that would have taken place if no such project had been undertaken, and/or if there was no CDM scheme present to begin with. The task of setting the baseline inherently requires a sensible method of spelling out scenarios for a counterfactual world. The incentives for CDM on the part of the parties are significantly affected by the choice of such methods.

In this chapter, we develop two models incorporating the uncertainties faced by parties involved in CDM. The first model is microscopic, and focuses on the

* Kyoto Institute of Economic Research, Kyoto University, Yoshidahonmachi, Sakyo-ku, Kyoto 606-8501, Japan
† Graduate School of Economics and Management, Tohoku University, Kawauchi, Aoba-ku, Sendai 980-8576, Japan

question of short-run incentives for CDM investment under alternative ways to set the CDM baseline. The second model is macroscopic, and focuses on the question of the longer-run "evolution" of the entire Kyoto mechanism. That is, we address questions pertaining to the decisions made by nations as to their Annex I or non-Annex I status, and show that the choice between the alternative baseline schemes does matter for either question.

Ex-Ante Versus Ex-Post Baseline Schemes

The Set-Up: the Source of Uncertainty

One of the proposed methods to determine the baseline emissions level is called the "rate-based" baseline method. That is, we may set the baseline level of emissions in terms of its proportion to the level of economic or production activity to be realized in the future. For instance, suppose that the energy efficiency of a factory is improved by replacing its old equipment. This improvement results in a reduction in GHG emissions, say from e_0 per unit of output to e. Suppose that the total production in this factory is x units. Then we say that the baseline emissions level for this investment project (of replacing the production equipment) equals e_0x, while the corresponding net reduction in emissions equals $(e_0 - e)x$.

Nevertheless, it seems that most of the uncertainties stem from the output level x, which is subject to demand (and production) uncertainty.[1]

While e_0 could also be uncertain, perhaps owing to some unexpected discovery of a more efficient way of utilizing the old equipment, this can safely be dismissed as a rather unlikely theoretical possibility. Thus, let us suppose it is only the output x that is uncertain.

Ex-Ante Baseline Versus Ex-Post Baseline

When such uncertainty is present, the rate-based baseline need not coincide with the predicted value of the emission level with the old equipment, i.e., $E[e_0x] = e_0E[x]$. We refer to this latter baseline as the "ex-ante" baseline, whereas the former rate-based baseline is the "ex-post" baseline, emphasizing the different times at which the baseline emission levels are determined. The timing matters because an investment decision must be made prior to the revelation of the uncertain demand level.

Let p stand for the market price of one unit of emissions reduction. Assuming that the investing parties wish to maximize the private benefit accruing to them from the project, their maximand is written as

[1] The degree of the output uncertainty may vary from one case to another. In cases where equipment is always operated around its normal capacity, and/or the facility in question is just one among many facilities operated by the firm, the output, x, may not fluctuate as much as the total demand

$$E[u(p\max\{b-ex,0\}+\Pi(x)-I)]$$

where e is the actual emissions per unit of output, b denotes the baseline emissions level, Π denotes the profit from the project without the CDM credits, I denotes the investment costs (with adequate discounting), and $u(.)$ is the utility function defined over the monetary wealth, and reflects the investors' attitudes toward risk. The max operator in the expression ensures that the value of the CDM emissions reduction credit is nonnegative even in the event that the emissions reduction turns out to be negative.

Note that $b = e_0E[x]$ with the ex-ante baseline, and $b = e_0x$ with the ex-post baseline.

Thus, the emissions reduction realized from the ex-ante baseline is $e_0E[x] - ex$, which could be negative, while that from the ex-post baseline is $(e_0 - e)x$, which is always nonnegative provided that $e_0 > e$ and $x \geq 0$.

Given this set-up, we shall now compare the two baseline schemes in relation to the incentives they provide vis-a-vis the investors. At first glance, it may appear that the ex-post baseline should absorb more of the relevant risks, thereby favoring investors. Indeed, the ex-post baseline has been favored by many in view of its technical ease, which avoids the difficult task of handling the randomness in x, as well as that of agreeing on an appropriate estimate of $E[x]$. Furthermore, the conventional view seems to hold that there is no readily foreseeable adverse incentive effect caused by the adoption of the ex-post baseline scheme.

In what follows, we shall argue that the presumption in favor of the ex-post baseline, while not necessarily invalid on most counts, is subject to a certain set of qualifications which are often violated in reality.

In particular, we shall argue that the ex-post baseline scheme does not really absorb the risk faced by the CDM investors, but instead it only shifts the burden and/or the cost of the risk to the global environment. Granted that the actual size of emissions reduction from CDM activities is expected to be much smaller in magnitude relative to the total emissions from Annex I parties, nevertheless, it remains true that the emissions reduction credits generated through CDM activities constitute genuine additions to the total emissions amount assigned to the Annex I parities. Thus, the variability of CDM credits translates into the variability of the total amount assigned, including CDM credits, although the translation may not be one-to-one because of possible emissions "banking."[2] If the preference of the world community assumed risk-neutrality with respect to the Annex I total emissions level, then the increased variance in the emissions level

[2]The addition of the CDM credits may not necessarily lead to an increased variability in the Annex I total emissions level by the same magnitude during the same commitment period, when such an addition affects the emissions banking activities of the Annex I parties as well. (In an extreme case, it is even conceivable that some particularly environmentally concerned parties may choose to destroy some of their emissions permits.) However, as long as there is a close relationship between the variability of the total amount assigned and the actual emissions level, the risk-shifting inherent in the ex-post baseline scheme may well induce a change in the risk of the total emissions level

(holding its expected value intact) would have been immaterial. However, this assumption of risk-neutrality is clearly unwarranted. Furthermore, we note that the total amount of emissions assigned to the Annex I parties could have been certain were it not for the uncertain addition of the CDM credits. Hence, given that the global community is actually risk-averse, and that the CDM investment projects are inherently uncertain, we may conclude that there is a case against the CDM scheme on this account.

In the following analyses of the consequences of such risks, we shall resort to a simple mean–variance approach to characterize the decision-maker's attitude toward risks.[3] Parties are assumed to prefer a project with a high mean of returns given a fixed variance level of returns, and to prefer low variance given a fixed mean.[4] Equipped with this simplifying assumption, our task of evaluating the risk consequences under different baseline schemes is reduced to a simple matter of calculating the variances of the project returns under the alternative sets of assumptions.

CDM Investment Risks Under the Alternative Baseline Schemes

As stated above, the only source of uncertainty in the present setting is the output level x. First, let us consider the simplest case, i.e., when Π is 0 (or constant and/or independent of x), and p is certain. Let us further distinguish the following two cases:

$$\text{(i) } e_0E[x] - ex^+ \geq 0, \qquad \text{and} \qquad \text{(ii) } e_0E[x] - ex^+ < 0$$

where x^+ denotes the upper limit of the potential value of x given the probability distribution representing the uncertainty in the demand level.

Case (i) arises when the ex-ante baseline emissions level exceeds (or equals) the maximum possible actual emissions level ex^+. In this case, we obtain the most straightforward result, as stated in the following proposition, depending on whether the emissions reduction effect of the CDM project is drastic ($e_0/2 > e$) or nondrastic ($e_0 > e > e_0/2$).

Proposition 1: *Suppose* $\Pi = 0$, $\text{Var}[p] = 0$, $x^+ \leq e_0E[x]/e$, $e < e_0$. *Then the mean of the CDM project return is the same under the two baseline schemes. On the other hand, its variance is higher (lower) with the ex-ante (ex-post) baseline if the reduction is nondrastic, while the variance is lower (higher) with the ex-ante (ex-post) baseline if the reduction is drastic.*

---

[3] This is primarily for the sake of simplicity. In fact, many of the arguments developed below may carry over, even if we drop the mean–variance approach and instead evaluate the extent of risks in accordance with the stochastic dominance criterion, an approach that is better received in the economic literature

[4] This implies that they have "quadratic" risk preferences

Proof: Let the value of the project under the ex-ante baseline be denoted by V_A, and that under the ex-post baseline by V_P. Under the assumption of this proposition, we have $e < e_0$ and $e_0E[x] - ex \geq e_0E[x] - ex^+ \geq 0$. Thus,

$$V_A = p(e_0E[x] - ex) - I$$
$$V_P = p(e_0 - e)x - I$$

Hence, we find

$$E[V_A] = E(V_P)$$

As to the variance,

$$\text{Var}[V_A] = (pe)^2 \text{Var}[x]$$
$$\text{Var}[V_P] = p^2(e_0 - e)^2 \text{Var}[x]$$

Thus, since $e < e_0$, we find that $\text{Var}[V_A] - \text{Var}[V_P] > 0$ if and only if $e_0 > e > e_0/2$, i.e., when the reduction is non-drastic. ■

The proposition indicates the different ways through which the demand uncertainty is translated into a credit level under the two baseline schemes. Under the ex-ante baseline scheme, a variability in the credit equals that of the actual emissions level since the baseline level is predetermined. Under the ex-post baseline, on the other hand, the credit amount equals the output demand times the reduction rate. Then the result follows from noting that the variance of the reduction is greater than the variance of the emission level if the reduction is drastic, which in turn implies that the credit under the ex-post baseline has a higher variance.

Incentives to Manipulate Output Levels Under the Alternative Baseline Schemes

We also note that the formulae for V_A and V_P indicate that V_A is decreasing in output realization x given the ex-ante baseline $b = e_0E[x]$, whereas V_P is increasing in x, provided that $e \leq e_0$. This implies that if there is any room for the parties to affect the level of x, the investment incentives provided by the two baseline methods could be diametrically opposite.

For example, suppose that a party engaged in CDM actually owns two factories, X and Y, where only factory X is considered for CDM. Denote their respective output levels by x and y. Let the two factories share an identical linear marginal production cost function $MC_X(x) = x$, $MC_Y(y) = y$. Furthermore, assume that the total output $x + y = z$ is subject to the demand shocks analogous to the ones considered previously.

In the absence of CDM, the party would divide the realization of z into x and y so that $MC_X(x) = MC_Y(y)$, which would yield the optimal allocation: $x^*(z) = y^*(z) = z/2$ given z. If a CDM operation entity (OE) takes a naive view that the ex-ante baseline should merely be based on what would happen without a CDM, then the ex-ante baseline would be $b = e_0 E[x^*(z)] = \frac{1}{2} e_0 E[z]$, while the ex-post CDM baseline would simply be $b = e_0 x$ again.

However, unlike previously, the party now has the freedom to manipulate the output x for a given realization of z by adjusting y. Thus, under the ex-post baseline scheme, the optional choice of output, given z, is the solution of $MC_X(x) = MC_Y(y) + p(e_0 - e)$, $x + y = z$, yielding $x^*(z) = z/2 + \frac{1}{2} p(e_0 - e) > z/2$. That is, the party has an incentive to manipulate x upward. The expected value of certified emission reductions (CERs) generated in this case is then

$$E[p(e_0 - e)x^*(z)] = \frac{1}{2}(e_0 - e)E[z] + \frac{1}{2} p^2 (e_0 - e)^2$$

which is greater than the amount $\frac{1}{2}(e_0 - e)E[z]$ which would have resulted without the output-level manipulation. While the generation of such additional CERs is not favored from the viewpoint of the global environment, there is apparently no good way to cope with this increase in CERs as long as we stick to the ex-post baseline.

On the other hand, if the ex-ante baseline is naively set at $b = \frac{1}{2} e_0 E[z]$, then the party would choose x and y for a given z, so that $MC_X(x) = MC_Y(y) - pe$, $x + y = z$, which yields $x^*(z) = z/2 - \frac{1}{2} pe < z/2$. Hence, in this case, the party has an incentive to manipulate x downward, so that the expected value of CERs would be

$$E\left[p\left(\frac{1}{2} e_0 E[z] - ex^*(z)\right)\right] = \frac{1}{2}(e_0 - e)E[z] + \frac{1}{2} p^2 e^2$$

which is also greater than the amount $\frac{1}{2}(e_0 - e)E[z]$ which would have resulted if the output level was not manipulated.

Note, however, that in the case of an ex-ante baseline, a sophisticated OE could set the ex-ante baseline at $b = \frac{1}{2} e_0 E[z] - \frac{1}{2} pe^2$ instead of the naive baseline $b = \frac{1}{2} e_0 E[z]$ in order to offset the effect of the output manipulation on the generation of CERs. It may not be always practical to expect the CDM–OE to set such a sophisticated ex-ante baseline, because this would require that the OE was aware of the existence of the second factory Y, which is not subject to the CDM, and of the extent to which such output manipulation would occur.

That V_A is decreasing in x also has a bearing on a provision contained in the so-called Bonn agreement (2001), which stipulates that parties experiencing an unexpected slump should not be granted extra CDM permits because of the slowdown in their economies. The analysis above indicates that it is more or less inevitable that the ex-ante baseline scheme will grant extra credits to the parties under such circumstances.

Shifting the Risk Burden to the World Environment Under the Ex-Post Baseline Scheme

Consider a particular emission source to which a CDM project is applicable, and suppose that the Annex I parties use up all the CDM credits generated from this project without either destroying or "banking" their emissions permits.[5] Then the total quantity of emissions associated with this particular project equals the sum of the resultant emissions from this particular source itself on the one hand, and the amount $b - ex$ of CDM credits, which allows for emissions elsewhere, on the other hand, i.e.,

$$ex + (b - ex) = b$$

That is, the total emission level equals the baseline emission level b, and it is clear that the variance of b is greater under the ex-post baseline scheme than under the ex-ante scheme. Therefore, to the extent that CDM credits are fully used by Annex I parties for their emissions, the ex-post baseline contributes to an increase in the variance of the contemporaneous global emission level. The stronger the correlation among output demands faced by each party, the more pronounced this contribution would be.[6]

Proposition 2: *Given the assumption of this section, and the assumption that credits are fully consumed by the Annex I parties, the variance of the contemporaneous global emission level is higher under the ex-post baseline than under the ex-ante baseline. This effect becomes more pronounced when the correlation among the output demands faced by each CDM project is stronger.*

The Benefit of More CDM Activities

So far we have argued that the ex-post baseline scheme, as opposed to the ex-ante baseline scheme, may lessen the risks faced by Annex I parties that are contemplating investments in "nondrastic" CDM projects (Proposition 1), but only at the cost of added risks to the global environment (Proposition 2). While it is

[5] The Annex I parties may not choose to use their CDM credits right away, but may instead conduct "banking" thereof. Note, however, that even in that case, as long as the credits are not destroyed, but are to be used eventually, the above conclusion essentially remains valid in that the ex post baseline scheme leads to higher levels of cumulative emissions, which in fact is even more relevant to the state of the global climate

[6] Note that for many products, demands tend to move in the same direction, which would aggravate this tendency

the latter effect that has a direct bearing on the problems of the global environment per se, the former effect nevertheless has its own virtue, to which we shall now turn.

Suppose there are many investment CDM projects which are at the margin of adoption, and recall our assumption that the investors are risk-averse.[7] Then as far as the nondrastic projects are concerned, more of them would be adopted for the CDM. Recalling that the total emissions level associated with each single project equals the baseline emissions level b when the project is actually adopted and there is no banking, and that the ex-post baseline implies $b = e_0x$, it is clear that there being more CDM projects adopted induces no net reduction in the overall global emissions level. However, some proponents of CDM have argued that more CDM projects are valuable in their own right. That is, even if there is no net reduction in the expected global emissions level, the argument says that a higher level of economic/CDM activities may be regarded as being beneficial. Should we subscribe to this argument, this additional benefit of having more CDM activities may partly counterbalance the above-mentioned downside of the ex-post baseline scheme.

Note, however, that this benefit is limited to the case of nondrastic CDM projects. For the "drastic" projects, the project return variance, as well as the total emissions level variance, should become smaller under the ex-ante baseline. In this case, the ex-post baseline scheme would have no advantage over the ex-ante scheme on any of the accounts discussed above.

Finally, when the CDM projects have much larger risks, i.e., when the maximum possible output level is high enough to violate our previous assumption, we would have to take into account the possibility that the actual emissions level may end up exceeding the ex-ante baseline level. In this case, since CDM, while rewarding emissions reduction below the baseline, fails to punish emissions above the baseline, the formula $V_A = p(e_0E[x] - ex) - I$ now needs to be replaced by $V_A = p(\max\{e_0E[x] - ex, 0\}) - I$, so that we find that

$$E[V_A] > E[p(e_0E[x] - ex) - I] = p(e_0 - e)E[x] - I = E[V_P]$$

$$\text{Var}\,[V_A] < \text{Var}\,[p(e_0E[x] - ex) - I] = p^2(e_0 - e)^2\,\text{Var}\,[x]$$

implying that our previous formulae for the expected value and the variance of the project return understates the attractiveness of the project under the ex-ante baseline. Thus, the above argument concerning the comparison between the ex-ante and ex-post baseline schemes needs to be modified accordingly.

Possible Extensions

In concluding this section, we briefly sketch a few of the possible extensions of the setting employed in this section, namely (a) when the market price of emis-

[7] In the present mean–variance approach, this implies that the investor's utility is decreasing with the variance of the investment return

sions reduction is subject to uncertainty, and (b) when the profit from the CDM project itself is uncertain.

Additive Noise

First, consider in the case where profit from the project apart from the CDM credits is uncertain. As is well known, the variance of the sum of the two random variables exceeds the sum of the respective variances if and only if the covariance of the two variables is positive.[8] Furthermore, suppose that the demand level x is positively correlated with the profit from the project. Then under the ex-ante baseline scheme, the presence of the positive output–profit correlation has the favorable effect of reducing the overall project return variance. This is because the CDM credits amount is positively correlated with the output x under the ex-ante baseline, while the opposite is true under the ex-post baseline. This additional effect must be taken into account as we compare the project risks under the alternative baseline schemes. For example, suppose that the profit is approximately linear in x, so that we may write the profit as $\Pi(x) = A + Bx$, where A is a constant and B is a positive constant. We then find that

$$V_{\mathrm{A}} = p(e_0 E[x] - ex) + A + Bx - I$$

$$V_{\mathrm{P}} = p(e_0 - e)x + A + Bx - I$$

$$E[V_{\mathrm{A}}] = E[V_{\mathrm{P}}]$$

$$\mathrm{Var}[V_{\mathrm{A}}] = \mathrm{Var}[p(e_0 E[x] - ex) + A + Bx] = (B - pe)^2 \mathrm{Var}[x]$$

$$\mathrm{Var}[V_{\mathrm{P}}] = \mathrm{Var}[p(e_0 - e)x + A + Bx] = (B + p(e_0 - e))^2 \mathrm{Var}[x]$$

If B is so large that $B - pe \geq 0$, then we find $\mathrm{Var}[V_{\mathrm{A}}] < \mathrm{Var}[V_{\mathrm{P}}]$ unambiguously. That is, when B is large relative to pe, then the variance, or the total return from the project under the ex-ante baseline, is smaller than that under the ex-post baseline, regardless of whether the reduction is drastic or not. If B is relatively small, so that $B - pe < 0$, we find $\mathrm{Var}[V_{\mathrm{A}}] < \mathrm{Var}[V_{\mathrm{P}}]$ if and only if $e < B + e_0/2$. That is, the condition under which the ex-ante baseline yields a lower project return variance becomes less stringent, and even for a nondrastic project (i.e., a project with $e > e_0/2$), the ex-ante baseline leads to a smaller variance. Thus, the presence of a positive correlation between x and $\Pi(x)$ means that the ex-ante baseline encourages more uptake of the CDM projects.

Multiplicative Noise

Next, let us assume that the price p of emissions reduction credit is random. In this case, our analysis involves variances of random variables and their products, which is known to be fairly complex. Therefore, we will limit ourselves to a simple parametric example to explore the bottom line situations. Suppose that p is perfectly and positively correlated with x; i.e., that a world economic boom simulta-

[8] That is, $\mathrm{Var}[X + Y] = \mathrm{Var}[X] + \mathrm{Var}[Y] + 2\mathrm{Cov}[X,Y]$

neously boosts both the demand for the output of the project and the emission price determined in the world market. Let $p = cx$ with $c > 0$, so that

$$V_A = cx(e_0E[x] - ex) - I = c(-eX^2 + (e_0 - 2e)\bar{x}X + (e_0 - e)\bar{x}^2) - I$$
$$V_P = cx(e_0 - e)x - I = c(e_0 - e)(X^2 + 2\bar{x}X + \bar{x}^2) - I$$

where $X = x - E[x] = x - \bar{x}$ denotes the deviation of x from its own mean $E[x] = \bar{x}$.

Furthermore, for the sake of simplicity, let us assume that the distribution of x is symmetric around its mean. In this case, we find the following results:

$$E[V_A] = c(-e\text{Var}[X] + e_0\bar{x}^2) - I$$
$$E[V_P] = c(e_0 - e)\text{Var}[X] - I$$

so that

$$E[V_A] - E[V_P] > 0 \Leftrightarrow \bar{x}^2 - \text{Var}[X] = E[x]^2 - \text{Var}[x] > 0$$

Likewise,

$$\text{Var}[V_A] = c^2e^2\text{Var}[X^2] + c^2(e_0 - 2e)^2\bar{x}^2\text{Var}[x]$$
$$\text{Var}[V_P] = c^2(e_0 - e)^2\left(\text{Var}[X]^2 + 4\bar{x}^2\text{Var}[X]\right)$$

so that

$$\text{Var}[V_A] - \text{Var}[V_P] > 0 \Leftrightarrow (2e - e_0)\text{Var}[X^2] < (3e_0 - 4e)\bar{x}^2\text{Var}[X]$$

From these results, we first observe that the expected value of the project return is no longer equal for the two alternative baseline schemes once the emissions price is positively correlated with the output. Furthermore, since x is nonnegative and symmetrically distributed around the mean, its standard deviation cannot exceed the mean. Thus, it is unambiguously the ex-ante baseline scheme that yields the higher expected value of the project return.

Second, turning to the variance comparison, we recall our previous result (Proposition 1) that the variance is lower under the ex-ante baseline if the reduction is drastic. Indeed, if $e < e_0/2$, then $3e_0 - 4e > 0$ automatically holds, so that the above condition holds. However, unlike the previous case, the drastic reduction is no longer a necessary condition for the ex-ante baseline to prevail. For "intermediate" reductions satisfying $\frac{3}{4}e_0 > e > \frac{1}{2}e_0$, the condition becomes

$$\frac{\text{Var}[X^2]}{\bar{x}^2\,\text{Var}[X]} < \frac{3e_0 - 4e}{2e - e_0}(> 0),$$

which may be satisfied if the distribution is sufficiently thin-tailed. Finally, however, for very nondrastic reductions satisfying $e > \frac{3}{4}e_0$, the condition is an impossibility, and the ex-post baseline will then prevail.

Therefore, as far as this particular example could suggest, the overall effect of introducing the price uncertainty appears to favor the ex-ante baseline scheme on account of both the higher expected mean and the lower variance for the project return.

CDM and the Long-Run Evolution

In addition to the short-run incentive issues discussed above, there is another important factor that ought to be considered in designing a CDM or Kyoto mechanism, namely, the incentive toward the longer-term evolution of the mechanism, i.e., the extent to which currently non-Annex I (or B) parties are motivated to join the Annex I category.

Although this evolution question may well involve quite intricate dynamic aspects, we shall not pursue the avenue of developing a fully dynamic model, but instead explore the impact of uncertainties on a nation's decision for an Annex I/non-Annex I status within the template of a simple static model.

That such decisions are made at the national level motivates another methodological simplification that we make. In stark contrast to our microscopic analyses in the previous section, here we will employ an aggregate "macroscopic" model of the CDM, which is often found in the literature. That is, instead of focusing on individual decisions regarding the adoption of separate CDM projects, we consider a continuum of such projects, and a nation as a whole that can control its aggregate emissions level by subjecting such projects to CDM one after another, starting with lower-cost projects and proceeding to higher-cost projects. Let $k(.)$ be the aggregate emissions abatement cost function, so that the cost necessary to bring the aggregate emissions from its initial level e_0 down to a given level e is given by $k(e_0 - e)$. The marginal cost of emissions reduction is then $M(e_0 - e) \equiv k'(e_0 - e)$, and we assume that the marginal abatement cost is initially zero, i.e., $M(0) = 0$. Furthermore, for $p \geq 0$, we denote the inverse function of $M(.)$ by $M^{-1}(p)$, i.e., $M(M^{-1}(p)) = p$.

We consider a "small" nation H which currently has the non-Annex I status. If it chooses to retain the non-Annex I status, its CDM project opportunities are as characterized above. It then faces no direct obligation to cut back emissions, but is motivated to do so only because of the benefit that accrues via CDM. On the other hand, if it chooses to switch to Annex I status, it will assume an emissions reduction obligation, and join a "cap and trade" emissions quota scheme that operates among the Annex I parties.

In order to examine the effect of CDM on the incentive for nation H to join the Annex I camp rather than remaining in the non-Annex I group, we resort to a simple method of identifying the quota level at which H is equally drawn to the two options. This is in line with the work of Bohm and Carlen (2000). However, unlike their study, we assume that CDM is fully functional. Also, while their study emphasizes a comparison between the monetary inducements and quota adjustments applicable to a nation once it has joined the Annex I parties, we focus primarily on the latter policy device, for the exact choice of device is of little importance for our purpose.

Certainty Case: A Benchmark

First, let us consider the case without uncertainty, which will be regarded as the point of reference for our subsequent analyses. We will discuss the situation when nation H, as a non-Annex I party, stands as a CDM host country, and then an alternative hypothetical situation where it actually joins the Annex I camp and assumes its obligation to cut down emissions on its own.

Clean Development Mechanisms

Since the CDM host nation H is a small country, its decision does not affect the emission price, p, determined by other nations. Taking that as given, nation H chooses its aggregate emissions level e (by way of choosing CDM projects) to obtain credits of value $p(e_0 - e)$ at a cost of $k(e_0 - e)$.

Here, we assume that $e_0 = e(0)$ is the baseline level. Then e is determined as a solution to

$$\operatorname*{Max}_{e}\{p(e_0 - e) - k(e_0 - e)\}$$

and the first-order condition is satisfied when $e = e^* \equiv e_0 - M^{-1}(p)$, or equivalently when $p = M(e_0 - e)$.

Typically, the CDM involves parties or private agents who invest in Annex I belonging to those nations, so that the intake for the host nation H would only be a fraction of the total gain. The proportion s that actually accrues to the host nation H should reflect its bargaining power vis-a-vis the investing parties, and the benefit for H would equal $s\{p(e_0 - e^*) - k(e_0 - e^*)\}$.

When Nation H Joins Annex I

Now suppose that nation H joins the Annex I camp, and faces its own emissions reduction obligation. Under the cap and trade scheme, a transferable quota q is levied on H's emission level. If the actual emission level e is less than q, then H can sell the superfluous quota, while it has to purchase emission rights if q is less than e. The benefit for H is then $p(q - e) - k(e_0 - e)$, and a maximization yields the same first-order condition as before, i.e.,

$$p = M(e_0 - e)$$
$$e = e^* \equiv e_0 - M^{-1}(p)$$

Incentive for Nation H to Join Annex I

Given these results, we may now ask under what condition does nation H have an incentive to move from non-Annex I status to Annex I status. To be more specific, we wish to identify the quota level that would make nation H just indifferent between the two options, namely CDM and the cap and trade quota scheme.

Recalling that the net benefit that accrues to nation H under the CDM is $s\{p(e_0 - e^*) - k(e_0 - e^*)\}$, while that under the quota is $p(q - e^*) - k(e_0 - e^*)$, the quota level that equates them is the solution to

$$s\{p(e_0 - e^*) - k(e_0 - e^*)\} = p(q - e^*) - k(e_0 - e^*)$$

i.e.,

$$q = q^* \equiv \frac{(1-s)(pe^* + k(e_0 - e^*)) + spe_0}{p}$$

Since the CDM benefit is nonnegative, it is clear that q^* is increasing in s. When $s = 0$ we have $q^* = e^* + k(e_0 - e^*)/p$, and when $s = 1$ we have $q^* = e_0$.

Uncertainty

Consider a local uncertainty which only affects the marginal costs of H. As an example, we consider the case where the national population change, denoted by y, affects the general level of economic activities, and therefore the emission level. We make a simplifying assumption that this change in population affects only the status quo emissions level e_0, while keeping the marginal emissions reduction cost schedule $M(.)$ intact. To indicate this dependence explicitly, we shall henceforth denote the status quo level by $e_0(y)$.

Once $e_0(y)$ is determined, the abatement cost is given by $k(e_0(y) - e)$, where y affects the cost only through $e_0(y)$.

Furthermore, we assume for simplicity that $e_0(y)$ depends linearly on y. Combined with another assumption that $E[y] = 0$, this yields the result that the expected value of the status quo emissions level equals the status quo level given $y = 0$, i.e., $E[e_0(y)] = e_0(0)$.

CDM: Ex-Ante Versus Ex-Post Baseline Schemes

When there is no uncertainty, we could simply identify the CDM baseline emissions level with, for example, the "status quo" emissions level associated with zero marginal abatement cost. Now that the status quo level itself is subject to a random fluctuation, we ought to distinguish between the ex-ante and the ex-post baseline schemes in an analogous manner to the previous section, which, in fact, is in accordance with policy proposals made in reality.

In the current setting, the ex-ante baseline is the expected value $E[e_0(y)]$ of the status quo emissions level, capturing the idea that the baseline should be determined at the time of investment. However, since we have $E[e_0(y)] = e_0(0)$, the ex-ante baseline level equals $e_0(0)$, which is the status quo level associated with $y = 0$. On the other hand, the ex-post baseline is simply $e_0(y)$ itself, which is determined after the realized value of y is observed.

We shall assume that the actual choice of the emissions level e is made only after the realization of y is revealed. This is a strategic assumption that we make in order to focus our attention on the impact of uncertainty on the choice of national status (i.e., Annex I versus non-Annex I), ruling out its impact on the emissions reduction choices.

Under the ex-post baseline scheme, the CDM benefit that accrues to nation H given the share parameter s, as well as the realization of y, is

$$V_P(y) = s\{p(e_0(y) - e^*(y)) - k(e_0(y) - e^*(y))\}$$

where the optimal emissions level $e^*(y)$ is the solution to $p = M(e_0(y) - e^*(y))$, or equivalently $e_0(y) - e^*(y) = M^{-1}(p)$. Thus, it turns out that $V_P(y)$ is actually independent of the population change y. We therefore find that

$$E[V_P(y)] = s\{pM^{-1}(p) - k(M^{-1}(p))\}$$

$$\mathrm{Var}\,[V_P(y)] = 0$$

On the other hand, under the ex-ante baseline, the benefit given to s and y is

$$\begin{aligned} V_A(y) &= s\{p(e_0(0) - e^*(y)) - k(e_0(y) - e^*(y))\} \\ &= V_P(y) + sp(e_0(0) - e_0(y)) \end{aligned}$$

where $e^*(y)$ is the solution to the same condition $p = M(e_0(y) - e^*(y))$. Hence, given our assumption that $E[e_0(y)] = e_0(0)$, we find that

$$E[V_A(y)] = E[V_P(y)] = s\{pM^{-1}(p) - k(M^{-1}(p))\}$$

$$\mathrm{Var}\,[V_A(y)] = s^2p^2\mathrm{Var}\,[e_0(y)]$$

When Nation H Joins Annex I

Similarly, the benefit that accrues to nation H given the share parameter s, as well as the realization of y, is

$$\begin{aligned} V_Q(y) &= p(q - e^*(y) - k(e_0(y) - e^*(y))) \\ &= \frac{1}{s}V_P(y) + p(q - e_0(y)) \end{aligned}$$

where $e^*(y)$ is the solution to the same condition $p = M(e_0(y) - e^*(y))$. Hence, we find that

$$E[V_Q(y)] = \frac{1}{s}E[V_P(y)] + p(q - e_0(0))$$

$$\mathrm{Var}\,[V_Q(y)] = p^2\mathrm{Var}\,[e_0(y)]$$

Incentive for Nation H to Join Annex I

Given these results, we now turn to the question of identifying the quota level that would make nation H just indifferent between the CDM and the cap and trade quota scheme. As before we shall take the mean–variance approach for simplicity, i.e., we assume that the utility function $u(.)$ representing the risk attitude of nation H is quadratic.

Let the solutions to $E[u(V_Q(y))] = E[u(V_A(y))]$ and $E[u(V_Q(y))] = E[u(V_P(y))]$ be denoted by q_A and q_P, respectively. That is, if the quota is set at the level of q_A, nation H would benefit equally from the quota scheme (Annex I status) or the ex-ante CDM scheme (non-Annex I status). Likewise, if the quota level is set at q_P, nation H benefits equally from the quota scheme or the ex-post CDM scheme.

Now recall the critical quota level q^* under certainty which was identified previously:

$$q^* \equiv \frac{(1-s)(pe^* + k(e_0 - e^*)) + spe_0}{p}$$

Among the three critical quota levels q_A, q_P, and q^*, we get the following result.

Proposition 3: *The three critical quota levels q_A, q_P, and q^* satisfy*

$$q^* \leq q_A \leq q_P$$

Furthermore, all these quota levels are increasing in s.

Proof: First, it is clear that $E[u(V_P(y))] > E[u(V_A(y))]$ since

$$E[V_P(y)] = s\{pM^{-1}(p) - k(M^{-1}(p))\} = E[V_A(y)]$$

$$\mathrm{Var}\,[V_P(y)] = 0 < s^2p^2\mathrm{Var}\,[e_0(y)] = \mathrm{Var}\,[V_A(y)]$$

Thus, the quota level needs to be more lenient for the quota scheme to be able to match the ex-post CDM rather than the ex-ante CDM. Hence, $q_A \leq q_P$.

Second, it is straightforward to show that q^* actually solves $E[V_A(y)] = E[V_Q(y)]$. On the other hand, we have shown that

$$\mathrm{Var}\,[V_Q(y)] = p^2\mathrm{Var}\,[e_0(y)] \geq s^2p^2\mathrm{Var}\,[e_0(y)] = \mathrm{Var}\,[V_A(y)]$$

Thus, if the quota is set at the level of q^*, then the quota scheme cannot match the ex-ante CDM scheme because of its larger variance. Thus, to restore the balance, the quota level must be made less stringent, i.e., $q^* \leq q_A$. ■

Discussions

The proposition above indicates that the critical quota level (i.e., the least lenient quota level that can induce nation H to switch to Annex I status from its current non-Annex I status) needs to be less stringent in the presence of uncertainty.

The presence of uncertainty also calls for a distinction between the ex-ante and the ex-post baseline schemes. It was then found that the effect of uncertainty on the critical quota level is more pronounced if the CDM employs the ex-post baseline rather than the ex-ante baseline. In the previous section, we argued that the ex-post baseline scheme, while mitigating the risk faced by the CDM investors, would in fact simply translate it to another risk faced by the global environment. The case is more or less the same in the present analyses. In addition to this, the present finding indicates that the ex-post baseline is likely to weaken the incentive perceived by nations currently under non-Annex I status to switch to Annex I status. Thus, so long as we value the evolution perspective, this can be construed as an additional downside of the ex-post baseline scheme.

The present analysis also has significant bearing on some of the most topical CDM-related issues, namely that of so-called "unilateral" CDM, and that of so-called CDM "low hanging fruits." Unilateral CDM allows non-Annex I parties to conduct CDM among themselves, which is likely to increase the overall CDM opportunities, and thus the likelihood of forcing the quota scheme to set more lenient quotas in order to match such expanded CDM opportunities. Likewise, if the CDM low hanging fruits argument works in favor of a higher share of CDM benefits accruing to the host nations, the analysis above implies that this would again lead to a more lenient quota level in order to match the increased CDM opportunities.

One final remark is in order regarding the assumption made in the above analyses. With risk-averse preferences, it is possible that the investing party and the CDM host nation H may come up with a contractual arrangement which is not necessarily a fixed-share arrangement. This would particularly be the case if risk attitudes differ among parties. For instance, if the investing party is risk-neutral (as assumed in Bohm and Carlen 2000), then full insurance would be given to risk-averse nation H. In this case, $q_A = q_P$ should hold.

Further Extensions

We now list some possible extensions of the above analyses. As in Bohm and Carlen (2000), we could lift the "small country" assumption to make the general equilibrium effect more important. As in their analysis also, we could compare quota versus other policy instruments in terms of the costs imposed on the rest of the world. Finally, we could consider other sources of uncertainty. These are left for future research.

Conclusion

After the Marrakech Accord (in 2001), the likelihood of the ratification of the Kyoto Protocol, as well as of the enactment of Kyoto mechanisms, has significantly increased. While the bottom-line principles and general guidelines have already been laid out, there remain numerous practical details to be settled. While the analyses conducted in this chapter need not be construed as being negative to the adoption of the CDM scheme itself, they do point to the important consequences of such practical details in dealing with the impact of uncertainties. It is important for the administrative authorities, e.g., the CDM executive board who are in charge of the actual practice of CDM, properly to address the risk and/or incentive issues discussed above in order to ensure the functionality of the CDM, as well as the "evolution" of the entire Kyoto mechanism.

Acknowledgments. Thanks are due to Mayumi Horie-Nakagawa, Mitsuru Nakagawa, and Kuninori Nakagawa for their comments.

Reference

Bohm P, Carlen B (2000) Cost-effective approaches to attracting low-income countries to international emissions trading: theory and experiments. Mimeograph (available from authors)

Feasibility Study on a CDM Project and an Investigation into an Effective Institution to Make CDM Projects Viable

RYUJI MATSUHASHI

Summary. The clean development mechanism (CDM) is expected to facilitate technology transfers from developed to developing countries, as well as to reduce greenhouse gas emissions economically. In the first part of this chapter we evaluate a CDM project utilizing photovoltaic (PV) systems. For this, we first estimated life-cycle CO_2 emissions from photovoltaic systems and other power generation systems based on input–output tables for environmental analysis. The cost of the photovoltaic systems was also evaluated, taking the effect of mass-production into consideration. Next we developed two dynamic scenarios to disseminate PV systems. One is a scenario for domestic dissemination only, while the other includes a CDM project utilizing PV systems as well as domestic dissemination. As a result, the internal rate of return (IRR) on the CDM was estimated to be very low. We proposed some measures to improve the IRR of the CDM utilizing photovoltaic systems.

In the second part, we explored some efficient institutions which could make CDM projects viable. For this purpose, we estimated IRR and other indicators of profitability for 42 CDM projects, taking account of volatilities in the price of certified emissions reductions (CER) and other costs. As a result of Monte Carlo simulations, the expected values and standard deviations in the IRR of the projects were quantitatively found. We also evaluated the distributions of return on equity (ROE) and debt-recovery rate to clarify the impacts of the debt/equity (D/E) ratio on project risks. In particular, CDM projects depending strongly on the CER value are exposed to higher risks than ordinary overseas projects. We concluded that it is effective to reduce the D/E ratio to some extent to suppress financial risks, so that CDM projects will be viable. Finally, we referred to partial security measures for CDM finance to realize the above-mentioned modifications.

Key words. Clean development mechanism (CDM) project, Photovoltaic systems for houses, Life-cycle analysis, Project finance, Risks in CDM, Debt/equity ratio, Monte Carlo simulation, Internal rate of return (IRR)

Institute of Environmental Studies, Graduate School of Frontier Sciences, University of Tokyo, 7-3-1 Hongo, Bunkyo-ku, Tokyo 113-8656, Japan

Introduction

Degradation of the global environment and the depletion of resources are becoming serious threats to the sustainable development of humankind. In particular, there has been growing concern about climate change caused by increases in greenhouse gases. Although Annex I countries in the United Nations Framework Convention on Climate Change (UNFCCC) have to control their greenhouse gas emissions according to the Kyoto Protocol, greenhouse gas emissions such as CO_2 are increasing in most countries. We therefore need to explore efficient and fair ways of internationally reducing greenhouse gas emissions. Under these circumstances, the clean development mechanism (CDM) is expected to provide powerful options to suppress the differences between the North and the South, as well as to reduce greenhouse gas emissions economically.

In this chapter, we first evaluate a CDM project utilizing photovoltaic (PV) systems. We then evaluate the profitability of various CDM projects to explore efficient institutions which will make CDM projects viable.

Life-Cycle Analysis of PV Systems

Life-Cycle CO_2 Emissions by PV Systems

Life-cycle inventories of PV systems are generally classified into four processes, i.e., manufacturing silicon wafers, PV cells, and PV modules, and fabricating the system, including peripheral equipment such as an electric inverter. The input materials and energy required in the above processes are shown in Tables 1–3. These data are quoted from Kato et al. (1999).

We input these values into the corresponding items in $\boldsymbol{f}$ in Eq. 1, so that life-cycle inventories for PV systems are obtained from Eq. 2.

$$\boldsymbol{X} = (I - A)^{-1}\boldsymbol{f} \tag{1}$$

$$C = \boldsymbol{eX} \tag{2}$$

where $\boldsymbol{X}$ is the production vector, $\boldsymbol{A}$ is the input–output matrix, $\boldsymbol{f}$ is the final demand vector, C is the estimated CO_2 emissions, and $\boldsymbol{e}$ is the CO_2 emissions vector. The environmental database developed by Ikeda et al. (1996) was used as the source for vector $\boldsymbol{e}$. The life-cycle CO_2 emissions of PV and other power generation systems were then estimated based on this concept. Figure 1 shows the estimated results. This figure indicates that the life-cycle CO_2 of PV systems is much less than that of fossil-fuel power generation. These values were used to estimate the certified emission reductions (CER) acquired by a CDM project using PV systems.

Cost of PV Systems

The cost of the PV system was also estimated using input–output price data, and was estimated to be 225 000 yen/kW. Price data from the input–output table are

TABLE 1. Input materials and energy needed to manufacture silicon wafers (10 MW)

			Input	Unit
High-purity silica process	Raw material	Silica sand	462	t
	Water glass	Soda lime	272	t
		Crude oil	172.8	kl
	Refinery	Acids	1 051	t
		Limestone	959.7	t
		Electricity	525.6	MWh
		Water	22 987	kl
	Carbon pellets	Acetylene	161	t
		Resorcinol	97	t
		Electricity	4 028	MWh
		Cl_2	36.6	t
		NaOH	22	t
Solar-grade silicon process	Silica reduction	Argon	16.4	$1000\,m^3$
		Graphite nozzle	1.8	t
		Graphite electrode	9.1	t
		Electricity	2 548	MWh
	Decarbonization filter	Electricity	309.4	MWh
	One-way coagulation	Brick	4 830	kg
		Electricity	250	MWh
	Slice	SiC	231	t
		Piano wire	37	t
		Electricity	1 750	MWh

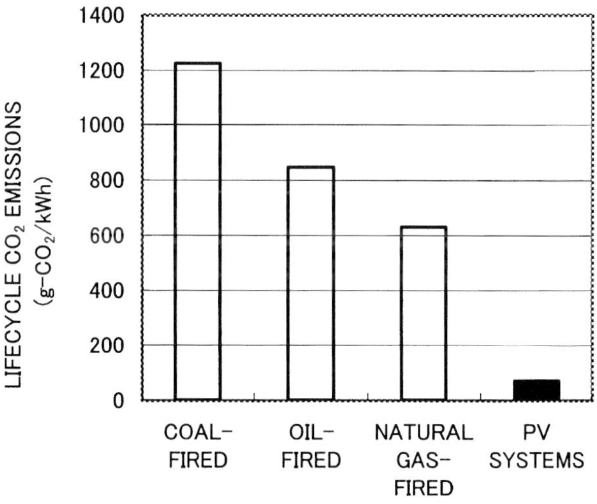

FIG. 1. Life-cycle CO_2 of photovoltaic (PV) systems and other power generation technologies. *Black bar*, capital production; *open bars*, operation and maintenance

TABLE 2. Input materials and energy needed to manufacture photovoltaic cells (10 MW)

		Input	Unit
Foundation surface processes (texture etching and cleaning)	KOH	11	t
	IPA	27	t
	Water	3666.7	t
	Teflon	0.38	t
	Electricity	80	MWh
Diffused joint formation (phosphorus diffusion)	$POCl_3$	0.59	t
	N_2/O_2	1.18	t
	Quartz pipe	0.25	g
	Quartz holder	0.05	g
	Water	3666.7	t
	Teflon	0.38	t
	Electricity	450	MWh
Protection film formation (TiO_2CVD)	TPT	0.7	t
	Electricity	280	MWh
Back-side etching	KOH	11	t
	IPA	27	t
	Water	3666.7	t
	Teflon	0.38	t
	Electricity	80	MWh
Back-side electrode formation	Screen	0.0073	Number
	Ag–Al paste	1.28	t
	N_2/O_2	200	t
	Electricity	420	MWh
Surface-side electrode formation	Screen	0.0073	Number
	Ag–Al paste	0.66	t
	N_2/O_2	400	t
	Electricity	420	MWh
Specific measurement	Xe lamp	Monetary	–
	Case	Monetary	–
	Electricity	15	MWh

TABLE 3. Input materials and energy needed to manufacture photovoltaic modules (10 MW)

	Input	Unit
Glass	84.5	1000 m^3
Aluminous frame	154	t
Copper wire	7.7	t
Filler (EVA)	92.2	t
Tedola (PVF)	12.7	t
Aluminous sheet	6.6	t
Back-side seal	19.3	t
Peripheral seal	9.3	t
Electricity	188	MWh

generally those of goods manufactured by mature processes. However, processes to manufacture PV systems are not very mature, so that the present cost does not coincide with the above values. We therefore interpret the estimated value as the ultimate cost, assuming that the present cost of producing PV systems will approach the ultimate cost as the annual production scale increases. The relationships between costs and production in the past were then used to identify parameters in a learning curve. In sum, we introduced the learning curve as Eq. 3, taking the effect of mass production into consideration.

$$\mathrm{Ln}\{\mathrm{Cost}(t) - 22.5\} = 6.02 - 0.20 \times \mathrm{Ln}\{\mathrm{Production}(t)\} \quad (3)$$

where Cost(t) is the capital cost of PV systems in a year $\boldsymbol{t}$, and Production(t) is the annual production of PV systems in a year $\boldsymbol{t}$.

Regional Differences for CDM Projects

In order to determine which regions are appropriate for CDM projects, we summarized the relevant data, as shown in Table 4. In CDM projects, PV systems are disseminated into regions, where the following conditions are satisfied.

1. Air pollution is so serious that the replacement of fossil power generation by PV systems could bring about environmental benefits. Information on air pollution is shown in the 5th column in Table 4.
2. Sufficient solar radiation can be received to improve the economics and CO_2 reduction potential of PV systems. Information on solar radiation is shown in the 4th column in Table 4.
3. The number of households is large enough to disseminate PV systems. This information is shown in the 3rd column in Table 4.

These conditions led us to adopt the province of Liaoning as a model for the CDM in this analysis.

Evaluation of the CDM Project Using PV Systems

Dynamic Scenarios to Disseminate PV Systems

Next, we investigated some dynamic scenarios for the CDM project, as well as its marketability. The Japanese government is currently subsidizing approximately one-third of the investment costs to houses installing PV systems on their roofs. Future policy on this subsidy will affect the marketability of the CDM project as well as domestic dissemination. Therefore the effects of mass production and the subsidy were both taken into consideration in this analysis. We thus assumed the two following scenarios. Case A simulates the domestic dissemination of PV systems without the CDM, while Case B simulates the domestic dissemination of PV systems along with the CDM.

TABLE 4. Number of households, solar radiation, and air pollution in provinces in China

	Province	Number of households	Equivalent annual solar radiation (h)	Number of air pollution incidents
1	Liaoning	6074538	1438	82
2	Guizhou	1791641	725	50
3	Tianjin	1721214	1428	27
4	Gansu	1666880	1694	26
5	Hebei	4030703	1253	26
6	Sichuan	4886504	680	25
7	Yunnan	2348060	1558	20
8	Shanxi	2456062	1090	12
9	Guangxi	2325513	1080	12
10	Jilin	4152994	1583	11
11	Shandong	7262846	1476	10
12	Qinghai	485322	1664	6
13	Anhui	4331175	1359	6
14	Zhejiang	3279579	1236	5
15	Neimenggu	3242198	1669	3
16	Ningxia	495825	1677	2
17	Henan	5011124	1307	2
18	Shanxi	2544667	1653	1
19	Jiangsu	5907776	1466	1
20	Heilongjiang	6655164	1406	1
21	Xinjiang	2358863	1398	1
22	Jiangxi	3679714	1335	1
23	Guangdong	4014048	1084	1
24	Fujian	2638379	945	1
25	Xizang	109371	1929	0
26	Beijing	2853211	1459	0
27	Hubei	6636000	1419	0
28	Hainan	663252	1417	0
29	Shanghai	3470314	1208	0
30	Hunan	3989493	1048	0
31	Chongqing	2036328	684	0

— Case A: the Government is assumed to continue the subsidy for PV systems of one-third of their investment cost until 2020.
— Case B: the Government is assumed to continue the subsidy for PV systems of one-third of their investment cost until 2020. The CDM will be initiated around 2010 as a project. In this project, PV systems will be disseminated in an appropriate region in China.

In order to estimate domestic demand for PV systems in the future, the logit share function identified by Okushima et al. (2000) was used in this analysis. We introduced Eq. 4, taking the effect of the subsidy into consideration.

$$\mathrm{Prob}(t) = \frac{\exp[42.03 - 3.110 \times \mathrm{Ln}\{\mathrm{Cost}(t) - \mathrm{Subsidy}(t)\}]}{1 + \exp[42.03 - 3.110 \times \mathrm{Ln}\{\mathrm{Cost}(t) - \mathrm{Subsidy}(t)\}]} \tag{4}$$

FIG. 2. Annual production capacity of PV systems in case A in Japan (kW)

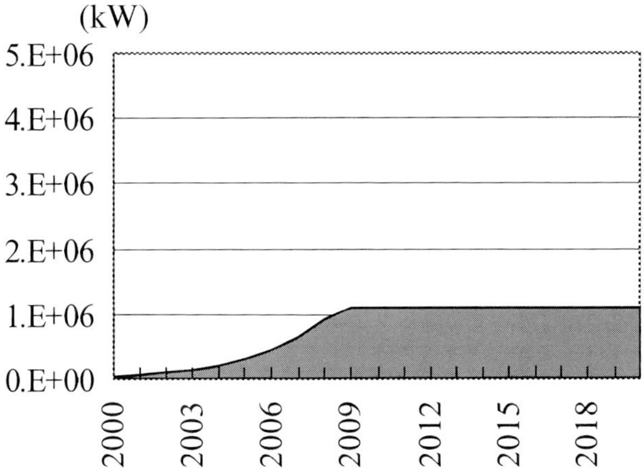

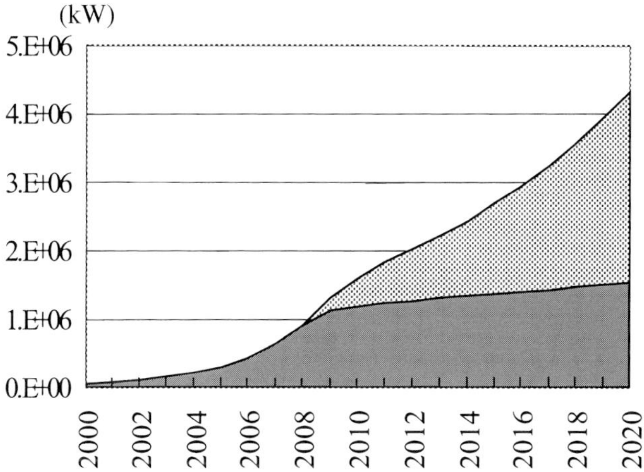

FIG. 3. Annual production capacity of PV systems in case B. *Black*, increased capacity in Japan (kW); *gray*, PV systems transferred to China (kW)

where Prob(t) is the rate of adoption if all households take a system.

and

$$\text{Potential demand for roof-top PV systems} = \text{Prob}(t) \times \text{annual number of household constructions} \times \text{PV capacity per household} \quad (5)$$

The evaluated results are shown in Figs. 2–5. Figures 2 and 3 indicate that domestic demand for PV systems will be saturated around 2010. On the other hand, the introduction of the CDM project will bring about a steady increase in annual PV system production. Figures 4 and 5 indicate that CO_2 reduction by introducing

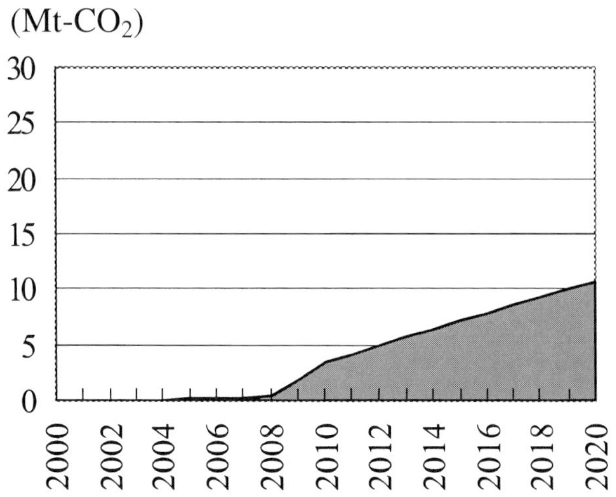

Fig. 4. CO_2 reduction by disseminating PV systems in case A in Japan

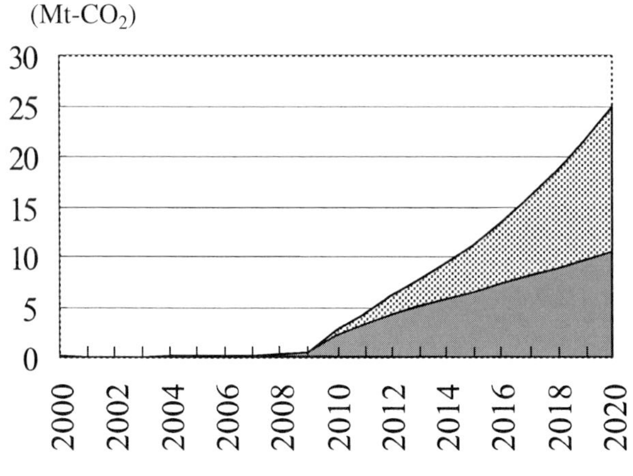

Fig. 5. CO_2 reduction by disseminating PV systems in case B. *Black*, Japan; *gray*, China certified emissions reductions

PV systems is not remarkable until 2010, regardless of the CDM project. In other words, the introduction of PV systems cannot be expected to make a significant contribution to CO_2 reductions in the first commitment period of the Kyoto Protocol. However, it can contribute to considerable CO_2 reductions between 2010 and 2020, including the second and the third commitment periods. Thus, it takes a long time for novel energy technologies to help to improve the global environment.

Internal Rate of Return of the CDM

Next we estimated the internal rate of return (IRR) of the CDM project in Case B, based on the following conditions.

1. The project is assumed to be initiated in 2010. The depreciation time is assumed to be 20 years.
2. Maintenance and transaction costs are assumed to be 3.0% and 30% of the investment costs, respectively. We set a relatively high value for the transaction costs, since negotiations to acquire verification and certification from operational entities and approval from host countries inevitably lead to cost escalation. The trade-in price of the project after 20 years is assumed to be 10% of the investment cost.
3. Certified emission reductions (CER) by the CDM project are estimated assuming that PV systems are substituted for the average electricity mix in Liaoning province.
4. The revenue of the project will be from sales of electricity and CERs.
5. The present price of CER is assumed to be 4 US\$/t-CO_2 based on current data.
6. According to an estimation using a global energy model, DNE (Akimoto et al. 1998), the price of emissions permits will rise to 22 US\$/t-CO_2, which corresponds to 21% of the annual escalation. However, this was a theoretical case in which the USA ratified the Kyoto Protocol. From present circumstances, possibility of the USA's participation is very slight. Therefore, we set a more conservative annual escalation of the CER price, i.e., 10%. After 2010, the price is assumed to increase at the same constant rate as between 2001 and 2010.
7. Electricity is supplied to household sectors in the target region by the project. The price of electricity is assumed to increase from the present value at a constant rate of 3.0%.
8. The annual tax rate for revenue was assumed to be 30%.

As a result, the IRR was estimated to be –0.5%. In ordinary market conditions, it would be impossible to introduce the CDM project in China. If some of the above assumptions were improved, the IRR could also be improved. For instance, if the transaction cost is reduced from 30% to 10%, the IRR increases to 0.7%, which is still very low. If the PV systems were produced not in Japan but in China, the investment cost could be further decreased, leading to a higher IRR.

We should also be aware of benefits other than profitability from this CDM project. In particular, the mitigation of air pollution will improve the acceptability of this project.

Schemes to Make CDM Projects Viable

In this chapter, we describe how to make CDM projects more attractive to the market.

Present Situation on CDM Projects

The institution of CDMs has been intensively discussed based on the Kyoto Protocol. In particular, issues of "additionality" are significant to certify projects such as CDMs. From past interpretations on investment additionality, operational entities would not certify a project as a CDM if it were profitable, even without revenue from sales of CERs. The interpretation of investment additionality was changed in the COP-7 meeting in Morocco. According to the agreements in COP-7, a project could be certified as a CDM if it was profitable without CER revenue. Nevertheless, the past interpretation of investment additionality seems still to be alive in some developing countries.

Under such circumstances, the profitability of CDM projects must be evaluated cautiously, so that the projects can be carried out. We therefore investigated the overall profitability of CDM projects, and the feasibility studies were performed under the sponsorship of New Energy and Industrial Technology Development Organization (NEDO). Table 5 shows the lists of the projects about which we were able to obtain the necessary information to evaluate the IRR. Data on initial investment costs, annual running costs, annual production of main commodities such as electricity, and annual reduction of equivalent CO_2 emissions were acquired from NEDO's reports on each project in Table 5.

Table 5 shows various types of CDM project, including a fuel switch from coal to natural gas-efficiency improvements in industrial sectors, the recovery of methane from coal mining, and energy conservation in commercial and residential sectors. These projects have different characteristics, especially in their revenue structure. If a CDM project is on high-efficiency power generation, the revenue is from sales of electricity and CER. Most projects rely on main products such as electricity, as well as CER, for revenue. However, the share of CER value in the total revenue is different, depending on the characteristics of each project. For instance, shares of CER sales are generally high in projects which recover methane from coal mining and utilize the gas for power generation, town gas production, or methanol synthesis. In such projects, the recovered methane is converted into equivalent CO_2 reductions by multiplying by 21 as its global warming potential. Thus, equivalent CO_2 emissions become large amounts, so that revenue from CER sales often exceeds sales of the main products. On the other hand, the share of the CER value would be relatively small in projects covering efficiency improvements in fossil-fuel power generation. The distribution of initial shares of CER sales in the total annual revenue from 42 projects is shown in Fig. 6. This figure shows the differences in revenue structure.

Monte Carlo Simulations to Estimate Risks Accompanying CDM Projects

Assumptions of Monte Carlo Simulations

The volatility of future CER value is generally higher than that of the value of the main product. Therefore, a CDM project which relies mainly on CER value

TABLE 5. Examples of feasibility studies on CDM projects sponsored by NEDO

	Project	Host country
1	Project to recover coal mine methane in China	China
2	Power generation utilizing waste heat in cement production in China	China
3	Feasibility study on a project to reduce CO_2 emissions through power generation by residual oil in petroleum refinery	China
4	Development of combined-cycle power generation systems utilizing blast furnace gas for iron and steel company in China	China
5	Rehabilitation of 300 MW coal-fired power generation in China	China
6	Survey of dissemination situation of various kinds of kiln in cement industry in China	China
7	Projects of energy savings in the Shengyang iron and steel company in Liaoning Province	China
8	Evaluation of efficiency improvement in conventional fossil-fuel power plants in China	China
9	Rationalization of energy consumption in Chinese iron and steel industries	China
10	Improvement of low-quality limestone to reduce CO_2 emissions in Chongqing city in China	China
11	Energy conservation projects in Bak and Novobak petroleum refinery plants	Azerbaijan
12	Energy conservation projects in Khabarovsk petroleum refinery plants	Russia
13	Investigation of improved waste disposal system in factories of producing starch from tapioca	Indonesia
14	Evaluation of energy-saving options in a commercial building in Indonesia	Indonesia
15	Rehabilitation of fossil-fuel power stations in Lyazanskaya city	Russia
16	Reduction of losses in electric power generation, transmission, and consumption in Myanmar	Myanmar
17	Replacement of coal-fired power stations by combined-cycle power generation in Poland	Poland
18	Improvement of heating reactors in three petroleum refinery plants in Quivishef, Shizranniand, Novoquibishef in Russia	Russia
19	Fuel switch from conventional coal-fired power plants to natural-gas-fired power plants in Sakhalin Province	Russia
20	Rehabilitation of fossil-fuel power stations in Khabarovsk	Russia
21	Rehabilitation of fossil-fuel power stations in Chekingskaya city	Russia
22	Fuel-switch project in the first and ninth combined heat and power plants in Irkutsk	Russia
23	Combined-cycle projects in Konakovo power stations in Russia	Russia
24	Fuel switching from Amursk coal-fired power plants to natural-gas-fired power plants in Russia	Russia
25	Fuel-switch project in power plants in Ignovaskaya	Russia
26	Repowering of three gas-fired power stations	Russia
27	Project of generic rehabilitation in fossil-fired power plants in provinces along the oceans	Russia

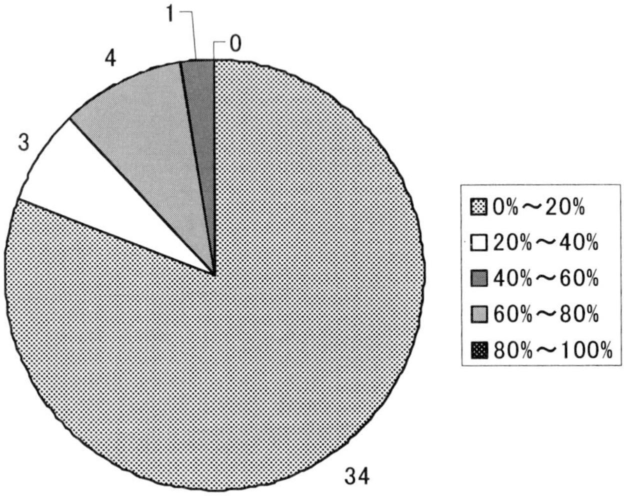

FIG. 6. Distribution of initial shares of certified emissions reductions (CER) sales in total annual revenue from the CDM projects

is exposed to higher risks. If the price of CER did not rise as anticipated, the project would not bring in sufficient revenue to the sponsors of the project. Conversely, the sponsors would receive more revenue with higher prices for CERs. In particular, the participation of the USA would promptly raise the price and increase the revenue of the project.

In order to quantify these risks, we performed Monte Carlo simulations on the profitability of CDM projects. We estimated the IRR of 42 projects by Monte Carlo simulations. Economic data were acquired from NEDO's reports.

The major assumptions of the estimations are same as those described above. Other necessary assumptions for these simulations are listed below.

1. We estimated an implied volatility in CER price from present data on call options by the Black–Scholes equation. Based on this estimation, the volatility was assumed to be 23%.
2. The annual escalation of main products such as electricity was assumed to be 3%. The volatility in prices of the main products was assumed 1.5%.
3. We estimated the IRR of each CDM project by generating 100 000 random numbers according to the above assumptions.

Figure 7 shows the computed results of the Monte Carlo simulations. It depicts the expected values and standard deviations of the IRR of the 42 CDM projects with and without CER revenues. This figure shows how risks and returns in CDM projects increase by including revenues from CER sales. Although Fig. 7 tempts us to make portfolios of various CDM projects, it would difficult to lower the risks accompanying investment in CDM projects owing to the high correlation among revenues by the projects. The risks accompanying CDM projects are generally classified as follows.

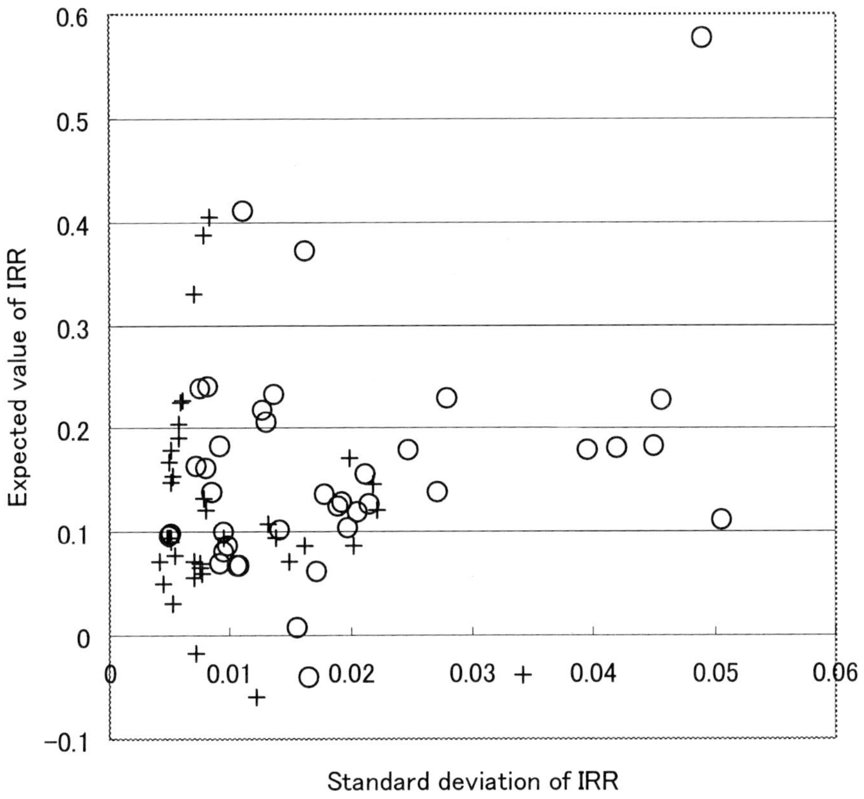

FIG. 7. Expected values and standard deviations in the internal rate of return (IRR) for the 42 CDM projects with and without CER revenues

1. There is certainly a significant risk that volatility in the CER price will change the revenue from CDM projects. This is already taken into consideration in the above Monte Carlo simulation. Here the risk is called the "CER risk."
2. Another risk which is also crucial is whether or not a project could be certified as a CDM. If the project were not certified by operational entities, sponsors of the project could not acquire CER. This risk is called the "certification risk."
3. We should also pay attention to the risk that the amount of CER will change owing to future modified baselines. We call this the "baseline risk."
4. We should also be cautious about whether or not a CDM project is approved by the host countries.
5. A CDM project is also accompanied by country risks owing to politic or economic instability in host countries. Here, we call risks 4 and 5 "country risks."

Among the above-mentioned risks, certification risk and baseline risk strongly depend upon the project type. If technologies to recover and combust methane from coal mining in host countries could be improved, the baseline in projects recovering methane would be lowered, leading to a reduction of CER units. In the worst case, operational entities might not certify the project as a CDM. To suppress these risks, we should invest in different types of project, rather than in similar projects.

On the other hand, the risks in points 4 and 5 depend upon political and economic stability in host countries. In order to lower these risks, we should invest in projects in different countries, rather than in a single country.

The CER risk is difficult to avoid by portfolios of CDM projects, since it is similar to systematic risks in the securities market. It might be effective to have portfolios of CDM projects as well as stocks of energy-intensive companies such as electric utilities, or iron and steel companies. This is because the price of CERs and the stock of energy-intensive companies might be negatively correlated.

In sum, further investigation is needed in order to compile effective portfolios which include CDM projects in order to lower the above-mentioned risks.

Alteration of the Debt/Equity Ratio

We next investigated the more detailed structure of profitability. In an ordinary project, the initial cost of the project will be covered by loans from financial organizations or by equity. The ratio of debt to equity (D/E) is significant when evaluating the economic feasibility of projects. Therefore, we investigated the influences of the D/E ratio on economic indicators, and focused on a CDM project of introducing an integrated coal gasification combined-cycle plant in a developing country.

Figure 8 shows the result of 100000 Monte Carlo simulations. This figure depicts the distribution of the return on equity (ROE) of the CDM project with three values of the D/E ratio. In the case of $D/E = 9/1$, the ROE is broadly distributed in this project, which implies a high risk. Risks in ROE are reduced with lower D/E ratios, although the expected value of ROE also decreases slightly. In particular, the ROE in the case of $D/E = 5/5$ would be allowable from an investor's viewpoint if the expected value of the ROE were a little better.

While Fig. 8 shows the probabilistic distribution of ROE, Fig. 9 depicts the distribution of the debt-recovery rate in the project. If the debt-recovery rate is less than 100%, the project is said to fall into default. In the case where $D/E = 9/1$, the project falls into default more than 1600 times in 100000 simulations. In the case where $D/E = 7/3$, the number of defaults decreases to 50, and no default case is observed when $D/E = 5/5$. An index of value at risk (VaR) is often used to estimate risks in financial engineering. When we sort data on the debt-recovery rate from the worst case to the best case, the VaR is about 1000th (1%) out of 100000 cases. We estimated VaR when $D/E = 9/1$, 7/3, and 5/5 to be 92%, 100%, and 100%, respectively. Thus, the case where $D/E = 9/1$ is too risky from the viewpoint of financial organizations.

Schemes to Make CDM Projects Viable

The observations in the last section indicate that we need to reduce the D/E ratio when financing CDM projects up to a certain value in order to lessen the risks

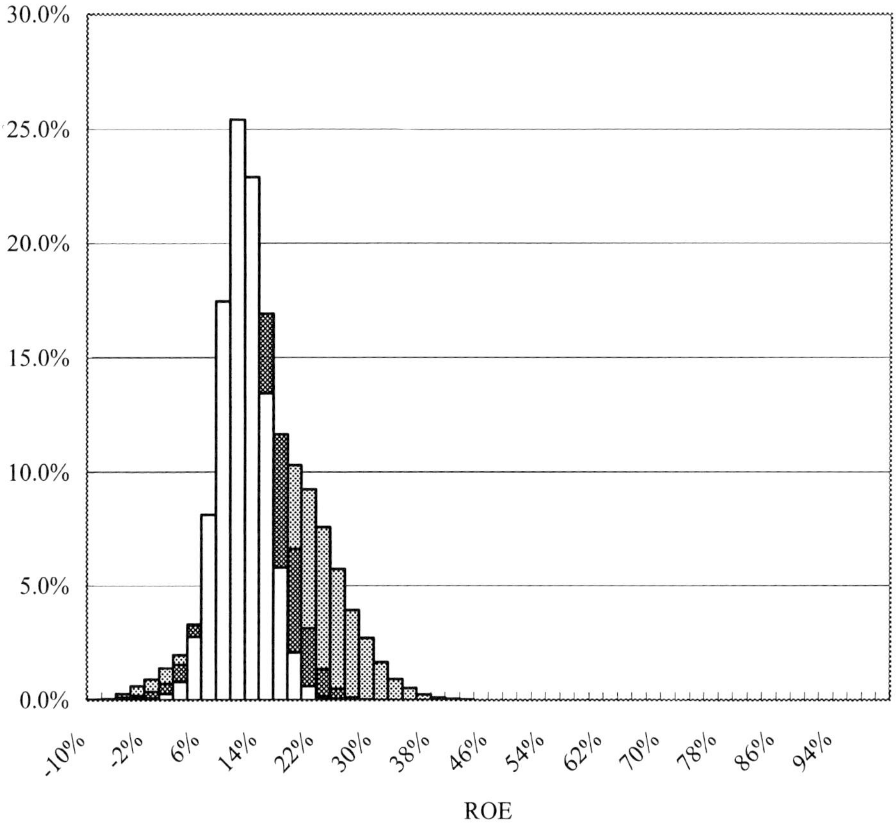

FIG. 8. Probabilistic distribution of the return on equity (ROE) of a CDM project estimated by Monte Carlo simulations. *Light shading*, *D/E* = 9/1; *dark shading*, *D/E* = 7/3; *white*, *D/E* = 5/5

in ROE and in the debt service. At the same time, it is desirable for sponsors to diversify their investments into various projects in various countries in order to reduce baseline risk, certification risk, and country risks.

Various measures could be taken to realize the above scheme in CDM projects. One is a partial security system for equity. If sponsors include investment banks or security companies, they could evaluate the total risks in CDM projects, structure them, and develop various types of security system. Then they could sell these systems to general companies and to qualified institutional buyers (QIBs). Figure 10 shows a scheme of this concept.

Possible investors or QIBs would be companies which want to supplement their own efforts to reduce CO_2 emissions by the CER, companies which want to raise their environmental ranking, or companies which take it just as an investment business. From the standpoint of investors, the projects in which they prefer

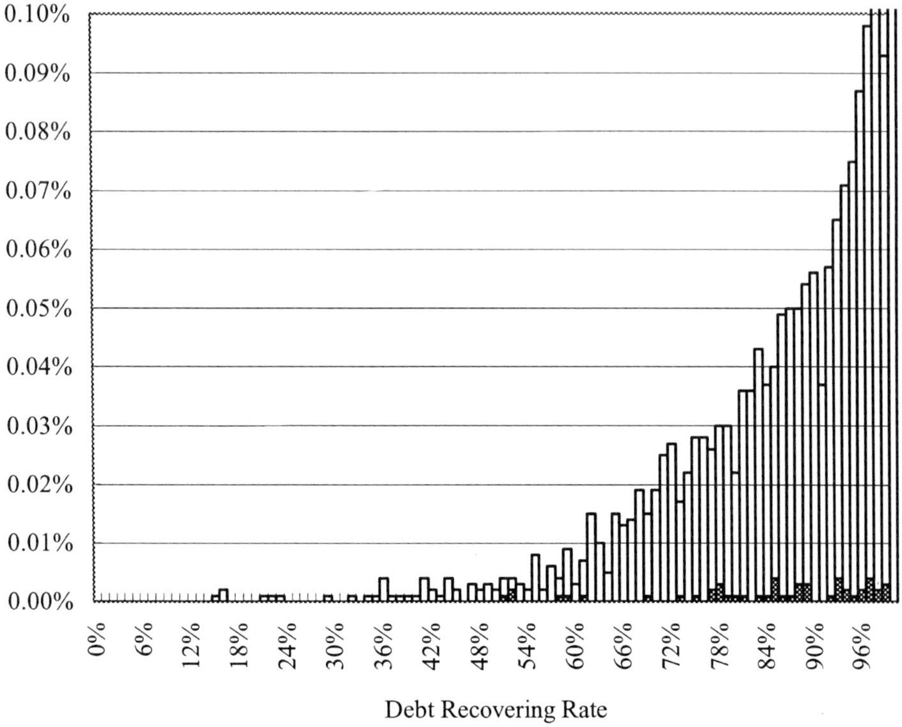

FIG. 9. Probabilistic distribution of debt-recovery rate of a CDM project estimated by Monte Carlo simulations. *White*, *D/E* = 9/1; *dark shading*, *D/E* = 7/3

to invest depend on their attitudes toward the risk. It also depends on the necessity for CERs in each company. Some companies take the investment opportunity to comply with the Voluntary Action Program, while others take it as just an environmental business or advertisement. If there are wide variations in the security systems being sold in the market, companies or QIBs are able to buy them according to their own attitude toward the risk and necessity of CERs. Thus, the concept of security systems for CDMs deserves further investigation. We will establish the institution more clearly and evaluate it in our future work.

In order to reduce the risks in CDMs, governments of donor countries could also play an important role. In this respect, the CER units procurement tender (CERUPT) initiated by The Netherlands's government deserves attention. In CERUPT, the government purchases CER units acquired from CDM projects where host countries and donor countries have ratified the Kyoto Protocol. In the future, other governments, including Japan, should also take the establishment of such a system into consideration. If a donor country's government could offer the lowest price for a CER unit in such a system, it could contribute to lowering CER risks.

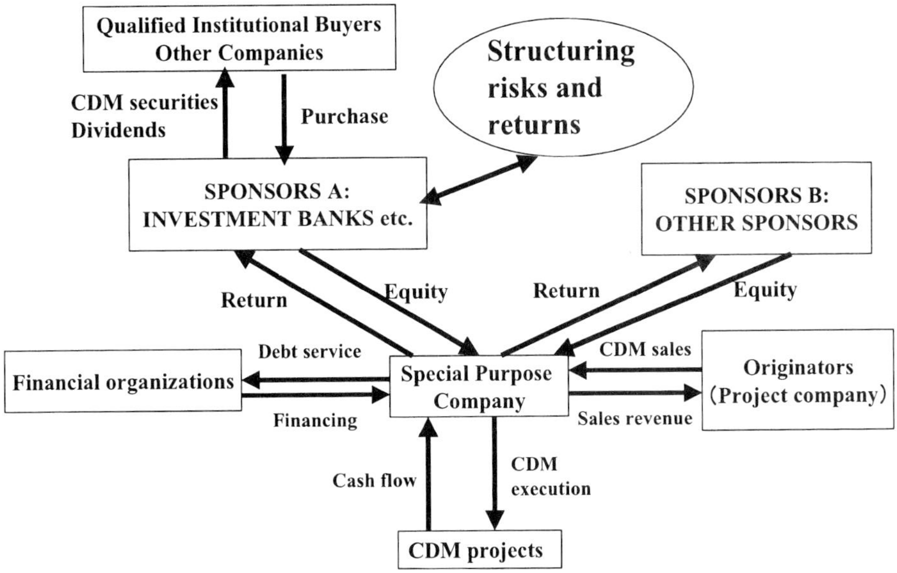

FIG. 10. Financial flow accompanying partial security systems for CDM projects.

Conclusions

In the first part of this chapter we evaluated a CDM project using photovoltaic systems. For this purpose, we first estimated life-cycle CO_2 emissions from PV and other power generation systems based on input–output tables for the environmental analysis. The cost of the PV systems was also evaluated, taking the effect of mass production into consideration. We then developed two dynamic scenarios to disseminate PV systems. One was a scenario for domestic dissemination only, while the other included a CDM project using PV systems as well as domestic dissemination. The scenario analysis indicated that a continuation of the present subsidies for domestic dissemination is imperative to increase demand, and consequently to reduce the cost of PV systems. A continuation of the subsidy was also shown to be effective in promoting CDM projects using PV systems. Nevertheless, the IRR of the CDM was estimated to be very low. We therefore proposed some measures to improve the IRR of a CDM using a photovoltaic system.

In the second part, we explored the establishment of efficient institutions to make CDM projects viable. For this, we evaluated some CDM projects on which feasibility studies were performed under NEDO's sponsorship. We estimated the IRR and other indicators of profitability for 42 projects, taking account of volatilities in the CER price and other costs. As a result of Monte Carlo simulations, the expected values and their standard deviations of the IRR of 42 CDM projects were found quantitatively. The risks accompanying CDM projects were classified into CER risk, certification risk, baseline risk, and country risk. Although it is difficult to suppress CER risk by diversifying investment into various CDM projects, certification risk, baseline risk, and country risk could effectively be lowered by compiling portfolios of various types of CDM project in various countries.

We also evaluated the distribution of ROE and debt recovery rate to clarify the impacts of the debt/equity ratio on project risk. In particular, CDM projects depending strongly on the CER value are exposed to higher risks than ordinary overseas projects. We concluded that it is effective to reduce the D/E ratio to some extent to suppress financial risks, so that CDM projects will be viable. Finally, we referred to partial security systems for CDMs, and a role for donor country governments to realize that these modifications could make more CDM projects viable.

Acknowledgments. The author is grateful to Haruo Imai, Hidenori Niizawa, Jiro Akita, Shunsuke Mori, Tatsuyoshi Saijo, Tetsuo Tezuka, Yuichi Soda, and Takamitsu Sawa for their useful comments in frequent discussions on the CDM.

References

Akimoto K, Matsunaga A, Fujii Y, Yamaji K (1998) Game theoretic analysis for carbon emission permits trading among multiple world regions with an optimizing global energy model (in Japanese). Trans Jpn Inst Electr Eng 118-C:1424–1431

Ikeda Y, Shinozaki M, Suga M, Hayami H, Fujiwara K, Yoshioka K (1996) Input–output table for environment analysis (in Japanese). Keio University Economic Observatory

Kato K, Weng K, Okajima K, Yamada K (1999) Life-cycle analysis and economy of CO_2 emissions reduction in photovoltaic systems for houses (in Japanese). Energy Resources 20:390–396

Okushima K, Tezuka T, Sawa T (2000) Introduction of photovoltaics and its effect on CO_2 reduction (in Japanese). Proceedings of the 19th Conference of the Japan Society of Energy and Resources, Japan Society of Energy and Resources, Osaka, pp. 255–260

Section II: China and International Cooperation in Global Warming

The Kyoto Protocol and China–Japan Cooperation to Reduce CO_2 Emissions: A Macroeconomic Analysis of Cooperation Potential

Mitsuo Ezaki*, Lin Sun*, and Morihiko Kinjo†

Summary. This chapter analyzes the potential of North–South cooperation in achieving the Kyoto targets by examining the case of China and Japan with a focus on their bilateral cooperation at the macroeconomic level. The potential, based on the framework of joint implementation (JI) between China and Japan, is quantitatively investigated. Simulation analysis based on computable general equilibrium models is used to investigate the possibility of joint implementation based on marginal cost curves and analyze its impact on the economy and society in both countries.

The analysis estimates that Japan will lose 3.1 trillion yen (0.3% of GDP) as a social deadweight loss if it achieves the Kyoto targets by domestic measures alone, with carbon tax of 53500 yen/ton. If the JI with China is accomplished, China will lose 2.1 billion yuan (0.01% of GDP) as a social deadweight loss, while Japan would lose only 30 billion yen (0.003% of GDP).

The cooperation potential between China and Japan depends on the perception of estimated costs and the allocation of benefits arising from the cooperation.

Key words. Kyoto Protocol, International cooperation, Clean development mechanism, Joint implementation, Computable general equilibrium model

Introduction

The Third Conference of the Parties (COP-3) to the UN Framework Convention on Climate Change (UN FCCC) was held at Kyoto, Japan, in December 1997. This conference adopted the Kyoto Protocol, which set legally binding numerical targets to reduce emissions of greenhouse gases (GHG) in industrialized countries for the commitment period 2008–2012, e.g., 6% below the 1990 level

* Graduate School of International Development, Nagoya University, Furo-cho, Chikusa-ku, Nagoya 464-8601, Japan
† School of Political Science and Economics, Tokai University, 1117 Kitakaname, Hiratsuka, Kanagawa 259-1292, Japan

for Japan, 7% below the 1990 level for the United States, 8% below the 1990 level for the European Union, and an average of 5.2% below the 1990 level for other industrialized countries. As supplements to domestic measures to attain the reduction targets, the COP-3 also agreed to introduce three measures of international flexibility called the Kyoto Mechanism; these were emissions trading (ET), joint implementation (JI), and a clean development mechanism (CDM). ET and JI cover international cooperation within industrialized countries which have reduction targets, while the CDM is concerned with international cooperation between these industrialized countries and developing countries which have no reduction targets. The Seventh Conference of the Parties (COP-7) met at Marrakesh, Morocco, in November 2001, and reached overall agreement on the rules governing the implementation of the Kyoto Protocol, but without the participation of the United States. The Kyoto Protocol is expected to come into force in 2002.

The participation of developing countries is essential in order to prevent global warming. The development and enhancement of the endogenous capacities and technologies of developing countries is one thing, but the implementation of the CDM between industrialized and developing countries in order to achieve the Kyoto targets of emissions reduction is another. The CDM is a measure of international cooperation based on joint projects at the microeconomic level in developing countries. However, the CDM also has a macroeconomic meaning for international cooperation when its individual joint microprojects are selected and aggregated, one after another, in accordance with their cost-effectiveness in reducing CO_2 emissions. It is therefore necessary to enquire to what extent international cooperation of this kind is possible to help achieve the Kyoto targets, i.e., what is the macroeconomic potential of North–South cooperation.

This chapter attempts to answer the question by examining the case of China and Japan, and focusing on their bilateral cooperation at the macroeconomic level. Looking at the CDM aggregation as bilateral cooperation within the macroeconomy, we investigated its potential quantitatively based on the framework of JI between China and Japan, i.e., China with a zero reduction target and Japan with a reduction target of 6% below the 1990 level. We represent GHG by carbon dioxide (CO_2), and take the case of introducing a carbon tax in both countries. The methodology used is basically the same as that of Weyant and Hill (1999). We first evaluated the costs (social deadweight loss) and benefits (reduction of CO_2 emissions) of introducing a carbon tax in each country by using simulation analysis based on computable general equilibrium (CGE) models. We then investigate the possibility of JI based on marginal cost curves, and analyzed its impacts on the economy and society in both countries.

The simulation period was 18 years from the benchmark year of the CGE model (1995) to the end of the commitment period of the Kyoto Protocol (2008–2012). Japan was committed to reduce GHG emissions during 2008–2012 to 6% below the 1990 level. The level of CO_2 emissions in Japan in 1990 was 291 million tons of carbon (MtC), so that 6% below the 1990 level is 274 MtC, which becomes the target level of CO_2 emissions during 2008–2012. How to achieve this target by JI with China, which has a zero numerical target, and how to distribute the benefits obtained therefrom will be the main content of this analysis.

The next section covers the analysis of the situation in China. We present a CGE model, a baseline scenario (business-as-usual, BAU), and the impacts of introducing a carbon tax. The following section covers a similar analysis of the situation in Japan. We then compare the marginal cost curves of China and Japan, and investigate the possibility of JI to reduce CO_2 emissions by introducing a carbon tax. We then present some concluding remarks.

Carbon Price and the Chinese Economy

CGE Model of the Chinese Economy

We constructed an environmental CGE model of China to analyze CO_2 emissions based on the 1995 input–output table. The model includes nine industries: (1) agriculture; (2) light manufacturing; (3) electricity; (4) oil and gas; (5) coal; (6) metals and chemicals; (7) machinery; (8) construction; (9) services.[1]

We have adopted the Armington hypothesis for each industry. This aggregates domestic and imported goods into composite goods in the domestic market by employing the constant elasticity of substitution (CES) aggregation function. The elasticity of substitution (σ_Q) is basically derived from GTAP.[2] We treated energy as composite goods in its intermediate and final uses, aggregating electricity, oil and gas, and coal by the CES function in order to allow substitutions between them. The elasticity of substitution (σ_E) is assumed to be 0.5. We employed the CES production function for each industry to allow substitution between energy, labor, and capital inputs, with Leontief technology (i.e., fixed coefficients) for the remaining intermediate inputs. The elasticity of substitution (σ_X) is assumed to be 0.5.[3] For households, we assumed the Cobb–Douglas utility function in order to derive consumption demand.

The CGE model used for nonenvironmental aspects is essentially based on the classic CGE model developed by Dervis et al. (1982, Chap. 7). However, our model has two important deviations from the classical one in addition to the treatment of energy as composite goods. The first is the labor market, for which our model does not assume equilibrium between demand and supply, but treats agricultural labor as a residual. The second is the balance between nominal savings (domestic plus foreign) and nominal investment, which is dropped as being redundant by the law of Walras. This leads to price determination at

---

[1] Note that item 4, oil and gas, covers not only oil and gas mining, but also their products; item 5, coal, covers coal mining and its products; and item 6, metals and chemicals, also includes related mining activities

[2] See Hartel (1997), Table 4.1 (p. 125). GTAP is the Global Trade Analysis Project

[3] The values for σ_E and σ_X assumed here are generally the same as those which Babiker and Rutherford (2001) assumed for their global general equilibrium model

absolute levels based on the unitary price of nominal savings selected as numeraire.

Our environmental CGE model includes, in addition to the above nonenvironmental system, the CO_2 emissions in proportion the use of energy, and the carbon tax in proportion to the quantity of CO_2 emissions. In our model, the carbon tax is levied on demand side, i.e., on industries for their use of energy as intermediate inputs, and on households for their use of energy as consumption,[4] while the revenues from the carbon tax are all used for investments in environmental protection without any real production effects.

The environmental CGE model for China is recursively dynamic, which means capital stocks by industry are given at the beginning of each year, but are partially adjusted for profit rates by the industry allowing for capital accumulation at the end of the year (i.e., at the beginning of the next year). In other words, this is a model in which market equilibrium is attained in each successive year in accordance with predetermined but changing levels of capital stocks.

Baseline Scenario (BAU) and Impacts of Carbon Tax

To establish the impacts of a carbon tax in the commitment period (2008–2012) of the Kyoto Protocol, we need a standard simulation of the Chinese economy for the period 1995–2012 in which no measures were taken to limit CO_2 emissions. This is the baseline scenario of BAU, and the results of the simulation are shown in the first two columns of Table 1 (column 1, average levels for 2008–2012; column 2, average growth rates for 1995–2012).

The baseline scenario is obtained by assigning proper and reasonable values to exogenous variables and parameters in order to be consistent with a priori anticipated rates of growth or levels for several key endogenous variables. In other words, we have assigned specific values to exogenous variables and parameters such as total labor supply (1% growth), nonagricultural wage (8% growth), real investment (8% growth), foreign exchange rate (fixed or 0% growth), etc., for the simulation period 1995–2012 to obtain a priori anticipated predictions on such key endogenous variables as real GDP (about 7% growth), deflator of GDP (about 2% growth), real wages (about 6% growth), average total factor productivity (about 3% growth), the current balance of payments (almost zero in 2008–2012), and so on.

Our baseline scenario in Table 1 predicts that the total quantity of CO_2 emissions in China around 2010 will be 1760 MtC. Zhang (1998) predicts that the quantity in 2010 will be 1430 MtC. Predictions made by others range from 937 MtC to 1960 MtC (Zhang, 1998, Table 15). Li (1999) also predicts that it will be 1430 MtC. The GREEN model of the Organization for Economic Cooperation and Development (OECD) (e.g., Koppel and Lee, 1996) gives a prediction in the

[4] See Ezaki et al. (1998) for data on the CO_2 emissions by industries and households in 1995

lower part of the range given above. Our baseline prediction for CO_2 emissions is in the upper part of the range.[5]

Our baseline prediction corresponds to the case of BAU, with no measures being taken against CO_2 emissions. In that case, what would happen to the emissions, as well as to the economy, if a carbon tax was introduced in China? Table 1 (column 4) indicates the impacts of a carbon tax in China by percentage deviations from BAU for the case where 40 yuan of carbon tax is levied in nominal terms per ton of carbon emissions (PCO_2 = 40 yuan/ton) for the period 2008–2012. In this case, the total quantity of CO_2 emissions decreases by 113 MtC per year on average. The carbon tax causes only a small decrease in real GDP (–0.1% in terms of percentage deviation from BAU) and also in the real output of most industries, especially the energy industries, and accelerates inflation at both aggregate and industry levels.[6] Needless to say, the impacts are larger (smaller) for higher (lower) rates of carbon tax.

Carbon Price and the Japanese Economy

CGE Model of the Japanese Economy

We constructed an environmental CGE model of Japan which is basically in accordance with that of China in terms of its framework, i.e., classification of industries, CES functions of aggregation and production together with substitution elasticity, Cobb–Douglas utility function, recursively dynamic system, CO_2 emissions in proportion to energy demands, and so on. It must be noted, however, that there are two important differences in the framework of the two models. First, the Japan model assumes that the labor market is competitive and in full employment through wage adjustments. Second, the Japan model uses the revenue from the carbon tax to refund households by income transfer.[7]

Baseline Scenario (BAU) and Impacts of Carbon Tax

The baseline simulation (BAU) for the Japanese economy is also presented in Table 1 in terms of levels (column 5) and growth rates (column 6) for the com-

---

[5] We derived our prediction of CO_2 emissions based on the emission coefficients for electricity, oil and gas, and coal fixed at the 1995 levels, with an allowance for substitution between the three kinds of energy. Our prediction may be an overestimation, judging from the declining trend of total CO_2 emissions in China which was observed in the late 1990s

[6] The model assumes that revenues from the carbon tax would be used for environmental investment, with demand but no supply effects. This is the reason for the positive impacts observed on the real output of the construction industry in Table 1

[7] Because full employment is assumed for Japan but surplus labor is assumed for China, the impacts of exogenous or policy shocks on real variables such as real GDP are generally small in Japan but large in China. A carbon tax in China has multiplier effects on real GDP since the revenue is treated as an environmental investment, while a carbon tax in Japan is nearly neutral on real GDP owing to the assumptions of redistribution and full employment

TABLE 1. Baseline scenario (business-as-usual, BAU) and impacts of a carbon tax (PCO_2): average for 2008–2012[a]

	China				Japan					
	Baseline (BAU)		PCO_2 = 40 yuan/ton		Baseline (BAU)		PCO_2 = 4000 yen/ton		PCO_2 = 53 500 yen/ton	
	Level (2008–2012)	Growth (%)	Level (2008–2012)	Deviation (%)	Level (2008–2012)	Growth (%)	Level (2008–2012)	Deviation (%)	Level (2008–2012)	Deviation (%)
Real output, agriculture (*XS1*)	32 142	3.1	32 168	0.1	121 321	−1.2	121 242	−0.1	120 399	−0.8
Real output, light manufacture (*XS2*)	77 628	6.2	77 519	−0.1	904 721	0.9	904 093	−0.1	897 440	−0.8
Real output, electricity (*XS3*)	7 889	6.5	7 695	−2.5	201 396	0.9	199 805	−0.8	184 736	−8.3
Real output, oil and gas (*XS4*)	11 269	7.2	11 143	−1.1	103 315	0.7	100 770	−2.5	79 713	−22.8
Real output, coal (*XS5*)	3 468	4.5	3 216	−7.3	13 918	0.4	13 239	−4.9	8 778	−36.9
Real output, metals and chemicals (*XS6*)	120 786	8.8	120 405	−0.3	1 205 272	1.8	1 202 184	−0.3	1 172 315	−2.7
Real output, machinery (*XS7*)	88 815	9.7	88 672	−0.2	1 817 186	2.4	1 815 933	−0.1	1 804 281	−0.7
Real output, construction (*XS8*)	42 576	8.0	42 911	0.8	1 206 651	2.1	1 206 527	0.0	1 205 289	−0.1
Real output, services (*XS9*)	91 249	7.2	91 076	−0.2	6 452 421	2.4	6 450 755	0.0	6 429 580	−0.4
Labor supply (*LS*)	72 438	1.0	72 438	0.0	7 497	1.0	7 497	0.0	7 497	0.0
Capital stock supply (*KS*)	557 519	10.4	557 519	0.0	28 480 868	1.8	28 480 868	0.0	28 480 868	0.0
Real GDP (*YR*)	169 500	7.0	169 347	−0.1	6 587 933	2.1	6 586 286	0.0	6 558 433	−0.4
Equivalent variation (*EV*)	80 662	7.2	80 350	−0.4	3 953 700	2.2	3 954 052	0.0	3 946 089	−0.2
Real private consumption (*CR*)	74 187	6.5	73 820	−0.5	3 969 884	2.2	3 970 628	0.0	3 967 887	−0.1
Real gov. consumption (*GR*)	19 677	7.1	19 654	−0.1	691 217	2.5	689 498	−0.2	675 235	−2.3
Real investment (*IR*)	76 191	8.0	76 191	0.0	2 017 432	2.4	2 017 432	0.0	2 017 432	0.0
Real exports (*ER*)	45 500	8.5	45 262	−0.5	763 120	3.2	761 987	−0.1	751 848	−1.5
Real imports (*MR*)	46 055	9.5	46 206	0.3	853 721	5.1	853 259	−0.1	853 970	0.0
Nominal GDP (*YN*)	194 663	8.1	195 164	0.3	9 924 539	5.0	9 932 097	0.1	9 981 546	0.6
Nominal investment (*IN*)	80 438	8.6	80 601	0.2	3 008 171	5.2	3 010 592	0.1	3 032 745	0.8
Nominal saving (*S*)	79 903	8.1	79 826	−0.1	3 151 880	5.1	3 154 458	0.1	3 171 587	0.6

Current BOP deficit (*F*)	537	0.0	777	44.8	–143713	2.0	–143871	0.1	–138847	–3.4
Current BOP in US$ (*F$*)	537	0.0	777	44.8	–106751	0.0	–106871	0.1	–103150	–3.4
Output price, agriculture (*PX1*)	2.60	6.3	2.60	–0.2	2.69	6.4	2.69	0.0	2.70	0.5
Output price, light manufacture (*PX2*)	1.43	2.5	1.43	0.1	1.73	3.7	1.73	0.1	1.75	1.0
Output price, electricity (*PX3*)	0.72	–1.8	0.74	3.5	1.96	4.4	1.98	1.1	2.20	12.2
Output price, oil and gas (*PX4*)	0.69	–2.1	0.69	0.1	1.68	3.4	1.67	–0.4	1.63	–2.9
Output price, coal (*PX5*)	1.19	1.5	1.20	0.1	1.70	3.5	1.71	0.2	1.81	6.3
Output price, metals and chemicals (*PX6*)	0.83	–0.9	0.84	0.6	1.62	3.2	1.62	0.3	1.67	3.5
Output price, machinery (*PX7*)	0.81	–1.1	0.81	0.3	1.60	3.2	1.60	0.1	1.61	0.8
Output price, construction (*PX8*)	0.96	0.0	0.96	0.3	1.46	2.7	1.46	0.1	1.48	1.0
Output price, services (*PX9*)	1.12	1.0	1.13	0.2	1.42	2.5	1.42	0.0	1.42	0.5
Nominal wage, average (*PL*)	1.92	9.9	1.92	–0.1	829	4.5	828	0.0	824	–0.6
Nominal rent, average (*PK*)	0.05	–8.1	0.05	0.0	0.11	2.2	0.11	–0.2	0.11	–1.1
Exchange rate (*REX*)	1.00	0.0	1.00	0.0	1.35	2.0	1.35	0.0	1.35	0.0
Private consumption deflator (*PC*)	1.25	1.5	1.25	0.4	1.48	2.8	1.48	0.1	1.50	1.0
Government consumption deflator (*PG*)	1.12	1.0	1.12	0.2	1.42	2.5	1.42	0.1	1.42	0.5
Investment deflator (*PI*)	1.05	0.5	1.05	0.2	1.49	2.8	1.49	0.1	1.50	0.8
Export deflator (*PE*)	1.00	0.2	1.00	0.3	1.57	3.1	1.57	0.1	1.59	1.1
Import deflator (*PM*)	1.00	0.0	1.00	0.0	1.35	2.0	1.35	0.0	1.35	0.0
GDP deflator (*PY*)	1.15	1.1	1.15	0.4	1.50	2.9	1.51	0.1	1.52	1.0
CO_2 emissions (100 MtC)	17.63	4.9	16.50	–6.4	3.95	1.6	3.79	–3.9	2.74	–30.6
CO_2 emissions by industries (100 MtC)	14.25	4.3	13.42	–5.8	3.58	1.7	3.44	–4.1	2.44	–31.8
CO_2 emissions by households (100 MtC)	3.38	7.8	3.08	–8.8	0.36	1.6	0.36	–1.8	0.30	–18.9
Carbon tax revenue (*$YGCO_2$*)	0.00		660		0.00		15174		146519	

[a] Units: price variables are 1995 = 1.0; level variables (except CO_2 emissions) are 100 million yuan for China and 100 million yen for Japan; CO_2 emissions = 100 MtC (million tons of carbon)

1 yuan = 100 yen in purchasing power parity for 2008–2012, so that PCO_2 = 40 yuan/ton = 4000 yen/ton

Growth (%): average for 1995–2012

Letters in Italics mean variables in the CCE models

mitment period of the Kyoto Protocol. The results of the simulation are obtained, as in the case of China, by assigning specific values to exogenous variables and parameters such as total labor supply (1% growth), real investment (2.4% growth), foreign exchange rate (2% depreciation per year), etc., to obtain a priori anticipated predictions of such key endogenous variables as real GDP (about 2% growth), deflator of GDP (about 3% growth), and so on.

The baseline scenario predicts that the total quantity of CO_2 emissions in Japan will increase from 313 MtC in 1995 to 394 MtC in 2008–2012 on average. Predictions for Japan in 2010, made by more than ten participating teams in the Energy Modeling Forum (EMF), are similar to one another with only a few exceptions, and generally range from 350 to 400 MtC (Weyant and Hill 1999; Kainuma et al. 1999). Our prediction is also close to those international values.

What are the impacts of introducing a carbon tax in Japan? Table 1 shows them for two different rates of carbon tax: PCO_2 = 4000 yen/ton and PCO_2 = 53 500 yen/ton in nominal terms.

We can see from the last two columns of Table 1 that the total quantity of CO_2 emissions in Japan can be reduced to 6% below the 1990 level (i.e., 274 MtC) by levying a carbon tax of 53 500 yen/ton for the commitment period 2008–2012. This is about 30% below the BAU level (i.e., 395 MtC). In this case, real GDP decreases by 0.4% from the BAU level.[8] Compared with other studies on Japan (Amano 1997, Table 7–5), our results lie in the group of high estimates of the required level of carbon tax, but in the group of low estimates of the negative impacts on real GDP. The loss of GDP needed to attain the Kyoto target is fairly small: 3.0 trillion yen in real terms at constant 1995 prices, 5.7 trillion yen in nominal terms, and 0.4% in terms of ratio. The corresponding impacts on the macroeconomy, prices, and the industrial structure are not large.

In Table 1, a carbon tax of 4000 yen/ton (PCO_2 = 4000 yen) for Japan is comparable to a carbon tax of 40 yuan/ton (PCO_2 = 40 yen) for China, since the exchange rate between the yen and the yuan is estimated to be 100 yen/yuan in terms of purchasing power parity for the commitment period. A combination of the two cases results in optimal cooperation between China and Japan in the light of the JI framework, as discussed in the next section.

Macroeconomic Analysis of China–Japan Cooperation

Carbon Tax and Joint Implementation (JI)

The carbon tax is equal to the marginal cost of reducing CO_2 emissions in equilibrium, where cost means a deadweight loss (i.e., a decrease in consumer surplus *plus* producer surplus *plus* tax revenue) caused by levying the carbon tax on the

[8] Note that a nominal 53 500 yen/ton = a nominal 400 dollars/ton = a real 35 200 yen/ton at 1995 prices = a real 390 dollars/ton at the 1995 rate of foreign exchange. This is due to the simulation results on prices and exchange rates shown in Table 1

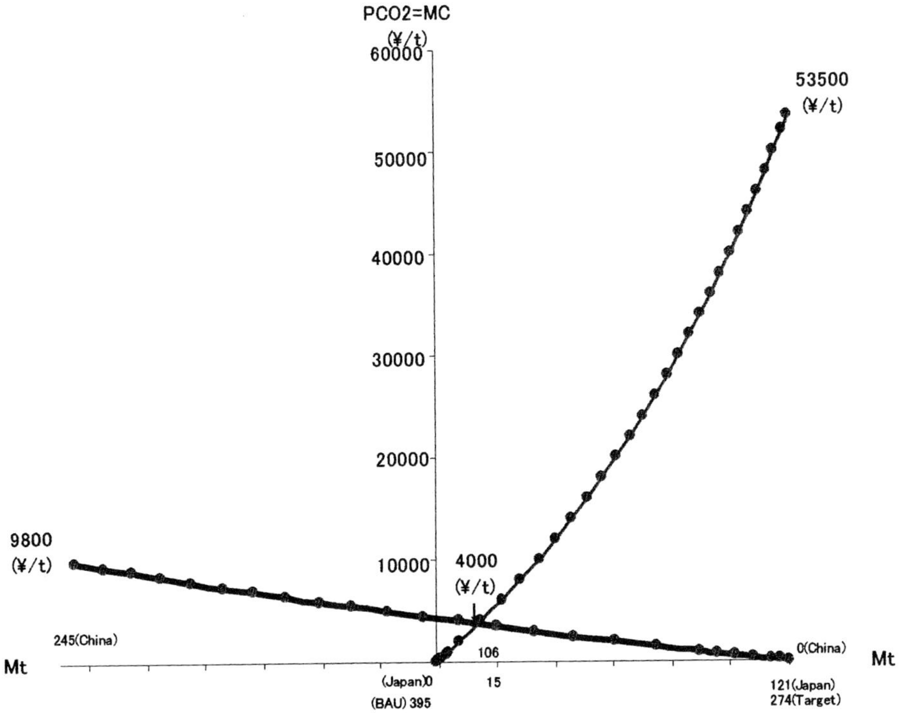

FIG. 1. Macroeconomic analysis of China–Japan cooperation. The horizontal axis represents the average CO_2 emissions in 2008–2012. China is plotted from right to left. The vertical axis represents the nominal carbon tax (PCO_2) or marginal cost (MC). Yuan are converted to yen (¥) by purchasing power parity (PPP)

energy industry (i.e., on CO_2 emissions in proportion to either energy production or energy input/consumption). Figure 1 shows the marginal cost curves of reducing CO_2 emissions for Japan and China. These are derived from the simulation results given in the previous two sections.

In Fig. 1, the vertical axis indicates the carbon tax (PCO_2) or marginal cost (MC), while the horizontal axis indicates the total amount of CO_2 reduction measured by deviations from the BAU level of CO_2 emissions. Note that in Fig. 1, the MC curve for China is drawn from right to left, starting from an origin located at the reduction target for Japan (121 MtC) along the horizontal axis of Japan. Note also that the marginal cost and carbon tax for China are indicated in terms of the Japanese yen based on a purchasing power parity of 100 yen/yuan for the period 2008–2012.[9]

[9] The exchange rate and purchasing power parity in 1995 are 11.55 and 98 yen/yuan, respectively. According to the baseline scenarios of Table 1, the GDP deflator increases for 1995–2012 at a rate of 1.2% in China and 1.5% in Japan, so that we have assumed a purchasing power parity in 2008–2012 approximately equal to 100 yen/yuan

Figure 1 provides an analytical framework for potential China–Japan cooperation to attain the Kyoto target in the light of JI between China (with zero reduction target) and Japan (with a reduction target of 121 MtC). We can see from Fig. 1 that the cost (i.e., deadweight loss) involved for Japan to achieve the target reduction of 121 MtC by levying a carbon tax of 53 500 yen/ton is the area under the MC curve for Japan from the origin to the point of 121 MtC, which will be about 3.07 trillion yen at current prices.[10] This amounts to 0.3% of nominal GDP (993 trillion yen). Real GDP, when compared with the BAU level, decreases by 0.4%, as shown in Table 1.

If the target of 121 MtC can be reached jointly with China by introducing a carbon tax in both countries, the optimal point will be the one at which the two marginal cost curves intersect, since the total cost for the two countries to reduce CO_2 emissions jointly by 121 MtC (i.e., the area under the two MC curves) is a minimum at this point. At this optimal point, China reduces emissions by 106 MtC annually by levying a carbon tax of about 40 yuan/ton, while Japan reduces emissions by 0.15 MtC annually by levying a carbon tax of about 4000 yen/ton.

The cost, or deadweight loss, to China from this joint implementation is about 210 billion yen (2.1 billion yuan) at current prices,[11] which is only 0.01% of nominal GDP. Real GDP decreases by 0.1% from the BAU level, as already shown in Table 1. On the other hand, the cost, or deadweight loss, to Japan is quite small. It is about 30 billion yen at current prices, which is 0.003% of nominal GDP, and real GDP decreases by 0.02% from the BAU level. Japan can lessen the cost of attaining the Kyoto target from 3.07 trillion yen to only 240 (= 210 + 30) billion yen by this joint implementation. It is necessary for Japan to share the benefits of 2830 (= 3070 – 240) billion yen with China through income transfer, in addition to the compensation paid for direct costs to China (210 billion yen).

The question here is what is the appropriate level of benefit sharing, and the most natural answer is to regard the marginal cost, or carbon tax, as the price of carbon from the point of view of the market economy. Then the amount which needs to be transferred from Japan to China (both costs and benefits) will be about 420 billion yen (= 4000 yen/ton × 106 MtC), or about 27 billion yuan by the exchange rate conversion.[12] In this case, China receives only 210 billion yen, or 8% of the benefits. Another answer which would reduce this contribution

---

[10] The area under the MC curve for Japan from the origin to the point of 121 MtC is approximated by two triangles and one rectangle, using the breakpoint at which the two MC curves intersect: the first triangle is from the origin to the breakpoint (0.03 trillion yen), the second triangle is from the breakpoint to the point of 121 MtC (2.62 trillion yen), and the rectangle is below the second triangle (0.42 trillion yen). This approximation totals 3.07 trillion yen

[11] This again is an approximation from the triangle under the MC curve. The same is true for the discussions that follow

[12] The rate of exchange in about 2010 is estimated to be 15.6 yen/yuan (= 1.35 × 11.55 yen/yuan). Conversion from yen to yuan in income transfer may be made by a combination of the exchange rate and purchasing power parity, e.g., the exchange rate for benefits allocated to China, and purchasing power parity for the direct costs to China

would be to allocate the benefits in proportion to the size of CO_2 emission reductions in each country. In this case, China would receive 2480 billion yen, or 88% (= 1.06/1.21) of the benefits. The most feasible solution politically seems to be an average of the two, thereby allocating a payment to China of 1350 billion yen, or 48% of the benefits.

Extension of the Analysis

The analysis above depends crucially on Fig. 1, in which the cost of reducing CO_2 emissions is measured by the deadweight loss caused by introducing a carbon tax on energy industry, i.e., a decrease in social welfare (consumer surplus *plus* producer surplus *plus* tax revenues). Alternatively, the cost of reducing CO_2 emissions can be defined by the decrease in real GDP or equivalent variation (EV). Marginal cost-curves based on such alternative cost measures are presented in Fig. 2, and show similar patterns to the curve based on the deadweight loss. Almost the same analysis may be made on the basis of alternative cost measures.

Another extension of the analysis is the regional breakdown of national marginal cost curves in order to investigate the cooperation potential between Japan and individual regions of China. For Fig. 3, China is divided into Sichuan Province and others. The optimal amount of CO_2 reduction in China (106 MtC) is distributed to the two regions at a carbon tax of 4000 yen/ton if the JI is done simultaneously with the two regions.[13] However, JI with Sichuan alone requires a very high rate of carbon tax (33 000 yen/ton), which would have a serious impact on the Sichuan economy, as shown clearly in Fig. 3.[14] Similar results will be obtained for JI with Thailand, or Indonesia, or some other countries in Southeast Asia. It would be necessary for Japan to have JI with a group of countries if it considered cooperation with Southeast Asia. The same is true for JI with individual provinces in China if such cooperation was possible.

Concluding Remarks

We have analyzed the potential for cooperation between China and Japan at the macroeconomic level on the basis of the JI framework, and regarding a carbon tax as the shadow price of carbon for the total economy. Our findings are summarized below.

If Japan achieved its reduction target from the Kyoto Protocol by domestic measures alone, it would lose 3.1 trillion yen in nominal terms (0.3% of nominal GDP) as a social deadweight loss in the commitment period. The carbon tax required would be about 53 500 yen/ton, or 400 dollars/ton in nominal terms. Real GDP would decrease by 0.4% compared with BAU. Whether the loss would be perceived as large or small is the problem. If it was considered to be large, then

[13] This is based on the link CGE model, with China consisting of two regions only
[14] This is based on the individual CGE model for Sichuan Province

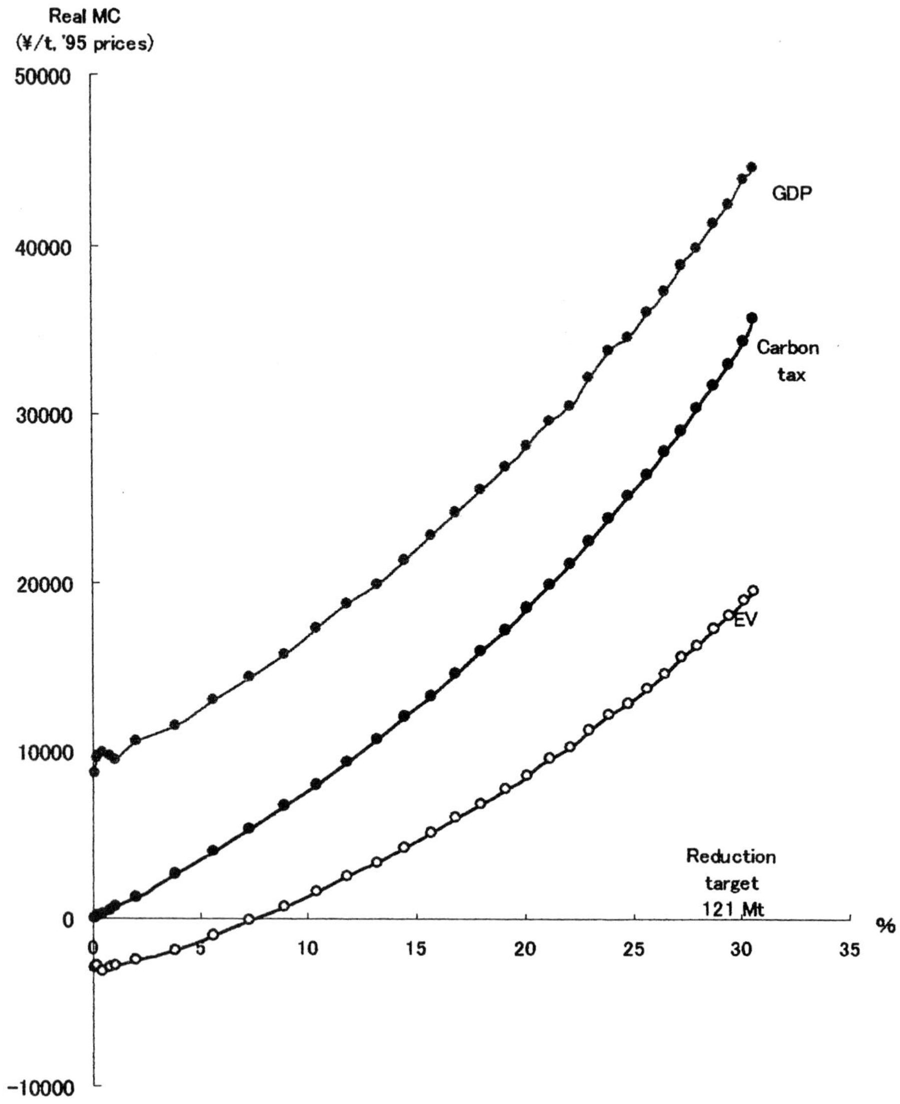

FIG. 2. Comparison of real marginal costs. The horizontal axis represents the CO_2 emissions averaged for 2008–2010. The real MC measured in GDP = $\Delta GDP/\Delta CO_2$, and real MC measured in equivalent variation (*EV*) = $\Delta EV/\Delta CO_2$, where ΔGDP = decrease in real GDP relative to the baseline (BAU) level averaged for 2008–2012, and ΔEV = decrease in equivalent variation relative to the baseline level averaged for 2008–2012

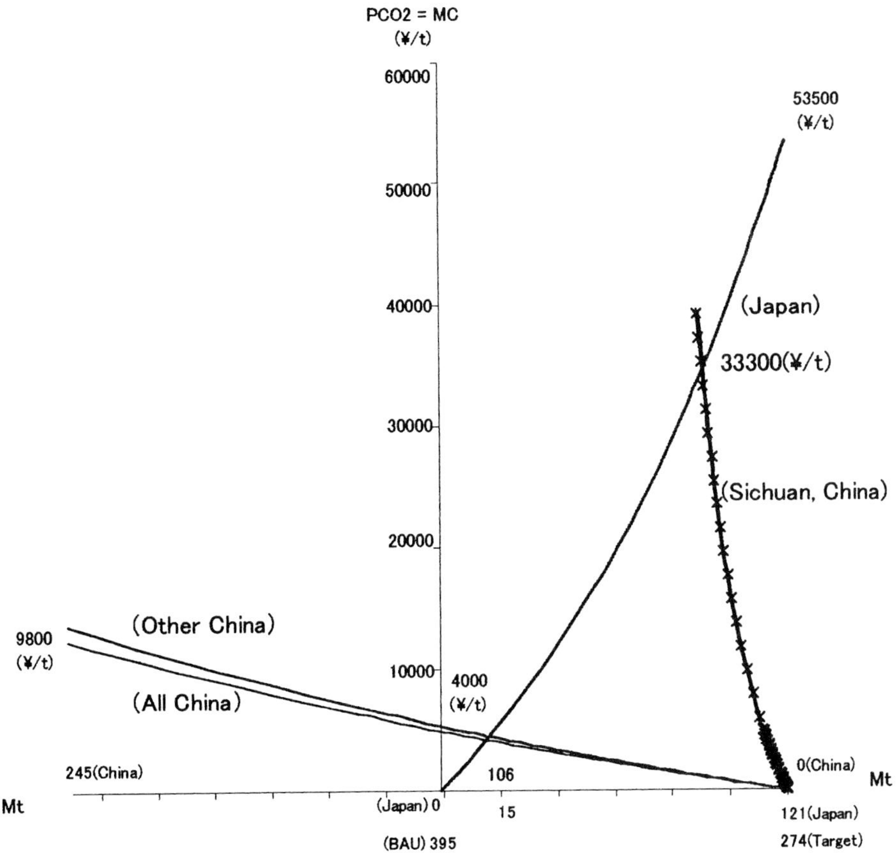

FIG. 3. Macroeconmic analysis of China–Japan cooperation: two-region link computable general equilibrium model for China. The horizontal axis represents the average CO_2 emissions in 2008–2012. China is plotted from right to left. The vertical axis represents the nominal carbon tax (PCO_2) or MC. Yuan are converted to yen (¥) by PPP

cooperation with China would be beneficial for Japan. If JI with China was carried to the point where the marginal cost curve of China crosses that of Japan, China would have a social deadweight loss of 2.1 billion yuan in nominal terms (0.01% of nominal GDP), which is trivial. The problem here is how to allocate the benefits of JI (2.83 trillion yen in total). The natural allocation would be to regard the carbon tax of 4000 yen/ton as the market price, which would require an income transfer of 0.42 trillion yen (27 billion yuan) from Japan to China, although there may be several other feasible methods of allocation. The cooperation could be extended to the countries of Southeast Asia by JI with a group of countries, but not with individual ones, because their size must be taken into consideration.

References

Amano A (1997) Economics of global warming (in Japanese). Nihon Keizai Shinbun Sha

Babiker MH, Rutherford TF (2001) The economic effects of border measures in subglobal climate agreements. Paper presented at the CREST Workshop on the Kyoto Mechanism, Kyoto, Japan, September 27–28

Dervis K, de Melo J, Robinson S (1982) General equilibrium models for development policy. Cambridge University Press, Cambridge

Ezaki M, Kinjo M, Gu L, Qi S (1998) Reducing CO_2/SO_2 emissions by environmental tax in China: a CGE simulation analysis. Working Paper No. 11. Towards an optimal framework for the preservation of the global environment. Realization of environmentally friendly society. CREST (Core Research for Evolutionary Science and Technology) of the Japan Science and Technology Corporation

Hartel TW (1997) Global trade analysis: modeling and applications. Cambridge University Press, Cambridge

Kainuma M, Matsuoka Y, Morita T (1999) Analysis of post-Kyoto scenarios: the Asian–Pacific integrated model. Energy J, Kyoto Special Issue, pp. 207–220

Koppel J, Lee H (1996) The framework convention and climate change policy in Asia. In Mendelsohn R, Shaw D (eds) The economics of pollution control in the Asia Pacific. Edward Elgar, Brookfield, Vt. pp. 26–59

Li Z (1999) Environmental protection system in China (in Japanese). Toyo Keizai

Weyant JP, Hill JN (1999) Introduction and overview. Energy J (Special Issue. The costs of the Kyoto Protocol: a multi-model evaluation), pp. vii–xliv

Zhang ZX (1998) Macroeconomic effects of CO_2 emission limits: a computable general equilibrium analysis for China. J Policy Modeling 20:213–250

A Socioeconomic Analysis of International Cooperation Between Japan and China to Mitigate CO_2 Emissions

Noriyuki Goto

Summary. This study attempts to inquire into the socioeconomic feasibility of international cooperation between Japan and China to mitigate CO_2 emissions jointly from a long-term perspective. The approach taken here is a scenario-based simulation analysis employing a dynamic macroeconomic energy equilibrium model. The simulation results show that collaborative scenarios between the two countries could be over ten times more efficient economically than when Japan carries out its Kyoto commitments independently. Also, collaboration may create large net economic benefits that can be shared even if the reduction targets are more stringent than those of the current Kyoto scenario.

Key words. Global warming, International cooperation, Scenario-based simulation analysis

Introduction

The issue of global warming is now widely recognized as a very serious problem requiring urgent international attention. As far as the political debate is concerned, we could postulate that international cooperation to cope with this issue was in its first stage at the Third Conference of Parties (COP-3) of the United Nations Framework Convention on Climate Change (FCCC), which was held in Kyoto in late 1997, and at which the Kyoto Protocol was adopted. The Kyoto Protocol obliged developed countries (more precisely, Annex I parties) to reduce overall emissions of greenhouse gases (GHGs) to around 5% below the 1990 level in the period 2008–2012, with specific targets for each country or region.

Although realization of the Protocol remains uncertain until it has been ratified by all member countries, it is clear that the next step forward is to find ways to construct a more effective longer-term regime of global cooperation which includes developing regions, and in particular countries with large populations

Graduate School of Arts and Sciences, University of Tokyo, Komaba, 3-8-1 Komaba, Meguro-ku, Tokyo 153-8902, Japan and/or rapid economic growth, such as China and India. At present, the Kyoto Protocol imposes no explicit obligations on non-Annex I countries. It is widely assumed that the growing economies of developing countries will increase the total emissions of carbon dioxide (CO_2) at a much higher rate than the reduction scenario presumed in the Kyoto agreement. This may significantly offset the effectiveness of the efforts of the more developed countries. Therefore, the next step will undoubtedly be to establish the role that should be played by developing countries in making certain commitments to reduce their CO_2 emissions and/or taking part in some of the supplementary mechanisms proposed in the Protocol. Hopefully, this would lead to the realization of a sustainable global system in the long run.

This study attempts to inquire into the socioeconomic feasibility of long-term scenarios of international cooperation between Japan and China to mitigate CO_2 emissions jointly. I first address the possibility of such collaborative scenarios, and then investigate quantitatively the potential economic benefits to be expected. The approach taken here is a scenario-based simulation analysis. It is not a prediction, but a scenario analysis. In the following sections, therefore, I assign a relatively large space to envisioning future scenarios in my analytical framework. I then present the simulation model to be used for numerical examinations, and discuss the computational results and their policy implications.

Analytical Framework

Feasibility and Possibility of Cooperation

The principal objective of this study is to evaluate, from a long-term point of view, the potential benefits that could be created by cooperation, rather than to make a practical assessment of what are called the "Kyoto mechanisms," such as the trading of emissions permits or the clean development mechanism (CDM), that are expected to be introduced in the near future. I start my discussion by recognizing the positional differences in the current status quo between Japan and China, i.e., that Japan has already made a commitment in the Kyoto agreements to reduce CO_2 emissions, while China has not yet officially accepted any obligations. It seems reasonable to postulate that China might require larger socioeconomic and even political benefits if it participated in the regime than the economic gains to be expected by adhering to the Kyoto mechanisms. In fact, China has been claiming, at least officially, that the developed countries should be fully responsible for, and hence must initiate, all action taken to lessen the threat of global warming. Also, China has insisted that developing countries have the right to their economic growth, and should not accept any obligations until they have grown to a certain economic level.

Taking into account the existence of various socioeconomic and political hurdles which must be overcome, and the magnitude of future uncertainties, the cooperative scenarios examined here are rather qualitative in nature. They span a long time horizon (50 years, i.e., 2000–2050), and are designed to allow for considerable flexibility in terms of mutual policy coordination from a macroscopic viewpoint.

However, the design of these scenarios is conceptually close to what is called "joint implementation" (JI). JI is simply a mechanism to encourage joint efforts which allow for free exchanges of policy implementation and cost bearing among several countries with existing commitments, provided that they achieve an overall emissions reduction target as a group. It is recognized, at least theoretically, that JI is an economically efficient policy measure to reduce GHG emissions regionally as well as globally.

It is evident that the total costs (covering both direct and opportunistic costs) of attaining a certain level of reduction can be minimized by taking up the option of trading in countries where materials and labor costs are cheaper.

Under the assumption of economic rationality in behavior, JI is a way to realize a minimum-cost scenario within a group of countries, and can be mutually beneficial to all participants. The many relevant studies of this subject are reviewed in, for example, IPCC (1995, Chap. 11).

However, the discussion is clearly based on the presumption that each country has already made a certain commitment and can therefore expect some degree of economic advantage from participating in JI cooperation. In reality, however, almost all of the developing countries (including China) are reluctant to agree to take any policy measures that might impede their economic growth, claiming that developed countries are fully responsible for the causes of global warming. Although this claim can be understood as reasonable in one sense, it should also be emphasized that cooperation could be beneficial to all participants. Developing countries may boost their own economic development by effectively utilizing funds provided by developed countries, and may also expect some improvements in their domestic environmental quality. In addition, there are estimates that a doubling of the atmospheric CO_2 concentration may result in greater damage in developing regions, e.g., damage costing more than 6.0% of GDP in China (Frankhauser 1992).

This case study addresses the possibility of international cooperation between Japan and China, implicitly taking into account the stages of negotiation and agreement that could be achieved if some economic incentives were offered by the former to the latter. The question to be examined is, "If Japan offered (or asked) to collaborate with China for a joint reduction of CO_2 emissions provided that the former should make funds available to the latter in return for its collaboration, then under what conditions might China accept the offer, and at what level could such cooperation be mutually beneficial?" After characterizing a feasible (or Pareto optimal) solution to the situation outlined above, I attempt to evaluate quantitatively the economic consequences and potential net benefits to be achieved through cooperation by applying the simulation model to some cooperative scenarios.

Socioeconomic Model of Cooperation

I briefly discuss the feasible cooperation scenarios by employing a simple mathematical representation. Let A (Japan) and B (China) be two countries that are candidates for cooperation. Table 1 gives the notation and briefly describes the analytical framework of the discussion.

Initial States

Let us presume, as initial conditions, that Country A has already made a certain commitment to reduce CO_2 emissions, but that Country B has not.

Feasibility of Cooperation

A cooperative scenario can be feasible when it results in a greater (at least an equal) reduction in CO_2 emissions than would be the case with noncooperation, and brings nonnegative economic benefits to both countries.

Negotiation

The negotiation between the two countries may be described as follows: (i) Country A offers Country B a combination of reduction target (joint target $r_{AB} = r_A + r_B$, or individual target r_B) and monetary funds transfer (TR) in return for acceptance, and then (ii) Country B agrees, or declines the offer.

Classification of Feasible Scenarios

It is convenient for our discussion to classify the feasible solutions into four regimes. (I) Only Country A can enjoy positive net benefits with minimum funds being transferred to partner B. (II) Only Country B can enjoy positive net benefits by receiving the maximum available funds transferred from partner A. (III) Both countries can enjoy positive net benefits by reallocating the net economic benefits created by cooperation in some way. (IV) The earth can enjoy maximum benefits if all of the net economic benefits produced are put into a further reduction of CO_2 emissions.

Structure of Model

It is easily shown that a feasible solution exists in the game structure outlined above. Clearly, the feasible region for the set of (TR, r_A, r_B) must satisfy

$$C_B\left(r_A^{\,c} - r_A\right) \le C_B(r_B) \le \mathrm{TR} \le C_A\left(r_A^{\,c}\right) - C_A(r_A)$$

Under the normal assumptions made about the functions $C_A(r_A)$ and $C_B(r_B)$, there exists a positive set of (TR, r_A, $r_B > r_A^{\,c} - r_A$) since we can find a positive value of $r_B (> r_A^{\,c} - r_A)$ for any r_A in the small neighborhood of $(r_A^{\,c} - r_A)$, and then a TR that satisfies the above inequality.

Figure 1 visualizes the discussion a little more precisely in the context of economic theory. By relaxing the level of reduction from $r_A^{\,c}$ (the initial commitment)

TABLE 1. Conceptual representation of the model

Countries

Country A (= Japan)

Country B (= China)

Initial states

$r_A^0 = r_A^c (> 0)$

$r_B^0 = 0$ (no commitment by B) (1)

Feasibility of cooperation

(1) Meaningfulness (overall reduction is not decreased by cooperation)

$$r_{AB} = r_A + r_B \geq r_A^c \qquad (2)$$

(2) Conditions for agreement (neither country is worse off by cooperation)

$$C_A(r_A) + \mathrm{TR} \leq C_A(r_A^c)$$

$$C_B(r_B) - \mathrm{TR} \leq 0 \qquad (3)$$

Negotiation

(1) Country A offers to Country B a combination of (TR, r_{AB}), or (TR, r_B)

(2) Country B agrees or declines

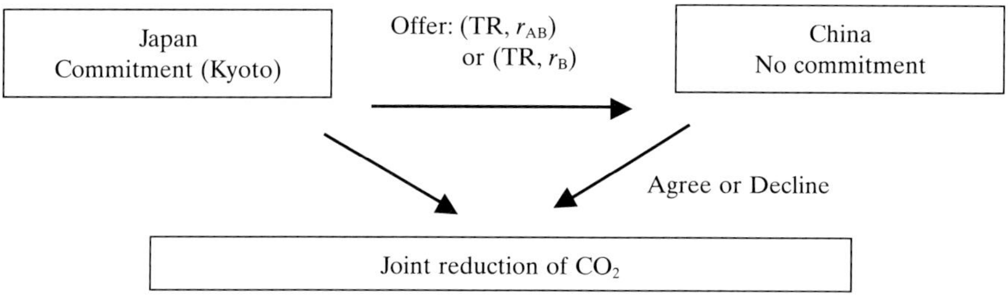

Classification of the feasible scenarios

(I) Only Country A can enjoy positive net economic benefits

$$C_A(r_A) + \mathrm{TR} < C_A(r_A^c) \text{ and } C_B(r_B) - \mathrm{TR} = 0 \qquad (4)$$

(II) Only Country B can enjoy positive net economic benefits

$$C_A(r_A) + \mathrm{TR} = C_A(r_A^c) \text{ and } C_B(r_B) - \mathrm{TR} < 0 \qquad (5)$$

(III) Both countries can enjoy positive net economic benefits

$$C_A(r_A) + \mathrm{TR} < C_A(r_A^c) \text{ and } C_B(r_B) - \mathrm{TR} < 0 \qquad (6)$$

(IV) The Earth can enjoy the maximum benefits

$$C_A(r_A^c) - C_A(r_A) - \mathrm{TR} = C_B(r_B) - \mathrm{TR} = 0 \text{ and Max } \{r_A + r_B\} \qquad (7)$$

Notations

r_A = the commitment to reduce CO_2 emissions (relative to BAU) by Country A

r_B = the commitment to reduce CO_2 emissions (relative to BAU) by Country B

$C_A(r_A)$ = total direct and opportunity costs of achieving r_A

$C_B(r_B)$ = total direct and opportunity costs of achieving r_B

TR = the amount of monetary funds transferred from Country A to Country B, where it is normally assumed for the functions $C_A(r_A)$ and $C_B(r_B)$ that $C_A(0) = C_B(0) = 0$, $dC_A(r_A)/dr_A > 0$, $dC_B(r_B)/dr_B > 0$, $d^2C_A(r_A)/dr_A^2 > 0$, and $d^2C_B(r_B)/dr_B^2 > 0$.

BAU, business-as-usual

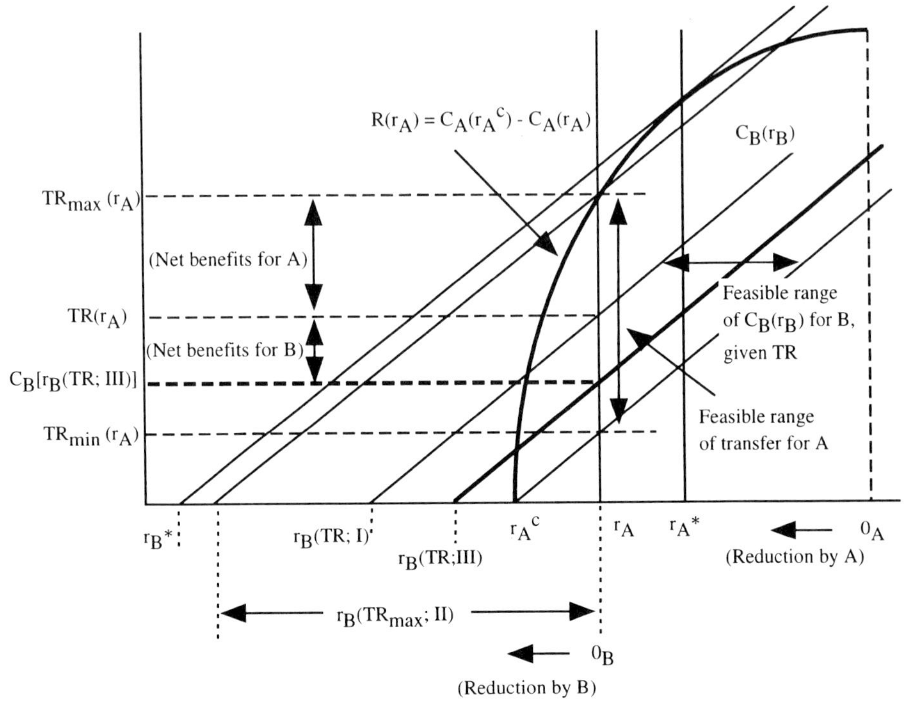

FIG. 1. The economic feasibility of cooperation

to r_A, Country A can reduce the total abatement costs by $R(r_A)[= C_A(r_A{}^c) - C_A(r_A)]$, and hence can offer Country B an amount of funds transfer somewhere between a maximum of $TR_{max}(r_A)[= R(r_A)]$ and a minimum of $TR_{min}(r_A)[= C_B(r_A{}^c - r_A)$, determined by the feasibility condition] in return for a commitment to the joint reduction. This is depicted as the "feasible range of transfer for A" in the figure. It is clear from Fig. 1 that as the total costs-saving curve, $R(r_A)$, becomes steeper, the feasible region becomes wider, and hence a larger incentive as well as flexibility in negotiation are given to Country A.

Here, the total costs of reducing carbon emissions in Country B are shown by straight lines, which makes the drawing easier, although they are normally assumed to be sloping upward and increasing. The curve (line) is located such that, given the amount of funds transferred from Country A, say $TR(r_A)$, the level of reduction by Country B, i.e., r_B, is shown by the length between its intercept with the horizontal axis and 0_B. The horizontal movement of the line describes the reduction level decided on by Country B. Let r_B be measured by the distance from point 0_B to the left. If $TR(r_A)$ is received from Country A, Country B may choose its reduction level between the maximum of $r_B(TR; I)$, in which all funds are put into its own reduction efforts, and the minimum of $(r_A{}^c - r_A)$, that exactly makes up the waiver in Country A, and which is depicted as the "feasible range for B." It is easily inferred that more funds transferred from Country A and a less steep marginal costs curve will give a wider feasible region and more flexibility in determining its own reduction level, and hence create stronger incentives for Country B to accept the joint reduction.

In the general case (III), Country A can get net economic benefits by $TR_{max}(r_A) - TR(r_A)$ (> 0), while Country B can enjoy $TR(r_A) - C_B[r_B(TR; III)]$ (> 0). In this case, the reduction level is also raised by $r_B(TR; III) - (r_A^c - r_A)$ compared with the noncooperative case r_A^c.

Finally, the distance from 0_A to r_B^*, where r_B^* is the intersection of the horizontal axis and Country B's total costs curve tangential to the total cost-savings curve of Country A, shows the maximum level of reduction attainable through cooperation when the earth enjoys the maximum benefit. In this particular case, Countries A and B reduce their emissions by r_A^* and $r_B^* + (r_A - r_A^*)$, respectively, and the marginal costs are equalized in both countries, which is known to be a necessary condition for optimality.

It should be noted that in any of the regimes, our scenarios allow for the possibility of achieving more stringent reduction targets than the level of initial commitment in the noncooperative scenario.

I have classified the feasible scenarios into four regimes, but it is not clear which of the regimes is likely to be realized under the assumptions and conditions made in this study. Although these issues, i.e., to what extent we should aim at mitigation, and how we should reallocate the economic benefits between the two countries, are recognized as being serious and are implicitly taken into consideration throughout the study, a careful examination of each regime has to be postponed to the second stage of the research. It is too early to decide on the preferred scenarios, not only theoretically, but also because of our limited previous experience of international environmental negotiations and their consequences (Young 1992; Haas et al. 1993; Victor et al. 1998).

Economic Efficiency and Common Net Economic Benefits

Theoretically at least, a cooperative reduction of CO_2 emissions will always be more economically efficient than noncooperation in terms of the different marginal costs to participating countries. Let us suppose that the levels of joint reductions were given, and were equal to those of the initial commitments. Figure 2 describes the basic mechanism of gains in efficiency which will be brought about by mutual exchanges of reduction and cost-bearing in the two-country model. As in many similar figures, it is drawn so that if the commitment to reduction by Country A (E_0) is given, and is represented by the length of the horizontal axis, the reduction by Country A is measured from left to right, and that by Country B, with no initial commitments, is measured in the opposite direction from right to left. If a level of cooperation is agreed, any point on the line indicates the burden sharing that exists. The length of the line to the left of the point indicates the reduction required by Country A, and that right of the point is the reduction required by Country B, and the sum is always equal to the initial commitment by Country A. The marginal cost curves for the two countries are also drawn in opposite directions, and correspond to the level of reduction in each country. It

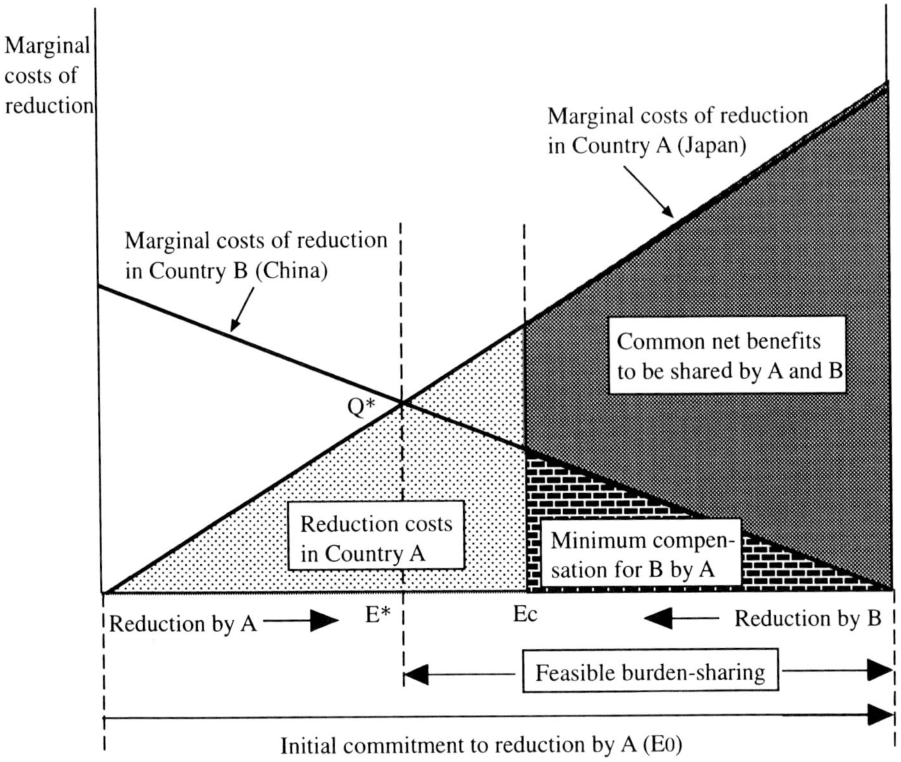

FIG. 2. The potential economic benefits of cooperation

is clear that the area (triangle) below the marginal cost curve shows the total costs. Country A can save costs by requesting Country B to take part of its obligation in exchange for bearing the costs or transferring funds, or even by itself carrying out reductions in Country B, whose marginal costs start from virtually zero.

Figure 2 also reflects our basic concerns about this issue. Most economists discuss it in the context of the market mechanism, and demonstrate the desirability as well as the efficiency of an equilibrium in which the marginal costs are equal for the two countries. The level of costs means the equilibrium price if we assume that such exchanges are dealt with in the market through the trading of emissions permits, for example. This equilibrium corresponds to Q^* (or E^* in terms of each share of the reduction) in Fig. 2.

I have taken a different view of the issue here, and have considered it not only from an economic, but also from a sociopolitical point of view. First, I do not pay particular attention to market equilibrium in the sense described above, since it is still controversial and we are uncertain whether or not the two countries concerned would agree to join any market system because of the differances in their status quo. In equilibrium, the cost-savings in the committed country and the transfer of profits in the partner country, which can be discussed in terms of what are known as "consumers' surplus" and "producers' surplus" to be mutually enjoyed from dealings in the markets, are determined so that much of the benefit appear to be allocated to Country A in our model. This is mainly because the marginal-costs curve in Japan is apparently far steeper than that in China. Furthermore, although the CDM is not yet clearly defined, the present discussion seems to suggest that the method of allocation ensures that the recipient (or host) country receives only the minimum compensation for its efforts, as shown in Fig. 2. It seems fairly reasonable, therefore, that the latter country may require more economic benefits than only market values or CDM values to be persuaded to participate in the collaboration.

Instead of focusing on the concept of market equilibrium, I presume that a certain level of joint reduction has been agreed (on E_c in Fig. 2, for example) within a feasible region for burden-sharing. I then regard the net economic benefits to be created by the cooperation, which are shown by the shaded area in Fig. 2, as "common net benefits," whose allocation should be open to negotiation between the two countries. It is evident from the figure, and also theoretically, that these common net economic benefits will be larger, and hence there may be stronger incentives for collaboration on both sides, when the differences in the steepness as well as height of the marginal cost curves are larger.

Before proceeding to the next section, it will be helpful to take a comparative view of the relationship between the levels of economic activities and CO_2 emissions in Japan and China. Figure 3 shows the levels of CO_2 emissions per real GDP in recent years, and shows that large differences exist in the marginal abatement costs (curves) of the two countries. Those of China are roughly ten times as high as those of Japan when evaluated in terms of the opportunity costs which will be incurred by reducing their CO_2 emissions in 1995.

In addition, Fig. 4 compares the levels of CO_2 emissions per capita in the two countries. This figure strongly supports the claim by China that cooperation

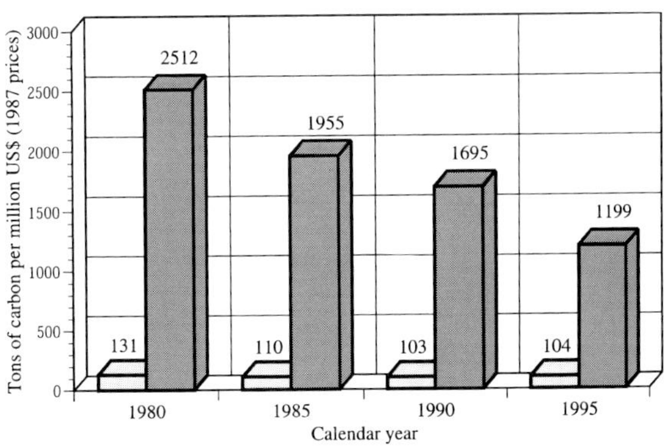

FIG. 3. CO_2 emissions per real GDP in Japan and China. *Open blocks*, Japan; *shaded blocks*, China

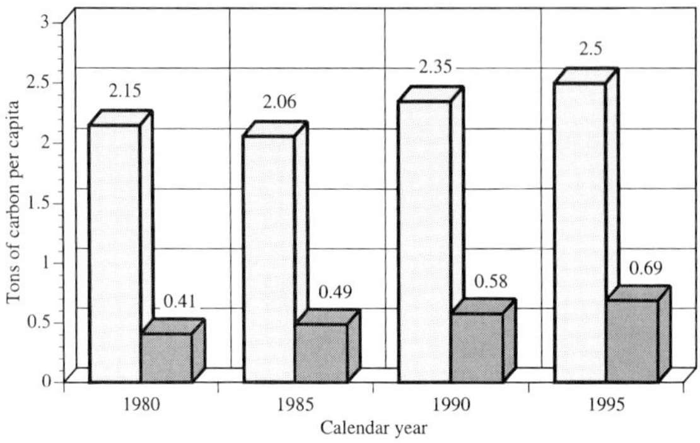

FIG. 4. CO_2 emissions per capita in Japan and China. *Open blocks*, Japan; *shaded blocks*, China

should be not only cost-effective, but also equitable. This principle is also stated in the United Nations FCCC (Article III, Clauses II and III).

Construction (and Extension) of a Simulation Model

Original Model (GDMEEM)

My previous work has concentrated on constructing and continuously extending an economic simulation model with the purpose of assessing various energy and/or environmental policies particularly related to the issue of global warming. The analytical model is a simplified general equilibrium model named Goto's dynamic macroeconomic energy equilibrium model (GDMEEM). GDMEEM is a hybrid integration of a bottom-up formulation for the behavioral characteristics of energy supply sectors and of a rather abstract top-down representation of those of industrial and residential sectors on the demand side. It is constructed as a mathematical programming model, which simulates a future dynamic equilibrium path under the assumption that markets for both primary and secondary energy resources (products) are efficient. The model is an integration of two underlying models: (1) a linearized nonlinear programming model that is designed to compute a cost-minimum supply of energy (secondary energy products) reflecting the technical and economic characteristics of both existing and explicitly specified future energy conversion technologies, as well as the availability of primary energy resources, and (2) a macroeconometric model that relates energy demand to macroeconomic activities, where the macroeconomy is disaggregated into 14 revenue-maximizing industrial sectors, that are further divided into 45 subsectors depending on the specific patterns of energy use (interfuel substitutions are rather limited among the subsectors), and one utility-maximizing residential sector that has five subsectors according to different patterns of energy use. The supply and demand behaviors are integrated in a competitive market that contains an explicit description of the energy network flow

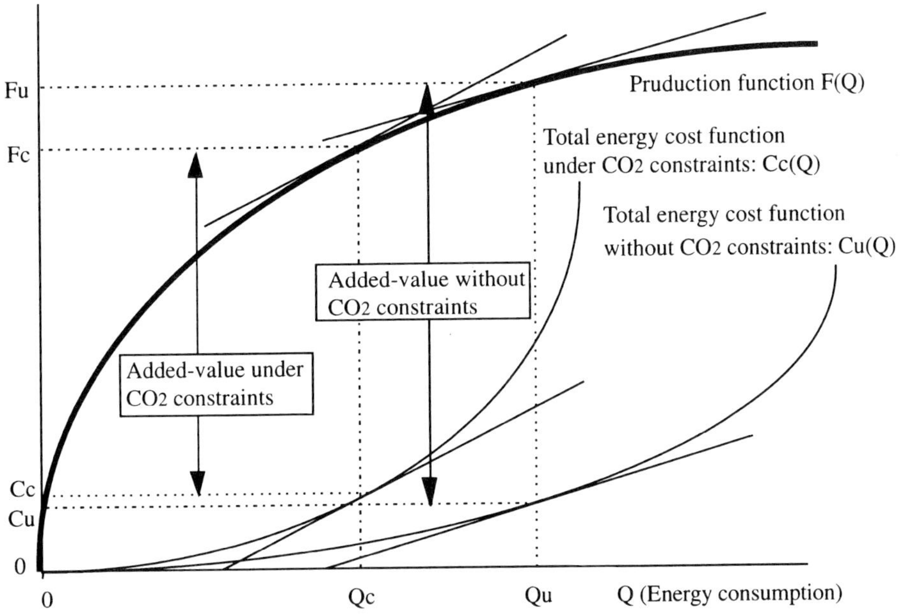

FIG. 5. Economic adjustments to the CO_2 constraints in the model

system. The network system specifies energy flows from six primary resources (i.e., crude oil, coal, natural gas, hydropower, nuclear power, and some renewables) to nine secondary energy forms (i.e., electricity, petroleum products, gaseous fuel, and solid fuel), and then to the 50 demand subsectors for final use. The equilibrium is solved so that the discounted sum of the producers' and consumers' surpluses over a planning horizon is maximized.

The quantity of CO_2 emissions can be computed directly if the consumption of each primary resource is determined. Environmental policies such as CO_2 emission controls are put into the model by appropriate constraints (e.g., quantity constraints or carbon taxes), which affect both supply and demand behaviors through interactions in the markets.

Figure 5 is a simplified version of the basic mechanism of economic adjustments to CO_2 emission constraints that is presumed in the model. The energy cost function shifts upward when CO_2 emissions are under control, and this is computed in terms of cost minimization by the bottom-up formulation of the energy supply sectors. Responding to the shifts in the energy cost function, nonenergy industrial sectors determine their energy demand in order to maximize their net economic profits, which are defined by the gross output in added value minus total energy costs. As a result, the total abatement costs can be evaluated as the sum of the opportunity costs in decreased outputs ($= F_u - F_c$) and increased energy costs ($= C_c - C_u$). The behavioral characteristics of residential sectors are almost same, with the production function $F(Q)$ in the figure as the utility function $U(Q)$ of energy consumption.

See Goto and Sawa (1992) and Goto (1995) for more details about the model, including its application to several assessments as well as the mathematical formulation and underlying methodology.

Extended Two-Country Model

The original model has been extended in the following ways.

1. The extended two-country model links the two national models for Japan and China with some simplifications of the latter module and the addition of a mechanism for the transfer of funds between them.
2. The two countries coordinate their policy options to reduce CO_2 emissions jointly in a cost-effective way. This is formulated mathematically by the maximization of the sum of two objective functions, which implies a utilitarian evaluation of the economic welfare of both countries.

Assumptions and Scenarios

Basic Assumptions

As was described in the last section, the simulation model takes account of the transfer of funds as the only explicit linkage between the two countries, assuming ceteris paribus for the other interactions and no hypothetical market mechanisms such as dealings in some projects through the CDM.

Another key assumption on the scenarios to be examined is that the funds to be transferred to China are financed by the revenue from carbon taxes in Japan. This assumption has been made for two basic reasons from a sociopolitical point of view. One is that the economic fruits of cooperation on the Japanese side are not real money obtained directly, but are savings in anticipated direct and/or opportunity costs, and hence they cannot be transferred freely. The explicit introduction of carbon taxes, and the use of tax revenues as the budget for transfers, may help us to address the practical operation of these scenarios. The other reason is that carbon taxation may be one valid form of response to the claim by China that developed countries should initiate action. From the analytical viewpoint, it means that the simulation output for the Japan module is characterized by an economic balance between a domestic reduction of CO_2 emissions through carbon taxes and a reliance on foreign efforts in return for the transfer of funds financed by tax revenues. Tax rates which are too high might seriously damage the domestic economy, but if they are too low (with lower tax revenues), they would not be sufficient to persuade a partner to be cooperative in return for the transfer of funds. The tax rates in Japan are fixed and are specified exogenously, without any relation to an economic equilibrium where the optimal carbon tax rates are determined by equalizing the marginal costs.

Business as Usual (BAU) Scenario

In this type of study, it is always difficult and troublesome to calibrate a business-as-usual (BAU) scenario which covers a long time-horizon. Although this work is still in the preliminary stage (and always will be in the strictest sense), I have constructed a base scenario which relies on a detailed examination of past trends, the current availability of energy resources and technologies, and the expected population and economic growth rates, including national programs and others, as well as a review of many relevant published reports (e.g., Clarke and Winters 1995; Imura et al. 1995; Zhang 1998; Goto 2001).

Table 2 gives a summary of the BAU scenarios for Japan and China. These are computer outputs of simulations under no CO_2 emission constraints, and hence also reflect all the input data and assumptions made in the model.

CO_2 Emissions Constraints

For the constraints on CO_2 emissions, I imposed the upper limits of the mass of CO_2 remaining in the atmosphere at the end of the time-horizon rather than those of the annual rate of emissions. This is based on the fact that it is not the rate of emissions but the atmospheric concentration which causes global warming, and that ideologically the latter should be the policy target to be achieved in the long run. A simplified relationship between the remaining mass at time t, $M(t)$, and the annual rate of emissions, $E(t)$, can be described by the formula

$$M(t) = 0.64E(t) + 0.992M(t-1)$$

which is an approximate estimate of the relationship between the two variables on a global scale. See Nordhaus (1991, 1994) for background details about this equation, and Goto and Sawa (1992) for some comparative analyses of its application to our simulation model.

Thus, the following formulation represents the policy targets, or the constraints on CO_2 emissions, that are employed in our simulation analyses:

$$M(2050) \leq M^{\mathrm{UL}}$$

Construction of Cooperative Scenarios

Four scenarios have been constructed for comparative examination. Table 3 gives a brief description of these scenarios. The first is a free scenario with no control of CO_2 emissions, which is the BAU scenario described previously, and is used as a yardstick to evaluate the other scenarios. The second, which I call the "Kyoto scenario," is virtually the base-case scenario, or the starting position in our considerations, i.e., Japan fulfils its commitment in the current Kyoto agreements on its own, while China has no obligations at all. Then, based on the computed value of M(2050; BAU) in the first scenario, i.e., the volume of CO_2 emitted from both countries and accumulated in the atmosphere at the end of the time-horizon in

TABLE 2. A brief summary of the BAU scenarios for Japan and China[a,b]

	Calendar year						Avg. growth rate (%/year)	Discounted sum[c] (10^{12} US$)
	2000	2010	2020	2030	2040	2050		
Japan								
GDP (10^{12} US$/year)	4.1	4.7	5.7	7.1	8.8	10.9	1.97	165.7
Energy demand (10^{12} GJ/year)	15.1	16.1	17.9	21.2	25.0	30.3	1.39	—
CO_2 emissions (10^9 tC/year)	0.32	0.33	0.38	0.44	0.50	0.60	1.26	—
China								
GDP (10^{12} US$/year)	1.0	2.1	3.8	6.4	9.6	13.4	5.17	132.2
Energy demand (10^{12} GJ/year)	38.5	55.0	77.4	95.7	127.1	164.7	2.91	—
CO_2 emissions (10^9 tC/year)	1.12	1.61	2.28	2.78	3.77	4.92	2.97	—
Japan + China								
CO_2 emissions (10^9 tC/year)	1.43	1.94	2.65	3.22	4.27	5.51	2.69	—
CO_2 accumulation (10^9 tC)	—	9.64	22.30	38.65	57.44	81.67	—	—

BAU, business-as-usual
[a] The time horizon is the 50 years from 2000 to 2050
[b] The exchange rates are fixed at a constant 15 Japanese yen for one Chinese yuan, and 120 Japanese yen for one US dollar
[c] The discount rate is assumed to be 5%/year for both countries and constant throughout the time-horizon. This is mainly for analytical simplicity

TABLE 3. Scenarios examined in this study

Scenarios to be examined

(1) BAU: Japan (BAU) + China (BAU)
(2) Kyoto: Japan (Kyoto) + China (BAU)[a]
(3) Coop.A: cooperation to achieve: $M(2050; \text{Coop.A}) \leq 0.9\ M(2050; \text{BAU})$
(4) Coop.B: cooperation to achieve: $M(2050; \text{Coop.B}) \leq 0.8\ M(2050; \text{BAU})$

Conditions for the cooperative scenarios

Japan: Japan introduces carbon taxes of 5000 yen (or US$42) per ton of carbon at present value throughout the planning horizon, the revenue of which can be used to finance the funds transferred to China.

China: In China, all costs, including direct and opportunity costs, are financed by funds transferred from Japan; more precisely, the total sum of discounted values of GDP, as well as that of consumers' utility enjoyed by energy consumption in residential sectors, are reimbursed so that they never decrease below the levels in the BAU scenario, although China is allowed to make intertemporal economic and energy-related technological adjustments

[a]The Kyoto scenario assumes the application of current Kyoto commitments (specified only for the period of 2008–2012) to the whole time-horizon

the BAU scenario, I have postulated two cooperative scenarios to implement a joint emissions reduction: the one called the "Coop.A scenario" achieves a 10% reduction from $M(2050; \text{BAU})$ by mutual policy coordination, and the other, the "Coop.B scenario," aims at a more stringent target which is 20% lower than $M(2050; \text{BAU})$.

Simulation Results

Potential Economic Benefits

Figure 6 summarizes the potential economic benefits to be expected from the cooperative scenarios. For Japan, while the opportunity costs of carrying out its Kyoto commitments on its own are estimated to be around US$937 billion (total sum over the time-horizon, 1990 prices, present values), which is losses of about 0.8% in terms of the discounted sum of real GDP in the BAU scenario, the Coop.A scenario costs only US$230 billion (around 75% saving), i.e., US$197 billion for domestic implementation and US$33 billion for the transfer of minimum funds to China, resulting in the creation of net economic benefits of US$707 billion (937 – 197 – 33 = 707). Since the atmospheric accumulation of CO_2 in 2050 with Japan following Kyoto and BAU in China is only 3.6% below the scenario for BAU in both countries (see later discussion, or Fig. 10), but is 10% below the BAU scenario with Coop.A, it is clear that there is great potential for improvements in economic efficiency.

Owing mainly to the rising marginal abatement costs in China, the Coop.B scenario forces Japan to raise the minimum transfer of funds to China to US$185 billion, together with US$309 billion for her domestic effort, and consequently

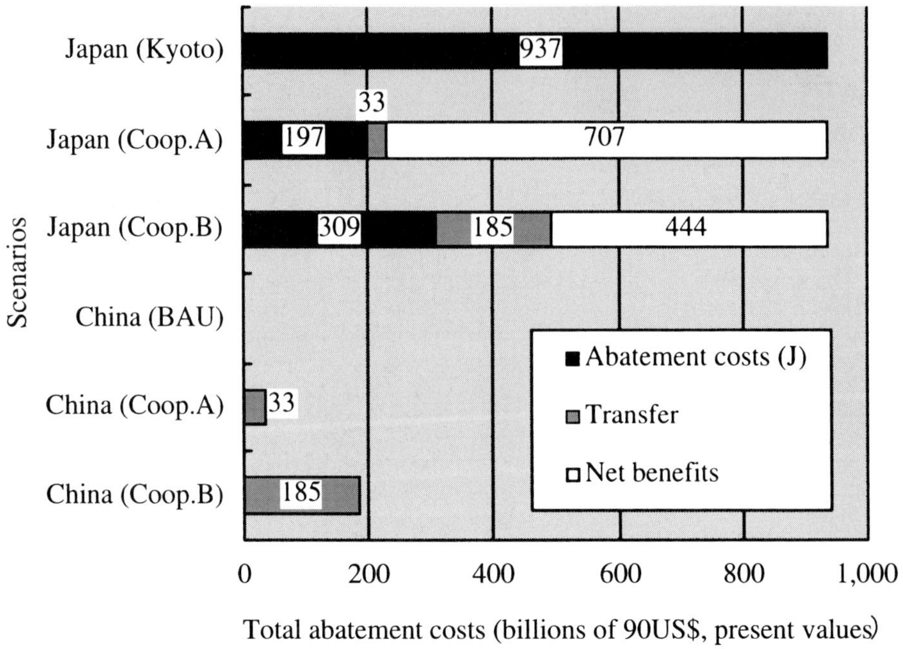

FIG. 6. Potential economic benefits in each scenario

the net benefits fall to US$444 billion (about 53% saving from Kyoto). However, this is still economically beneficial in spite of the 20% reduction in $M(2050)$ in this scenario.

In line with the arguments put forward throughout this study, I regard the estimated net economic benefits as common to both countries, and their allocation should be open to negotiation, although the figures shown in Fig. 6 appear to favor Japan. However, this issue is political as well as economic, and it is not at all clear how countries will share the net benefits in order to reach agreement. It should also be remembered that the net benefits estimated are not real money, but are potential savings of direct and opportunity costs in Japan, and hence their allocation has to be based on somewhat hypothetical expectations as well as on a willingness to share on both sides.

Notwithstanding these reservations, and even the wide margins of the estimates, the simulation results strongly suggest that we could expect potentially large net economic benefits to be mutually enjoyed through cooperation. They could be sufficiently large for both countries to see the attraction of engaging in the collaboration. China may expect further and/or different economic returns in addition to the minimum compensation for its efforts through negotiations, while a long-term contract for collaboration, if it was possible, might be especially beneficial for Japan, not only economically but also politically as a member of the Annex I parties in the FCCC.

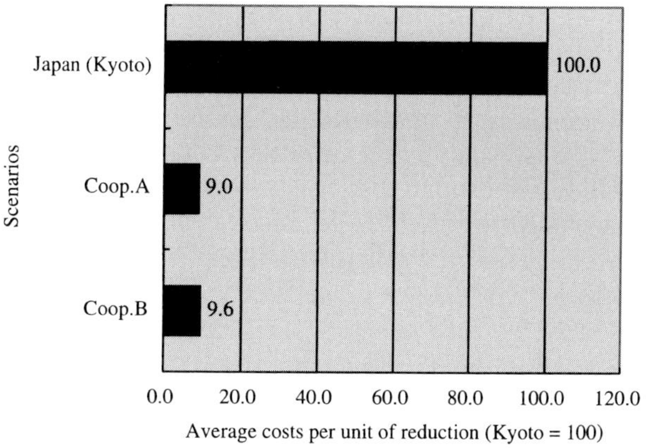

FIG. 7. Average costs of reducing CO_2 emissions in each scenario. The average costs used for the relative comparisons in the figure are calculated by the approximate total opportunity costs divided by the reduction in terms of $M(2050)$

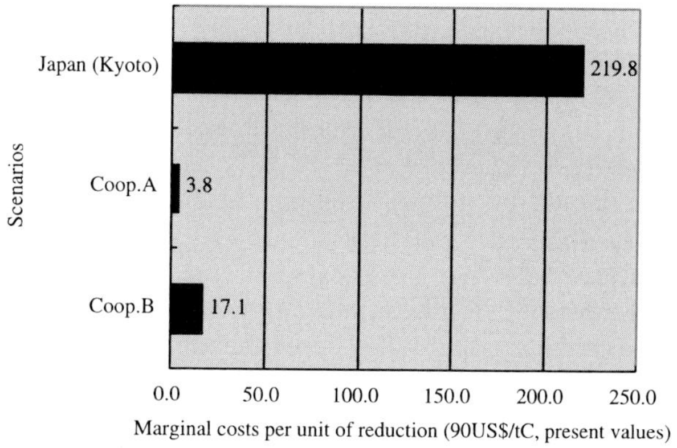

FIG. 8. Marginal costs of reducing CO_2 emissions in each scenario. The marginal costs in the figure are shadow prices associated with the upper-limit constraints on $M(2050)$ that are converted to units per emission

Economic Efficiency

Figures 7 and 8 compare the extents to which these cooperative scenarios can decrease the costs of reducting CO_2 emissions. Surprisingly enough (or perhaps reasonably enough), the simulation results are remarkable. According to Fig. 7, the average reduction costs per unit of CO_2 emission can be decreased by around 91% (in the Coop.A scenario) and 90.4% (in the Coop.B scenario) compared with the Kyoto scenario, which means that these collaborative reductions are almost ten times as more economically efficient than independent implementation by Japan in line with its Kyoto commitments. However, these low estimates are not really beyond our expectations if we remember that the ratio of CO_2 emissions per real GDP for the two countries was almost 10:1 in 1995, as was seen in Fig. 3.

Is has been, and is being, claimed by many analysts and others that the Kyoto scenario would be a heavy burden on the Japanese economy. According to professional discussions by economists, the Kyoto commitments are generally defective in the sense that they do not leave room for dynamic adjustments, e.g., later massive efforts by employing improved and expanded technological options in the future, or by following appropriate plans for capital depreciation and R&D for advanced technologies, and hence are economically inefficient. More practically, it is widely recognized that considering the high efficiency in energy utilization achieved in the past, in particular since the mid-1970s after the oil crises, it would be very costly to make further improvements in energy efficiency in Japan. By reflecting these views, and possibly some others, the simulation results also confirm that the marginal abatement costs for Japan to keep to the Kyoto scenario by its own domestic efforts might be extremely high, and are estimated to be around US$220 per ton of carbon (average over the time-horizon, at the present value). In our analytical framework, this figure equates exactly with the carbon tax rate that is required to implement the Kyoto agreements by relying only on tax policies.

Figure 8 shows that the estimated marginal costs drop sharply in the cooperative scenarios to US$3.8/tC in Coop.A and to US$17.1/tC in the even more aggressive Coop.B. Theoretically, these figures, which are still much lower than the carbon tax rate assumed in Japan (= US$42/tC), imply that there could be room for further improvements in economic efficiency than the levels reported here.

It should be noted, however, that these remarkable effects come from two causes inherent in the construction of our scenarios. One is clearly the large differences in the marginal costs (curves) between Japan and China, and the other is the flexibility of the intertemporal adjustments allowed in the model, although they cannot be quantified separately here.

CO_2 Emissions and Atmospheric Accumulation

We now look at the simulation results for the trajectories of annual CO_2 emissions and their atmospheric accumulations in each scenario. Figure 9 compares the annual CO_2 emissions from the two countries. We can see that much of the reduction is transferred from Japan to China, with multiplicative increases in volume in the cooperative scenarios, although the assumed imposition of carbon taxes also makes some contribution in Japan. This figure also shows the economic efficiency of dynamic adjustments, i.e., that of later reductions rather than earlier actions, which is theoretically reasonable, but may be arguable in another sense. The zigzag patterns observed in the figure are mostly due to the linearization of the model, but partly because of the vintage formulation of the treatment of large energy-conversion facilities, which is less important in these discussions.

Figure 10 shows the simulation results for CO_2 accumulation in the atmosphere in each case, which are regarded as the most appropriate criteria for evaluating the impacts of increasing CO_2 emissions on climate change. Hence, this study has paid more attention to these than to the annual rates of emissions. Although the figures depend heavily on the construction of our scenarios and the various assumptions made in the model, it can be seen that the implementation of the

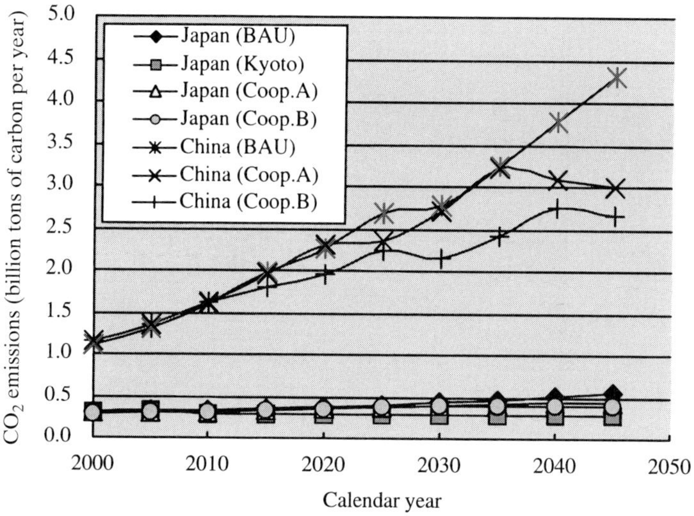

FIG. 9. Annual CO_2 emissions in each scenario

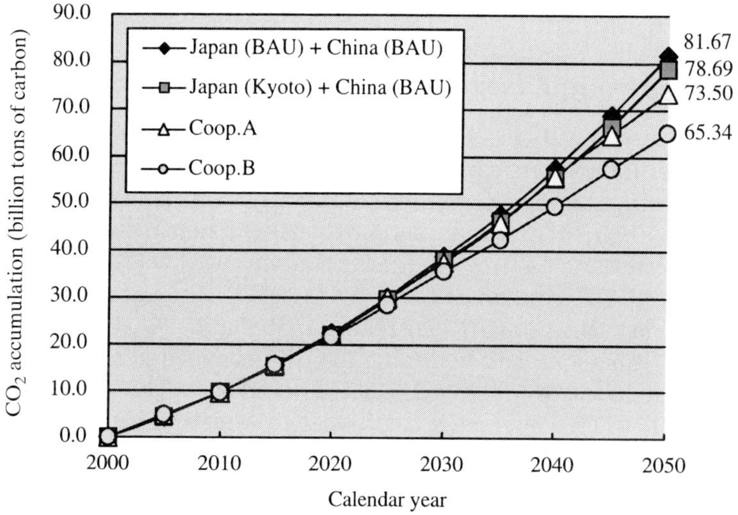

FIG. 10. Atmospheric accumulation of CO_2 in each scenario

Kyoto scenario by Japan alone may result in around 3 billion tons of carbon, or a reduction of only 3.6% from the BAU scenario in terms of the atmospheric accumulation in 2050. In this sense, our cooperative scenarios, in which the targets are set at –10% (Coop.A) and –20% (Coop.B) with respect to this criterion, are fairly mild, but can be regarded as aggressive compared with the Kyoto scenario for mitigating climate change.

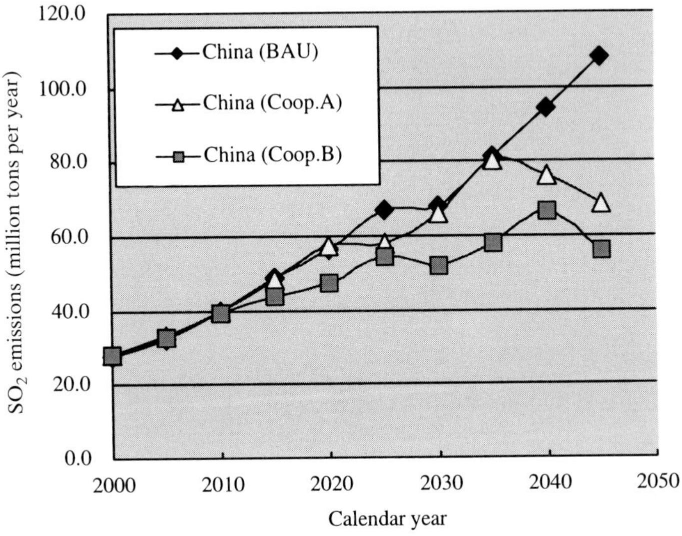

FIG. 11. Annual SO_2 emissions in each scenario

SO_2 Emissions in China

In the last section, I will include figures on the changes in the level of SO_2 emissions in China for each scenario, as a complement to our discussion. As is well known, local air pollution is a serious environmental issue in China. It was estimated by Li (1999), for example, that the damage to human health, as well as to land and/or forests, caused by local air pollution in China may correspond to 1%–7% of her GDP even today.

The model includes no particular policy or technological measures to reduce SO_2 (or NO_2) emissions, and hence the figures presented here are simply a by-product of taking measures to reduce CO_2 emissions. The objective of these observations is to show that cooperative scenarios may be necessary in order to mitigate local air pollution, and this is expected to make cooperation more attractive to China. Figure 11 shows the simulated SO_2 emissions in each scenario, which support our expectation that energy substitution above all, and also other policy measures related to various improvements in energy efficiency, could lead to considerable reductions in SO_2 emissions at the same time.

Concluding Remarks

The construction of a global system based on policy coordination among all countries is undoubtedly the key to creating a sustainable international society, and this inevitably requires cooperation between developed and developing countries. However, at present there seem to be large discrepancies in the views of these two groups about encouraging collaboration toward this brighter future. What is called the "North–South problem" has a long history of tragedy and suspicion. It will never be easy to overcome what happened in the past. In addition, it is not clear whether or not successful cooperation is realizable, when and under what conditions it might take place, and what types of cooperative scenario are likely to be achieved. However, the issue of global warming may force us to advance toward a better and equitable international community, since nobody can leave the earth.

For a better understand of this issue, I have attempted to examine, from a long-term perspective, the possibility of international cooperation between Japan and China to mitigate CO_2 emissions. By applying a simulation model to some cooperative scenarios between the two countries, I have shown that cooperation could be far more economically efficient than noncooperation, and that cooperation could be even more mutually beneficial if the net economic benefits were reallocated appropriately between the two countries. In other words, we could say that by collaboration we "kill two birds with one stone," or even make triple-win (Japan, China, Earth) scenarios.

The numerical results presented here are sensitive to the many assumptions made in the model, most of which are inevitably clouded with much uncertainty. Mainly for this reason, I have taken a long time-span and examined scenarios which allow considerable flexibility in many areas, leaving the vision of the future as rather qualitative and conceptual in nature. It should be noted, however, that many sensitive analyses have shown that the major policy implications derived in the study are rather robust.

While recognizing these reservations and analytical limitations, we conclude that the simulation results strongly demonstrate the importance of international cooperation between developed and developing countries, not only in terms of the remarkable gains in economic efficiency, but also because of the great potential for creating mutual economic benefits. Hopefully, continuing attempts to achieve successful cooperation on a global scale may lead to the realization of a sustainable international society in the long run.

Acknowledgments. The author would like to express thanks to all members of the CREST project for their useful comments and warm encouragement, and also to H. Yamashita for helping with the computations in the early stages of the research.

References

Clarke R, Winters LA (1995) Energy pricing for sustainable development in China. In: Goldin I, Winters LA (eds) The economics of sustainable development. OECD Development Centre, Cambridge University Press, Cambridge

Frankhauser S (1992) Global warming damage costs: some monetary estimates. CSERGE, Working Paper GEC 92-29, London

Goto N (1995) Macro-economic and sectoral impacts of carbon taxation: a case for the japanese economy. Energy econ 17:277–292
Goto N (2001) Empirical examination of the relationship between carbon emissions and economic development. Proceedings of the Department of Advanced Social and International Studies, Graduate School of Arts and Sciences, University of Tokyo, pp. 111–148
Goto N, Sawa T (1993) An analysis of the macro-economic costs of various CO_2 emission control policies in Japan. Energy J 14:83–110
Haas PM, Keohane RO (1993) Institutions for the Earth: sources of effective international environmental protection. MIT Press, Cambridge
Imura H, et al. (eds) (1995) Environmental problems in China (in Japanese). Toyo-keizai-shinpo-sha, Japan
IPCC (1995) Economic and social dimensions of climate change. Contribution of Working Group III to the Second Assessment Report of the Intergovernmental Panel on Climate Change
Li Z (1999) Environmental protection system in China (in Japanese). Toyo-keizai-shinpo-sha, Japan
Nordhaus WD (1991) The cost of slowing climate change: a survey. Energy J 12:37–65
Nordhaus WD (1994) Managing the global commons: the economics of climate change. MIT Press, Cambridge
Victor DG, Raustiala K (eds) (1998) The implementation and effectiveness of international environmental commitments: theory and practice. MIT Press, Cambridge
Young OR (1992) The effectiveness of international institutions: hard cases and critical variables. In: James NR, Czempiel EO (eds) Governance without government: order and change in world politics. Cambridge University Press, Cambridge, pp. 160–194
Zhang Z (1998) The economics of energy policy in China. Edward Elgar

An Econometric Study of China's Long-Term Economy, Energy, and Environment

Li Zhi Dong

Summary. An integrated econometric model, consisting of a macroeconomic submodel, an energy submodel, and an environment submodel, was developed and used to perform a long-term simulation study of China. In the next 30 years, the potential growth of GDP will be around 7% annually, and continuation of the rapid economic growth could result in insurmountable difficulties for energy security, air protection, and CO_2 emissions reductions. For sustainable development, more comprehensive measures should be adopted, including improvements in energy efficiency, more rapid energy switching from coal to natural gas and renewable energy sources, imposing a carbon tax, the development of clean coal technology, the establishment of strategic petroleum stockpiling, the enforcement of air protection, etc.

Key words. China, Economy, Energy, Environment, 3Es-model, Long-term simulation analysis

Introduction

China, characterized by the world's largest population, is one of the most rapidly growing countries, and is the second largest energy consumer just behind the United States. Its annual GDP growth rate has been about 10% during the last two decades, and the primary energy consumption in 1998 was estimated at between 810 and 880 million tons of oil equivalent (MTOE), which is equivalent to a fifth of the Organization for Economic Cooperation and Development (OECD) total and a tenth of the world total. At the same time, China also is one of the worst polluters, with the largest SO_x emissions and the second largest CO_2 emissions. Because of its huge potential impact, a study of China's economic development and related energy issues is vital not only for China itself, but also for the rest of the world.

Department of Management and Information System Science, Nagaoka University of Technology, 1603-1 Kami-Tomioka, Nagaoka, Niigata 940-2188, Japan

Many studies have been done on China's energy-related issues, including those conducted by IEA (2000), EIA/DOE/U.S.A (2000), the World Bank (1996a), Zhou and Zhou (1999), etc. However, there are few studies which focus on the relations between economy, energy, and environment by modeling these three components simultaneously, except for that of the World Bank. The World Bank's study focused on the analysis of issues and options in greenhouse gas (GHG) emissions control by introducing a so-called China GHG Model, which consists of a macroeconomic model, an input–output table, energy coefficients, and GHG emissions coefficients. In this chapter, an integrated econometric model consisting of a macroeconomic submodel, an energy submodel, and an environment submodel, named the 3Es-model, was developed for a simulation study to 2030.

The chapter is organized as follows. The next section describes the structure, data, and estimates of the 3Es-model. Then the assumptions and the results of the simulation are discussed in the next two sections. The final section gives the conclusions.

Overview of the 3Es-Model

Model Structure

The 3Es-model consists of a macroeconomic submodel, an energy submodel, and an environment submodel (Fig. 1). The macro submodel is designed to provide the indicators influencing energy supply/demand and the related pollutants emissions consistently. In this submodel, population indicators, fiscal policy indicators, and overseas economic indicators such as world trade, crude oil price, and exchange rates, are treated as exogenous variables. The endogenous variables solved in the results include: macroindicators such as the demand for goods and services, investment, import, and export, etc.; industry activity indicators such as the output of steel, cement, and other energy-intensive products, size of vehicle ownership, passenger-kilometers, and freight ton-kilometers, etc.; price indicators such as deflators for GDP and its components, WPI, CPI, and energy price indexes.

The energy submodel, which serves as the core of the 3Es-model, is designed to determine the energy flow from final energy consumption to primary energy consumption and energy trading position. First, energy demand is determined by sector and by energy source, based on the economic activity indicators and price indicators obtained by the macro submodel. Then, the routine representing the energy conversion sector calculates the required input for the output of transformed energy sources such as electricity and oil products. Finally, primary energy consumption can be obtained by aggregating the energy requirements from end-use sectors and transformation sectors, and the energy trading position can be identified by comparing the primary energy demand and domestic production. By now, the domestic production is exogenously given by a detailed survey.

The environment submodel is simply designed to generate the energy-related production matrices and emissions matrices of both SO_2 and CO_2, following the energy balance table. The production of SO_2 is estimated based on energy con-

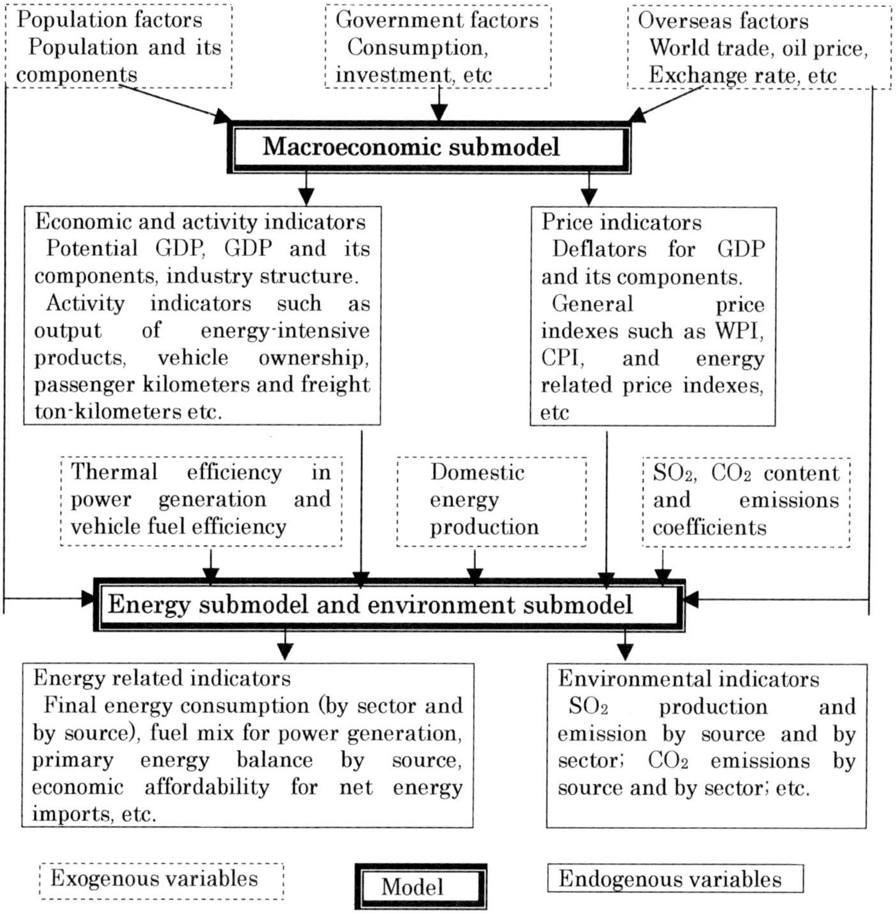

FIG. 1. Structure of the 3Es-model. *WPI*, wholesale price index; *CPI*, consumer price index

sumption obtained by the energy submodel and the sulfur content coefficient, and then emissions can be calculated based on the production and desulfurization coefficient. For CO_2, the model assumes that emissions are equal to production, based on the fact that no plausible CO_2 control technology is available commercially at present, and is not anticipated in the simulation period.

The 3Es-model is made up of 422 equations, with 83 for the macro submodel, and 339 for the energy and environment submodels.

Data for Model Estimation

Most of the macro indictors mentioned above were obtained from the *China Statistical Yearbook*, *China Fixed Asset Investment Statistical Yearbook*, *China Labor Statistical Yearbook*, *China Environment Yearbook*, *World Development*

Indicators, and other official statistics. The indicators, such as capital stock, operation rate, and energy price index, which were not available from any consistent sources, were estimated based on an extensive collection and analysis of the relevant materials, such as irregular statistical reports, research papers, etc. Energy supply and demand data were mainly collected from International Energy Agency (IEA) statistics. The pollutant-related indicators were calculated as mentioned above.

Because the results of model estimations and simulations are highly dependent on the data, careful attention and a lot of time were put into data collection. However, uncertainty about the reliability of official Chinese statistics could not yet be removed. Many studies, such as those by Maddison (1997), Rouen (1997), Meng and Wang (2000), found that the average annual GDP growth was overstated, and cast doubt on official GDP statistics. On the other hand, the studies carried out by The Energy Strategy Project Team of the China Engineering Academy (1997) and Yan and Yang (1999) found that energy production and consumption was understated in recent years. This highlighted the doubtful reliability of official energy-balance statistics. In this study, the understatement of energy is taken into account for simulation analyses, and the overstatement of GDP growth remains unchanged.

Example of Estimated Results and Their Implications

Except for defining equations, all of the structural equations are econometrically estimated by using the method of ordinary least squares (OLS), with the longest time series data available from 1951 to 1998 for macro indicators, and from 1971 to 1998 for energy-related indicators. Each equation for simulation analysis is chosen from three basic functional forms, linear, log-linear, and a mixture, by performing a coefficient test and a fitness test. Also, the performance of the simultaneous equations model is examined by a partial test, a total test, and a final test. Because the simulation results are highly dependent on the equation estimations as well as the data, it is meaningful to open the estimates instead of leaving them as a black box. Examples of estimated results and their implication are shown below.

GDP Production Function and Potential GDP

A Cobb–Douglas-type GDP production function is estimated as in Eq. 1.

$$\begin{aligned} &\ln((\mathrm{GDP})/(0.5L_{-1}+0.5L)) = -96.699 + 0.277\ \ln(\mathrm{OPR}(K+K_{-1})/(L+L_{-1})) \\ &(t\text{-statistics}) \qquad\qquad (-8.19) \quad (3.47) \\ &\qquad\qquad + 0.0467\,T \\ &\qquad\qquad (7.78) \qquad\qquad (1) \end{aligned}$$

The sample period is 1980–1998, $R^2 = 0.996$, and DW = 1.838, where GDP is the GDP in real terms, L is labor input, K is capital input, T is the time, and OPR is a proxy variable of operation rate calculated from iron production and capacity.

Table 1. Comprehensive comparison of GDP growth accounting (%)

	Estimates for China			Estimates for Japan Nakatani's study			
	This study	World Bank	EPA, Japan				
	1980–98	1985–94	1980–95	1955–61	1965–72	1973–80	1981–90
Annual growth rate	9.9	10.2	10.3	13.0	9.0	3.9	3.8
By capital input	2.9	6.6	3.2–3.6	2.9	5.2	2.7	1.8
By labor input	2.0	1.0	1.7	3.4	0.3	0.3	0.7
By technological progress	5.0	2.2	5.4–5.0	6.8	3.5	0.9	1.4

Sources. World Bank (1996b), The Chinese economy: fighting inflation, deepening reforms. EPA Japan (1997), China's future and the Asia Pacific economy. Nakatani (2000), Introduction to macroeconomics, 4th edition.

The estimates of capital elasticity, labor elasticity, and total factor productivity (technical progress rate) are 0.28, 0.72, and 4.7%, respectively. According to Nakatani (2000), capital elasticity during 1965–1990 is around 0.25 for the USA and 0.30 for Japan; the value for China is just between this two. For reference, Table 1 shows the growth accounting based on the estimation, compared with other studies.

The potential GDP is considered as the maximum GDP achieved at full capacity. The GDP is much lower than the potential from 1989 to 1991, and from 1998 to 1999, because of the recession due to the Tianmen crisis in 1989 and the Asian financial crisis from 1997.

Consumption Function and Investment Function

Private consumption is estimated as in Eq. 2. GDP is used as a proxy variable for disposable income owing to the unavailability of data, and the long-term marginal propensity to consume is estimated as 0.53. Also, the share of people aged 65 and over (aging rate) is considered to reflect the influence on consumption of the well-known birth-control policy, and the estimate is statistically significant.

$$\begin{array}{llllll} \mathrm{CP} = & -1209.03 + 0.242\ \mathrm{GDP} & + 474.778\ \mathrm{P65R} & + 0.456\ \mathrm{CP}_{-1} & & \\ (t\text{-statistics}) & (-1.78) \quad (4.70) & (2.49) & (3.45) & & (2) \end{array}$$

The sample period is 1953–1998, $R^2 = 0.996$, and DW = 1.241, where CP is private consumption, and P65R is the share of people aged 65 and over.

The private investment function is estimated as a function of GDP and K/GDPPA, the ratio of capital stock to potential GDP, shown in Eq. 3.

$$\begin{array}{lll} \mathrm{IP} = & 5871.25 + 0.102\ \mathrm{GDP} - 3564.4(K_{-1}/\mathrm{GDPPA}_{-1}) & \\ (t\text{-statistics}) & (2.38) \qquad (4.29) \qquad (-2.59) & \\ & + 0.493\ \mathrm{IP}_{-1} - 725.37\ \mathrm{DUM89} & \\ & (3.25) \qquad (-1.86) & (3) \end{array}$$

The sample period is 1980–1998, $R^2 = 0.990$, and DW = 2.167, where IP is private investment, K is capital stock, and DUM89 is a dummy with 1 for 1989 and 0 for the other years.

Export and Import Functions

The export of goods and services is estimated as a function of world trade and relative export price. As shown in Eq. 4, the long-run elasticity of world trade is 1.33 and the price elasticity is 1.05.

$$\begin{aligned}
\ln(\text{EX}) = \quad & -1.438 + 0.486\ln(\text{TWM}) - 0.384\ln(\text{EXDEF}/(\text{PEW} * \text{CEXR})) \\
(t\text{-statistics}) \quad & (-1.21) \quad (1.59) \quad (-2.47) \\
& + 0.635\ln(\text{EX}_{-1}) - 0.160\text{DUM8193} + 0.140\text{DUM87} \\
& (-4.86) \quad (-2.91) \quad (2.54)
\end{aligned} \tag{4}$$

The sample period is 1981–1998, $R^2 = 0.996$, and DW = 1.52, where EX is export, TWM is world trade, EXDEF is export deflator, PEW is world export deflator, CEXR is Yuan/$ exchange rate, DUM8193 is a dummy for 1981–1993, and DUM87 is a dummy for 1987.

The import of goods is estimated as a function of GDP and relative import price. Based on the estimate in Eq. 5, the long-run elasticity of income dropped from 2.18 in 1980 to 1.05 in 1995, and the long-run elasticity of price also went down from 1.42 to 0.97.

$$\begin{aligned}
\text{IM} = \quad & -42919.8 + 4789.83\ln(\text{GDP}) - 4417.62(\text{IMDEF}/\text{WPI}) \\
(t\text{-statistics}) \quad & (-7.74) \quad (7.84) \quad (-5.59) \\
& + 0.595\text{IM}_{-1} + 1339.12\text{DUM85} - 1115.50\text{DUM8990} \\
& (8.26) \quad (3.82) \quad (-4.34)
\end{aligned} \tag{5}$$

The sample period is 1981–1998, $R^2 = 0.992$, and DW = 2.12, where IM is import, IMDEF is an import deflator, WPI is wholesale price index, DUM85 is a dummy for 1985, and DUM8990 is a dummy for 1989 and 1990.

Energy Elasticity

Energy consumption by source and end-use sector is assumed to be a function of the relative energy price and activity factors such as GDP per capita, industrial output, etc. Table 2 shows the long-run elasticity of some main energy consumers based on the estimated equation. Generally, activity elasticity is higher than price elasticity; activity elasticity for clean energy sources is higher than that for less clean sources. These general trends are also observed in China.

Assumptions and Cases

Although many cases can be introduced for the simulation analysis, this study only covers a business-as-usual (BAU) case for macroeconomic simulation, and six cases for the energy-related simulation.

TABLE 2. Long-run activities and price elasticity estimated

Energy demand by sector	Activity factor		Price factor		Notes
	Variable	Elasticity	Variable	Elasticity	
Coal, steel	Steel	0.59	Coal price	0.31	In 1995
Coal, chemicals	Ammonia	1.17	Coal price	1.94	In 1995
Oil, households	GDP per capita	1.15	Oil price	0.07	Constant
Gas, services	GDP per capita	2.64	Gas price	1.02	Constant
Gas, households	GDP per capita	1.49			Constant
Electricity, services	GDP per capita	1.55	Electricity price	0.31	Constant
Electricity, households	GDP per capita	1.88	Electricity Price	0.39	Constant

TABLE 3. Key assumptions in a BAU case for macroeconomic simulation

Average annual growth rates	1980–98	1998–2000	2000–10	2010–20	2020–30
Government consumption (%)	9.9	5.0	6.0	5.0	4.0
Government investment (%)	10.9	9.0	6.0	5.0	4.0
World trade (%)	5.5	3.0	2.8	2.6	2.4
Absolute level	1998	2000	2010	2020	2030
Population (millions)	1248	1271	1369	1456	1501
Exchange rate (RMB yuan/$)	8.3	8.3	8.3	8.3	8.3
Crude oil price ($/BBL)	12.8	24.7	34.5	49.2	68.7

BAU, business as usual; RMB, Ren Min Bi, name of Chinese currency; BBL, barrel(s)

The BAU case for macroeconomic simulation assumes that the current trends influencing economic growth will continue to 2030. Key assumptions are shown in Table 3.

Based on the BAU case for macroeconomic simulation, a BAU case and five alternative cases for energy-related simulation are developed by changing the set of assumptions. The five alternative cases are an energy-saving case, a switch to non-fossil fuels case, a combination case taking energy saving and fuel switching together, a carbon tax case, and a comprehensive case taking these three alternative measures together. The relations among these cases are shown in Fig. 2, and the key assumptions are given in Table 4.

Results of the Simulation to 2030

Simulation Results of the Macroeconomy

On the basis of the assumptions made in Table 3, China is able to sustain the rapid economic growth started two decades ago to the year 2030. Its potential GDP will grow by some 7%, and GDP growth is expected to surpass 6%. The average annual growth rate will be 7.1% from 2000 to 2010, 6.1% from 2010 to 2020, and 5.1% from 2020 to 2030, with a one percentage point decline each decade. Although this growth rate is lower than the actual growth rate during the past two decades from 1980, it is still much higher than the predicted growth rate

TABLE 4. Key assumptions for energy-related cases

Case name	Key assumptions		
1. BAU case		1998	2030
	Thermal efficiency in thermal power plant (%)		
	Coal-fired	33	43
	Oil-fired	35	47
	Gas-fired	32	49
	Vehicle fuel efficiency (l/100 ton-kilometers)	9.5	4.1
	Non-fossil-fuel-fired capacity (GW)		
	Hydroelectric	70	250
	Nuclear	2.1	50
	Wind	0.04	20
	Geothermal, solar, and other renewables	–	100
2. Energy-saving case	Thermal efficiency will go up 2 points above that in BAU by 2030 Coal-fired 43% to 45%; oil-fired 47% to 49%; gas-fired 49% to 51% Vehicle fuel efficiency will be improved by 27% from that in BAU, to 3 l/100 ton-kilometers		
3. Switch to non-fossil fuels	Non-fossil-fuel-fired capacity in 2030 (GW)	This case	BAU
	Hydroelectric	300	250
	Nuclear	100	50
	Wind	128	20
	Geothermal, solar, and other renewables	200	100
4. Combination case	Energy saving measures in case 2 and fuel switch measures in case 3 will be introduced together		
5. Carbon-tax case	Carbon tax 10$/T-C will be imposed from 2011		
6. Comprehensive case	All measures adopted in cases 2, 3, and 5 will be conducted together (Energy-saving measures, fuel-switch measures, carbon-tax introduction from 2011)		

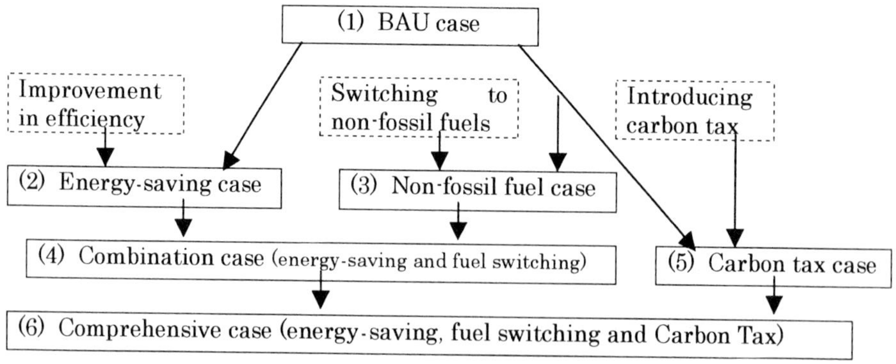

FIG. 2. Relations among energy-related cases. *BAU*, business as usual

TABLE 5. China's GDP growth accounting to 2030

	1980–98	1998–2000	2000–10	2010–20	2020–30
Potential GDP growth (%)	9.9	9.5	8.2	7.3	6.3
GDP growth (%)	9.9	7.6	7.1	6.1	5.1
By capital input	2.9	3.1	2.4	1.9	1.5
By labor input	2.0	1.7	0.8	0.3	−0.1
By technical progress	5.0	2.9	3.9	4.0	3.6
Contribution rate to growth (%)	100.0	100.0	100.0	100.0	100.0
By capital input	29.4	40.4	33.9	30.4	29.9
By labor input	20.4	21.9	11.2	4.7	−1.4
By technical progress	50.2	37.7	54.9	64.8	71.6

for the world. Comparing these figures with other forecasts of China's growth, the IEA assumed about 5% growth during 1997 to 2020 in its latest *World Energy Outlook* (IEA 2000), the Development Research Center (the State Council, China) predicted about 8% growth between 2000 and 2010, and about 6.5% from 2010 to 2020 (Li 2000), and the Econometric Research Institute (Academy of Social Science, China) forecast 8.1% growth between 2000 and 2010, and 6% during 2010 to 2030 (Wang 2000). Further, the Chinese government expected a higher than 7% growth during the first decade of new century in their Tenth 5-Year Plan. The growth rate obtained by this study is higher than the IEA assumption, lower than the two Chinese studies, and close to the government plan.

Table 5 shows the GDP growth accounting. The contribution of technical progress will go up and the other two factors will go down. In particular, the contribution of labor input will drop to a negative value, reflecting the decrease in labor input owing to the strict birth control measures.

The size of GDP in 2030 will reach US$6.1 trillion in exchange-rate terms and US$32.0 trillion in purchasing power parity terms (PPPs), both with constant 1995 prices. Owing to the high expansion of GDP and the low increase in population, per capita GDP in 2030 will go up rapidly, reaching US$4100 in exchange-rate terms and US$21 000 in PPPs. Taking account of the underestimation in the exchange-rate term and the overestimation by PPPs, a reasonable level should be somewhere between these two, e.g., around the average, US$13 000.

Regarding the macroeconomic structure, measured in nominal GDP and as shown in Fig. 3, little change is expected in the share of the secondary sector. The share of the primary sector will continue going down sharply to about 5% by 2030 from 19% in 1998, and by offsetting this decline, the share of the tertiary or services sector is expected to increase from 32% to 46%.

As shown in Table 6, the major activity indicators of energy-intensive industry, such as the output of steel, cement, vehicle ownership, etc., will increase rapidly. For example, by 2030, the output of steel is expected to surpass 200 million tons; vehicle ownership will reach 194 million, and the penetration rate into the population is expected to rise to 13%.

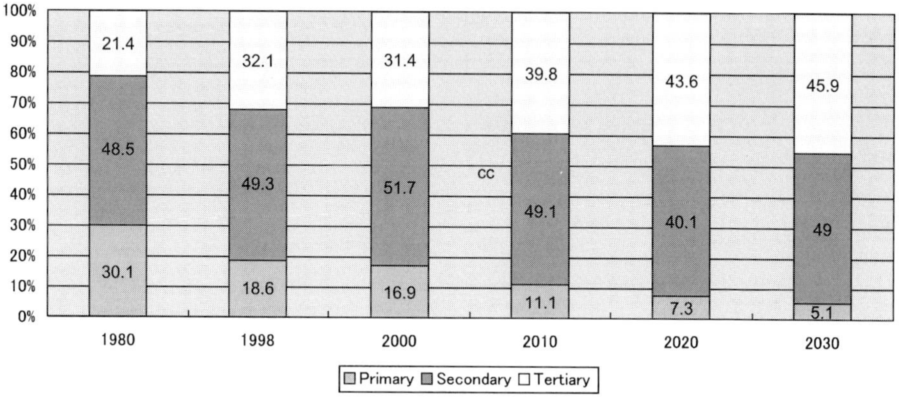

FIG. 3. The change in China's macroeconomic structrue to 2030

TABLE 6. The change in major industrial activity indicators to 2030 in China

	1980	1998	2000	2010	2020	2030
Steel production (million ton)	37	116	129	171	207	239
Cement production (million ton)	80	536	598	778	937	1073
Ethylene production (million ton)	0.5	3.8	4.4	7.9	13.8	22.0
Vehicle ownership (million)	1.8	13.2	16.3	43.6	97.0	194.0
Vehicle diffusion rate to population (per thousand)	1.8	10.6	12.8	31.8	66.6	129.2

Energy-Related Simulation Results in the BAU Case

Over the period from 1998 to 2030, the total primary energy consumption in the BAU case, as shown in Table 7, will grow by 4.1% per annum, compared with 3.9% from 1980 to 1998. It will reach 2990 MTOE in 2030, which is almost equivalent to the total energy consumption in North America plus Japan in 1998.

Energy-GDP intensity is expected to decline continually, but the annual rate of decline will slow down to 1.9%, which is much lower than the 5.5% in the past 18 years. This pushes up the energy-GDP elasticity to 0.67 from 0.39. It means that some new measures other than BAU should be adopted in order to achieve more improvements in energy efficiency.

The modernization of the energy structure will continue with respect to energy source switching and the change in consumption by sector. Regarding source-switching, the share of coal will drop from 71% in 1998 to 55% by 2030 in primary energy consumption, from 86% to 64% in power generation, and from 48% to 24% in final energy consumption. Offsetting this decline, the shares of other energy sources will go up. On primary consumption base, the share of gas will rise sharply from 2.4% to 12.3%, and the share of oil will go up to 26.1% from 23.6%. While expected growth in renewable energy and nuclear power will be a strong 15.5%

and 12.1% per annum, respectively, the share will still only reach 0.8% and 4.8%, respectively, in 2030. Gas is expected to be the main source to replace coal in power generation, and electricity is expected to be the main source to replace coal in final energy consumption. On the other hand, regarding the change in consumption by sector, the share of industry and nonenergy use in final energy consumption will decline sharply from 62.1% in 1998 to 41.2% by 2030. Reflecting the expected rapid improvements in living standards and the large increase in motorization, the share of transportation consumption will go up to 23.4% from 13.8%, and the share of consumption in other sectors will rise to 35.3% from 24.1%.

Three unacceptable issues will arise in the BAU case, as shown in Table 8. The first relates to energy security. Even assuming self-sufficiency in coal and an optimistic domestic production of crude oil and natural gas, the shortfall in the fossil fuel supply will increase to 842 MTOE in 2030, including oil 603 MTOE and natural gas 239 MTOE. All these have to be imported, resulting in the share of import-dependence rising to some 77% for oil and 65% for natural gas. Three points must be clarified. One is, who can supply this fuel to China. Russia and Central Asia look like the potential suppliers for natural gas, based on their vast reserves, but for oil, it seems very difficult to find a supplier based on the limited world reserves, with the ratio of reserves to production close to 41 years in 1998. Another point is whether China has the economic ability to import fuel. The ratio of payment for energy imports to total value of exports is estimated to be about 29% in 2030, meaning that it is not impossible economically, but the burden is likely to be too heavy. Even if these two points are resolved, whether China can maintain its infrastructure for energy imports, such as sea lines, pipelines etc., is also a critical question.

The second issue concerns air pollution related to energy. The SO_2 production will reach 71 million tons by 2030, which is more than three times the level in 1998. The actual level of SO_2 emissions in 2000 is estimated to be close to 20 million tons, and the government is planning to reduce it in the future. That means that the total desulfurization rate must be up to at least 72% by 2030. This should be not impossible, but very difficult.

The third issue relates to climate change. The CO_2 emissions from energy consumption in 2030 will increase to 2570 Mt-C, which is almost equivalent to the total emissions in North America and the OECD in Europe combined in 1998. Although the per capita emissions will rise to 1.7 t-C, which is much lower than the 3.0 t-C in the OECD in 1998, China's contribution to global GHG emissions will be unacceptably high. Unlike SO_2 reductions, nothing but reducing the consumption of fossil energy can contribute to CO_2 emissions reductions.

The above analysis means that the BAU case is unlikely to be sustainable, and substantial policy changes must be considered.

Energy-Related Simulation Results in Alternative Cases

Table 9 gives a summary of the energy-related simulation results in the five cases which are alternatives to BAU.

TABLE 7. Simulation results of China's energy consumption in the BAU case

	Level					Shares (%)					Average growth rate (%) or elasticity to GDP	
	1980	1998	2010	2020	2030	1980	1998	2010	2020	2030	1980–98	1998–2030
Total primary energy consumption (MTOE)	413	823	1341	1970	2990	100.0	100.0	100.0	100.0	100.0	3.9	4.1
Fossil fuels	408	801	1267	1818	2751	98.8	97.4	94.5	92.3	92.0	3.8	3.9
Coal	307	587	839	1099	1603	74.2	71.3	62.6	55.8	53.6	3.7	3.2
Oil	90	194	332	500	781	21.7	23.6	24.7	25.4	26.1	4.4	4.4
Gas	12	20	96	219	368	2.9	2.4	7.2	11.1	12.3	2.8	9.6
Nuclear power	0	4	37	89	143	0.0	0.4	2.8	4.5	4.8	–	12.1
Hydroelectricity	5	18	36	58	72	1.2	2.2	2.7	2.9	2.4	7.3	4.5
Other renewables	0	0	1	5	23	0.0	0.0	0.1	0.3	0.8	–	15.5
Power generation (input base, MTOE)	83	268	466	827	1467	100.0	100.0	100.0	100.0	100.0	6.7	5.5
Fossil fuels	78	247	392	675	1228	94.0	91.9	84.1	81.6	83.7	6.6	5.1
Coal	58	231	317	499	936	69.5	86.2	68.1	60.4	63.8	8.0	4.5
Oil	20	13	17	27	50	24.3	4.8	3.7	3.3	3.4	–2.5	4.4
Gas	0	3	57	149	242	0.2	1.0	12.3	18.0	16.5	16.1	15.3
Nuclear power	0	4	37	89	143	0.0	1.4	7.9	10.8	9.8	–	12.1
Hydroelectricity	5	18	36	58	72	6.0	6.7	7.8	7.0	4.9	7.3	4.5
Other renewables	0	0	1	5	23	0.0	0.1	0.2	0.6	1.6	–	15.5
Power generation (output base, TWH)	301	1170	2242	4238	7977	100.0	100.0	100.0	100.0	100.0	7.8	6.2

Thermal power	242	944	1673	3175	6332	80.6	80.7	74.6	74.9	79.4	7.8	6.1
Coal-fired	180	885	1343	2297	4678	59.9	75.6	59.9	54.2	58.6	9.2	5.3
Oil-fired	62	52	79	135	275	20.5	4.4	3.5	3.2	3.4	–0.9	5.3
Gas-fired	1	7	251	743	1379	0.2	0.6	11.2	17.5	17.3	13.9	17.8
Nuclear power	0	14	142	343	549	0.0	1.2	6.3	8.1	6.9	–	12.0
Hydroelectricity	58	211	421	670	842	19.4	18.0	18.8	15.8	10.6	7.4	4.4
Other renewables	0	1	6	50	254	0.0	0.1	0.3	1.2	3.2	–	19.4
Total final consumption (MTOE)	315	537	879	1249	1862	100.0	100.0	100.0	100.0	100.0	3.0	4.0
By source: fossil fuels	287	437	687	901	1228	90.9	81.5	78.2	72.1	66.0	2.4	3.3
Coal	220	259	367	405	442	69.7	48.2	41.8	32.5	23.7	0.9	1.7
Oil	60	164	288	439	682	19.1	30.5	32.8	35.1	36.7	5.7	4.6
Gas	7	15	32	57	104	2.2	2.7	3.6	4.5	5.6	4.4	6.3
Electricity	21	77	155	296	561	6.8	14.3	17.6	23.7	30.2	7.4	6.4
Heat	7	23	37	52	72	2.3	4.2	4.2	4.2	3.9	6.4	3.7
By sector: industry and non-energy use	196	333	519	635	767	62.3	62.1	59.0	50.8	41.2	3.0	2.6
Transportation	27	74	142	249	436	8.6	13.8	16.2	19.9	23.4	5.7	5.7
Other sectors	92	129	218	365	658	29.1	24.1	24.8	29.3	35.3	1.9	5.2
GDP (1995, trillion RMB yuan)	1.4	7.5	17.3	31.2	51.1						9.9	6.2
Energy/GDP intensity (TOE/million RMB)	302	109	77	63	58						–5.5	–1.9
Energy/GDP elasticity											0.39	0.67
Fossil/GDP intensity (TOE/million RMB)	299	107	73	58	54						–5.6	–2.1
Fossil/GDP elasticity											0.38	0.64

TWH, terawatt-hour = 10^9 kWh

TABLE 8. Simulation results on China's energy security and related environmental issues in the BAU case

	1980	1998	2010	2020	2030	Average growth rate (%) 1980–1998	1998–2030
Fossil energy consumption (MTOE)	408.2	801.1	1266.8	1818.3	2751.5	3.8	3.9
Coal	306.6	587.1	838.9	1098.8	1602.8	3.7	3.2
Oil	89.7	194.5	331.8	500.3	780.5	4.4	4.4
Gas	12.0	19.5	96.1	219.2	368.2	2.8	9.6
Fossil energy production (MTOE)	423.7	808.2	1090.1	1430.8	1909.5	3.7	2.7
Coal	303.9	624.6	838.9	1098.8	1602.8	4.1	3.0
Oil	107.9	161.0	177.8	212.4	177.8	2.3	0.3
Gas	12.0	22.6	73.4	119.6	128.8	3.6	5.6
Net imports of fossil fuels (MTOE)	–15.5	–7.1	176.7	387.5	842.0	–4.2	–
Coal	2.7	–37.5	0.0	0.0	0.0	–	–
Oil	–18.2	33.5	154.0	287.9	602.7	–	9.5
Gas	0.0	–3.1	22.7	99.6	239.3	–	–
Share of net imports to consumption (%)	–3.8	–0.9	13.9	21.3	30.6		
Coal (%)	0.9	–6.4	0.0	0.0	0.0		
Oil (%)	–20.3	17.2	46.4	57.5	77.2		
Gas (%)	0.0	–15.9	23.6	45.4	65.0		
Import price							
Coal ($/TOE)	89.8	63.2	105.4	137.0	179.5	–1.9	3.3
Oil ($/BBL)	34.6	12.8	34.5	49.2	68.7	–5.4	5.4
Gas ($/TOE)	221.9	114.6	223.9	301.2	404.6	–3.6	4.0

(A) Payment for energy net imports (billion $)	–4.4	0.4	44.1	133.8	400.3	–	23.9
Coal	0.2	–2.4	0.0	0.0	0.0	–	–
Oil	–4.6	3.1	39.0	103.8	303.5	–	15.4
Gas	0.0	–0.4	5.1	30.0	96.8	–	–
(B) Total value of exports (billion $)	22.8	207.6	435.2	780.3	1386.5	13.1	6.1
(A/B) Share of payment for energy net imports to total value of exports (%)	–19.2	0.2	10.1	17.1	28.9		
SO_2 production (million ton)	12	23	35	47	71	3.8	3.6
Share by fuel: Coal (%)	95.8	93.6	89.2	86.1	82.6		
Oil (%)	4.6	7.1	12.0	15.0	18.4		
Share by sector: Power generation (%)	19.1	37.3	34.5	40.2	49.7		
Industry (%)	44.3	32.0	33.9	29.0	21.9		
Transportation (%)	3.9	3.5	5.5	7.6	10.3		
Other Sectors (%)	23.2	11.0	8.7	6.9	5.4		
CO_2 emissions (Mt-C)	407	792	1212	1705	2570	3.8	3.7
Share by fuel: Coal (%)	81.4	80.0	74.8	69.6	67.4		
Oil (%)	16.7	18.4	20.1	22.2	23.5		
Gas (%)	1.9	1.6	5.1	8.2	9.2		
Share by sector: Power generation (%)	19.6	33.0	32.5	38.5	47.0		
Industry (%)	42.0	31.5	31.5	26.0	19.6		
Transportation (%)	6.1	7.8	9.8	12.1	14.0		
Other Sectors (%)	21.9	12.3	10.3	8.8	7.8		

TABLE 9. Comparison of energy-related simulation results for China: the BAU case and alternative cases

	Case 1 BAU	Case 2 Saving	Case 3 Switching	Case 4 Combination	Case 5 Carbon tax	Case 6 Comprehensive	Change ralative to BAU (BAU = 100)				
							Case 2	Case 3	Case 4	Case 5	Case 6
Total primary energy consumption (MTOE)	2990	2796	2914	2772	2725	2558	93.5	97.5	92.7	91.2	85.6
Fossil fuels	2751	2558	2474	2332	2487	2117	93.0	89.9	84.7	90.4	76.9
Coal	1603	1571	1398	1357	1417	1159	98.0	87.2	84.6	88.4	72.3
Oil	781	680	769	667	765	653	87.1	98.5	85.5	98.0	83.7
Gas	368	308	308	308	305	305	83.5	83.5	83.5	82.9	82.9
Nuclear power	143	143	286	286	143	286	100.0	200.0	200.0	100.0	200.0
Hydroelectricity	72	72	88	88	72	88	100.0	121.1	121.5	100.0	121.5
Other renewables	23	23	67	67	23	67	100.0	289.4	289.4	100.0	289.4
Shares (%)	100.0	100.0	100.0	100.0	100.0	100.0					
Fossil fuels	92.0	91.5	84.9	84.1	91.3	82.8					
Coal	53.6	56.2	48.0	48.9	52.0	45.3					
Oil	26.1	24.3	26.4	24.1	28.1	25.5					
Gas	12.3	11.0	10.6	11.1	11.2	11.9					
Nuclear power	4.8	5.1	9.8	10.3	5.3	11.2					
Hydroelectricity	2.4	2.6	3.0	3.2	2.7	3.4					
Other renewables	0.8	0.8	2.3	2.4	0.9	2.6					
Net imports of fossil fuels (MTOE)	842	673	762	661	757	645	80.0	90.6	78.5	89.9	76.6
Coal	0	0	0	0	0	0					
Oil	603	502	591	489	587	475	83.3	98.0	81.2	97.5	78.9
Gas	239	172	172	172	169	169	71.7	71.7	71.7	70.7	70.7
Payment for net imports of fossil fuels (billion $)	400	322	367	316	364	308	80.5	91.7	78.9	91.0	76.9
Share of the payment of fossil fuels net imports to total value of imports (%)	28.9	23.2	26.5	22.8	26.3	22.2					
SO_2 production (million ton)	72	69	65	61	65	53	95.5	89.2	84.3	90.1	73.8
CO_2 emissions (million T-C)	2570	2412	2300	2170	2316	1943	93.9	89.5	84.5	90.1	75.6

Case 2, the energy-saving case, assumes that thermal efficiency in power generation and vehicle fuel efficiency will be improved further than with BAU. Compared with the results in the BAU case, in 2030, the primary energy consumption will be 6.5% lower. Fuel imports are expected to decline from 842 MTOE with BAU to 673 MTOE, and the ratio of payments for energy imports will drop from 28.9% to 23.2%. SO_2 production and CO_2 emissions will be 4.5% and 6.1% lower, respectively. These results show both the effects and the limitations of improvements in the technical efficiency of energy use with respect to power generation and road transportation. Although this improvement in technical efficiency has the highest priority, it is not enough to resolve the BAU issues. The measures taken to improve technical efficiency should be extended to other energy-consumption sectors such as industry and households. Further, the measures taken to promote managerial efficiency, and to promote structural changes in the macroeconomic structure, the industrial structure, the product mix, the process and technology structure, the plant scale structure, etc., should all be adopted.

Case 3, the fuel-switch case, assumes that non-fossil-fuel power generation will be expanded further than in the BAU case. Compared with the results in the BAU case, in 2030, the primary energy consumption will be 2.5% lower, and the fossil fuel consumption will decrease by 10.1%. Fuel imports are expected to decline from 842 MTOE with BAU to 762 MTOE, and the ratio of payments for energy imports will drop from 28.9% to 26.5%. SO_2 production and CO_2 emissions will decrease by 10.8% and 10.5%, respectively. These results reflect both the effects and the limitations of the assumed fuel switching. In order to get greater effects, it is necessary to expand the exploitable reserves and improve the convertion efficiency of renewable energy.

Case 4 is a combination case based on Cases 2 and 3. Compared with the results in the BAU case, in 2030, the primary, energy consumption will decrease by 7.3%, and the fossil fuel consumption will decrease by 15.3%. Fuel imports are expected to decline from 842 MTOE with BAU to 661 MTOE, and the ratio of payments for energy imports will drop from 28.9% to 22.8%. SO_2 production and CO_2 emissions will decrease by 15.7% and 15.5%, respectively. Naturally, the effect is greater than in either Case 2 or Case 3. However, it is still not enough to resolve the sustainability issues.

Case 5, the carbon-tax case, assumes that a carbon tax of 10$/t-C will be imposed on energy consumption. Compared with the results in the BAU case, in 2030, primary energy consumption will decrease by 8.8%, and fossil fuel consumption will decrease by 9.6%. Fuel imports are expected to decline from 842 MTOE with BAU to 757 MTOE, and the ratio of payments for energy imports will drop from 28.9% to 26.3%. Both SO_2 production and CO_2 emissions will reduce by 9.9%. The Chinese government has started to impose a SO_2 discharge fee since the end of the 1990s, and some advanced regions such as Beijing and Shanghai introduced a new standard discharge fee which is higher than the general one. Although different from a carbon tax, the above result partially rationalizes the SO_2 discharge fee.

Case 6 is a comprehensive case based on Cases 2, 3, and 5. Compared with the results in the BAU case, in 2030, the primary energy consumption will decrease by 14.4%, and the fossil fuel consumption will decrease by 23.1%. Fuel imports are expected to decline from 842 MTOE with BAU to 645 MTOE, and the ratio of payments for energy imports will drop from 28.9% to 22.2%. In addition, SO_2 production and CO_2 emissions will decrease by 26.2% and 24.4%, respectively.

All the measures analyzed in these simulations are effective in reducing energy consumption and pollutant emissions, but the most effective way is to take all the measures together, as in the comprehensive case. However, even in the comprehensive case, issues relating to energy security, air protection, and CO_2 emissions reductions will remain. This means that it is not possible to find a way to resolve all the issues simultaneously and perfectly. In addition to the measures discussed above, the following measures should all be considered.

1. Switching to gas. Because most of the energy-related environmental issues come from coal consumption, and the energy security issue for China is essentially the issue of oil imports, switching from coal and oil to gas will benefit energy security and environmental protection.
2. Development of clean coal technologies. Although making only a limited contribution to CO_2 emissions reductions, traditional clean coal technologies such as coal washing, screening, and briquettes, can contribute considerably to local environment and energy security. As for the high-tech clean coal technologies such as IGCC (integrated coal gasification combined cycle), they can contribute a lot to CO_2 emissions reductions as well as to local environment and energy security.
3. Measures for oil security, such as the establishment of strategic oil reserves, the development of fuel cell vehicles, etc.
4. Measures for energy-related pollutant reductions, such as promoting desulfurization, improvements in environmental monitoring and management, etc.

Conclusions

1. China is likely to sustain a 6% economic growth in the next 30 years.
2. The expected continuation of rapid growth will lead to modernization by energy-source switching and changes in consumption by sector. On the other hand, under the assumptions of BAU, issues relating to energy security, air protection, and CO_2 emissions reductions will render this approach unsustainable.
3. It is not possible to find a way to resolve the sustainability issues simultaneously and perfectly. In addition to the measures analyzed by the simulation study, such as energy saving, switching to non-fossil fuel, and a carbon tax, additional measures, including switching to gas, the development of both traditional and high-tech clean coal technology, measures for oil security such as the establishment of strategic oil reserves, the development of fuel cell vehicles, and

measures for energy-related pollutant reductions such as promoting desulfurization, improvements in environmental monitoring and management, etc., should all be adopted together.

Acknowledgments. This study was funded by the Japan Ministry of Education and Science, and the Core Research for Evolutional Science and Technology (CREST) of the Japan Science and Technology Corporation. The author also thanks Koukiti Ito and Zhongyuan Shen of the Japan Institute of Energy Economics, Yasuhiro Murota of Syounan Econometrics, and Yande Dai of the China Energy Research Institute for their valuable comments and suggestions.

References

EIA/DOE/USA (2000) Annual energy outlook 2000. Washington, DC

EPA/Japan (1997) The China's future and Asia Pacific economy (in Japanese). Economic Research Institute, EPA, Tokyo

IEA (2000) World energy outlook. OECD/IEA, Paris

Li ST (2000) Prospects for China's economic development in the next 20 years. Summary report of the Workshop on Social/Economy/Energy Development and Carbon Emission Scenario Analysis, 25–27 May 2000, Beijing, China

Maddison A (1997) Measuring Chinese economic growth and levels of performance. OECD, Paris

Meng L, Wang XL (2000) Estimation of the reliability of the China's growth statistics (in Chinese). Econ Res J 10

Nakatani I (2000) Introduction to macroeconomic, 4th edition (in Japanese). Nippon-hyouronsha

Rouen R (1997) China's economic performance in an international perspective. OECD, Paris

The Energy Strategy Project Team of the China Engineering Academy (1997) Report on the sustainable energy strategy in China (draft) (in Chinese). China Engineering Academy

Wang TS (2000) Challenges and opportunities facing China in the 21st century. Summary report of the Workshop on Social/Economy/Energy Development and Carbon Emission Scenario Analysis, 25–27 May 2000, Beijing, China

World Bank (1996a) China: issues and options in greenhouse gas emissions control. World Bank discussion paper No. 339, World Bank, Washington, DC

World Bank (1996b) The Chinese economy: fighting inflation, deepening reforms. Washington, DC

Yan XC, Yang JM (1999) Report on China's oil industry (in Chinese). Management Press, China

Zhou FQ, Zhou DD (1999) Study on long-term energy development strategies (in Chinese). China's Energy Research Institute, China Plan Press

Section III: Assessments of Technology Strategies Toward Energy, Economic, and Environmental Issues

Energy and Technology Strategies in Long-Term Global Views: Simulations of the Integrated Assessment Model MARIA

Shunsuke Mori

Summary. This chapter describes an extended version of an integrated assessment model called MARIA (multiregional approach for resource and industry allocation), and how it was applied to develop global and regional greenhouse gas (GHG) emissions scenarios. The model has been developed to assess the potential contribution of fossil, biomass, nuclear, and other energy technologies and land-use changes to future GHG emissions. It also incorporates a simple carbon-cycle and climate-change model. Other extensions of the MARIA model include a higher degree of geographical disaggregation into eight world regions, and a more detailed nuclear fuel cycle. The chapter describes how the model was used in developing GHG emissions scenarios based on narrative storylines, and assesses mitigation strategies that would lead to the stabilization of atmospheric GHG concentrations. The results indicate that zero-carbon technologies such as fast-breeder reactors (FBR) and carbon sequestration technologies can make a significant contribution toward emissions mitigation, especially when drastic reductions are envisaged.

The MARIA model is then applied to evaluate a new hydrogen production process, the so-called sorption-enhanced reaction (SER) process, which is being developed by Air Products and Chemicals (Allentown, PA, USA), and which produces hydrogen through steam–methane reforming at a significantly lower temperature (300°–500°C) than that of conventional steam–methane reforming processes (around 800°C). One aim of this chapter is to assess this process as a liquid fuel supplier for long-term global warming strategies in comparison with other technological possibilities. The simulation results suggest that hydrogen with FBR could supply 5–8 billion tons of oil equivalent (GTOE) of hydrogen in the second half of the twenty-first century when a climate policy which stabilizes the atmospheric carbon concentration is introduced.

Biomass technology is often expected to be a major carbon-free energy source. Although biomass could not completely replace fossil energy sources, the simulations show that it effectively mitigates the marginal costs of carbon emission.

Department of Industrial Administration, Tokyo University of Science, 1-3 Kagurazaka, Shinjuku-ku, Tokyo 162-8601, Japan

Key words. Integrated assessment model, Climate change, Greenhouse gas (GHG) emission, Energy technology, Land-use change, Emissions mitigation, Hydrogen production, Sorption-enhanced reaction (SER) process

Introduction

It is now generally recognized that global warming issues will be a major barrier against world development, equity, and sustainability. Since these involve the natural sciences, technology developments, economic mechanisms, and policy measures, efforts toward the integration of "scientific knowledge" are substantial. The Intergovernmental Panel for Climate Change (IPCC) was established in 1988 to summarize and disseminate the latest scientific knowledge on this subject. The IPCC has published their Third Assessment Report (TAR), which touches on some scientific questions which are relevant to policies on the costs and benefits of global warming mitigation options. Integrated assessment models (IAM) have been developed to evaluate possible policy measures under the complex interrelationships among environment, energy, economy, technology, resources, and societal issues. EMF-14 began to organize IAM developers, and its successor EMF-19 and the activities of the IPCC contributed to a comparison of the models and investigations of "robust" scientific findings. Recent results from integrated assessment models are summarized and evaluated in a special issue of *The Energy Journal* edited by Weyant (1999).

In the procedures used to assess the policy and the technological instrumentation, the parameter settings for the "base-line" scenario are fundamental. The IPCC IS 92 emission scenarios (IPCC 1992, 1994) initiated these parameters, and they have since been used as the baselines for greenhouse gas (GHG) emissions. The IPCC started a new GHG emission scenario evaluation project in 1997, which was conducted by Nakicenovic. In 2000, a multimodel and multiscenario approach was employed. Six IAMs, i.e., AIM, ASF, IMAGE, multiregional approach for resource and industry allocation (MARIA), MESSAGE, and miniCAM, participated in this project. The outcome was published in 2000 (IPCC 2000), and an evaluation of atmospheric carbon concentration stabilization cases is summarized by Mori (2000b) and in Chap. 2 in IPCC–TAR (2001b).

The MARIA model used in this chapter was developed to assess future GHG reduction options, and participated in the IPCC emissions scenario investigations.

The research activities described above emphasize the importance of technology and its development and implementation strategies. Among many energy technology options, hydrogen has been expected to have a "major role" in the future because of its flexibility in terms of primary energy resources, and also as a source of carbon-free liquid fuel. Although many hydrogen production processes have been developed, its use is very limited in the area of energy systems owing to its high cost with respect to both production and end-use demand. However, new technological options for hydrogen are appearing. The first is demand-side fuel cells for automobiles and small residential buildings. The second is supply-side technologies. Recently, the "sorption-enhanced reaction" (SER) process, developed by Air Products and Chemicals, has been used in a new hydrogen production system. This is a steam–methane reforming process carried out at around 500°C, which is 300°C lower than the conventional process (Hufton et al. 1999, 2000). Shirasaki et al. (2001) have also reported a low-temperature hydrogen reforming process. Although these are not yet on the market, these lower-temperature processes provide new opportunities to produce hydrogen based on nuclear heat, waste thermal heat, etc.

The purpose of this chapter is to assess these new hydrogen processes based on an integrated assessment model MARIA. An extension of MARIA involving a carbon circulation model, Bern, is also described.

Background of the MARIA Model

MARIA was developed by the author, and aims at an integrated assessment of global warming issues. This is done by generating the international trade prices for fossil fuels, as well as the equilibrium prices for tradable carbon-emission permits under certain constraints. Land-use subsystems and food demand–supply equations are also included in older to evaluate the biomass energy resources under food supply constraints. The original MARIA incorporated four world regions (Mori and Takahashi 1999) and currently has eight world regions, as shown in Table 1. The basic structure of MARIA-8 follows that of MARIA-4, as shown in Fig. 1.

MARIA involves such "regrettable" technologies as carbon sequestration technologies as well as nuclear power technologies, e.g., once-through light water reactors (LWR), plutonium thermal reactors (LWR–Pu), and fast-breeder reactors (FBR).

MARIA is formulated as an intertemporal nonlinear optimization model including around 18000 variables and 15000 constraints. Detailed parameters and formulations are given elsewhere (Mori 2000a,b; Mori and Takahashi 1999).

TABLE 1. Regional Aggregation of MARIA-8

Region	Countries
NAM	USA, Canada
JPN	Japan
DC	Other OECD member countries in 1990
FSU	Former USSR and eastern European countries
ANS	Indonesia, Malaysia, Philippines, Singapore, South Korea, Thailand, Taiwan
CHN	China
SAS	India, Bangladesh, Pakistan, Sri Lanka
ROW	Other countries

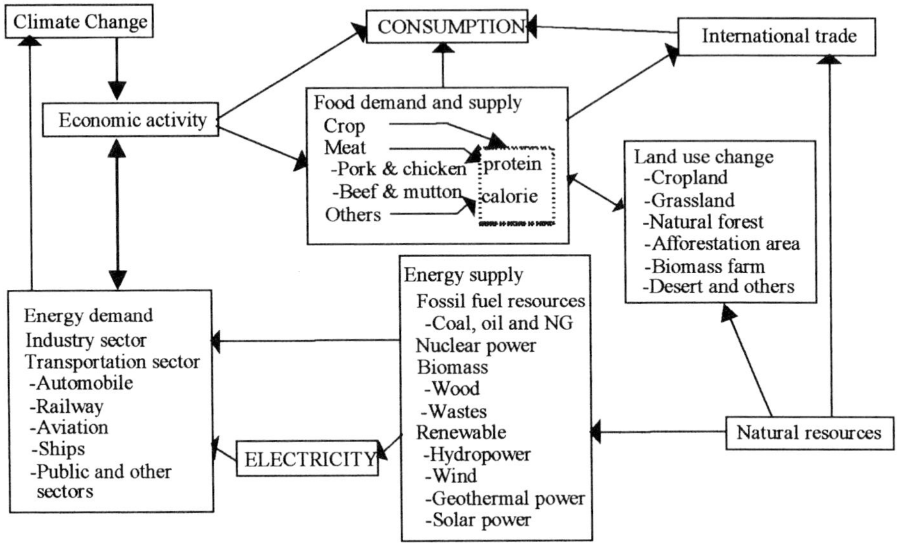

FIG. 1. Structure of the MARIA model. *NG*, natural gas

Expansion of MARIA, Incorporating the Bern Carbon Circulation Model

Since the first IPCC assessment report, investigations into climate change have progressed in both theory and model simulations, as well as with the expansion of computational capability. However, simulations of these detailed models require supercomputers and high costs. Although the simulation results are often available on a web site, they are produced by a very small group of experts. As the scale of climate models and society's interest in global warming issues grow larger, the need for "simple" climate models as policy evaluation tools becomes greater. The MAGICC model (Wigley 1993), which includes the radiative forces of carbon and noncarbon GHG, is a famous pioneering work in this field. The Bern carbon circulation (Bern-CC) model (Joos 1996) employed here was also developed for this purpose, and focuses on carbon emission and concentration processes. Bern consists of an atmospheric carbon circulation block, carbon absorption in the ocean, and emission and storage in the biosphere. The third feature enables us to evaluate the effects of fertilization and land-use changes. Although the formulation of the Bern model is simple, it also includes the results of large global climate models (IPCC 2001).

The Bern model consists of:

1. the box diffusion-type ocean carbon model used in the IPCC Second Assessment Report (SAR), including the effect of sea-surface warming on carbonate chemistry;

2. an impulse–response function which converts radiative energy into spatial patterns of changes in temperature, precipitation, and cloud cover on a global grid;
3. the terrestrial carbon LPJ model;
4. a radiative forcing module.

Figure 2 shows the basic structure of the Bern carbon circulation model.

Figure 3 shows a comparison of an existing version of MARIA and a new one based on the Special Report on Emission Scenarios (SRES) (IPCC 2000)-A1, B1, and B2 scenarios. By comparing the new version of MARIA and the existing one, it was found that (1) the behavior of the two models is basically the same, and (2) the new MARIA generates lower carbon concentrations than the old one, which reflects the explicit incorporation of fertilization effects. The second

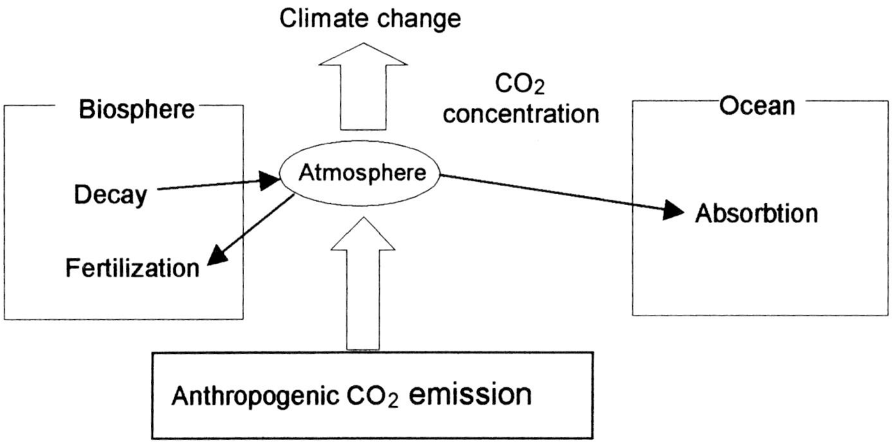

FIG. 2. Basic structure of the Bern model

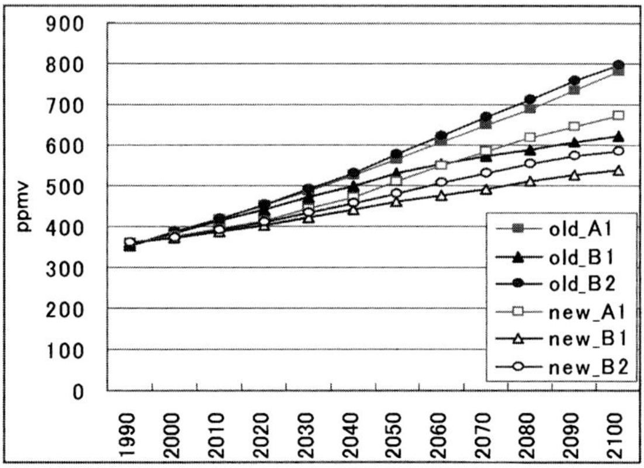

FIG. 3. Comparison of the existing and the new MARIA on SRES-A1, B1, and B2 carbon emission paths

point results in relatively cheaper mitigation costs in carbon concentration stabilization cases.

SER and the Palladium Membrane Process: Low-Temperature Hydrogen Processes

Hydrogen is expected to be a major liquid fuel source in the future because of its cleaness and the range of sources. Many industrial hydrogen production processes have already been developed.

These can be put into three basic categories: electrolysis, steam reforming, and hydrocarbon reforming. The first is simple but its energy efficiency is low, while the second and third require high temperatures.

Sorption-enhanced reaction (SER) is a new process which is being developed for the production of low-cost hydrogen through steam–methane reforming (Hufton et al. 1999, 2000). In this process, the reaction of methane with steam is carried out in the presence of a mixture of a catalyst and a selective adsorbent for CO_2. As a result, the reformation reaction occurs at a significantly lower temperature (300°–500°C) than that of conventional steam–methane reforming processes (around 800°C) while achieving the same conversion of methane to hydrogen. In addition, the product hydrogen is more than 99% pure from a SER reactor and only 70%–75% pure from a conventional reactors. Thus, the new process eliminates, or substantially reduces, the traditional downstream hydrogen-purification step. The SER process is expected to make small, low-cost plants more economical, and facilitate the growth of hydrogen delivery and service infrastructures. Figure 4 shows the average gas composition from SERs (Hufton et al. 1999). According to these authors, the yield of hydrogen is high at 450°C.

Shirasaki et al. (2001) have also reported a low-temperature hydrocarbon reforming process using a palladium membrane for the selective separation

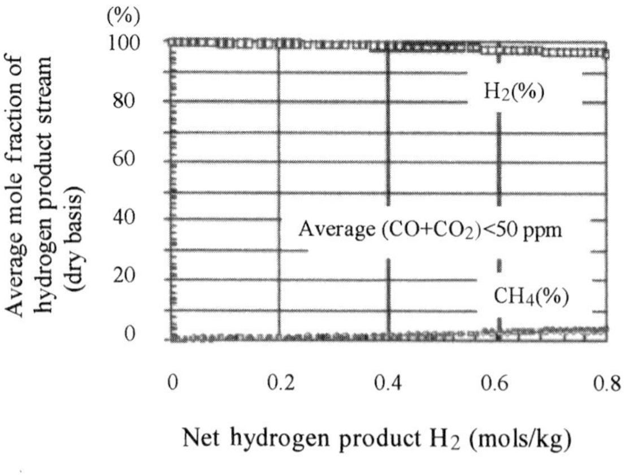

FIG. 4. Average product gas composition during a sorption–reaction step with initial H_2/steam pressurization as measured on a lab-scale SER#1 unit: 6:1 steam/carbon feed; 1:1 adsorbent (HTC)/catalyst; 55 psig; 45°C

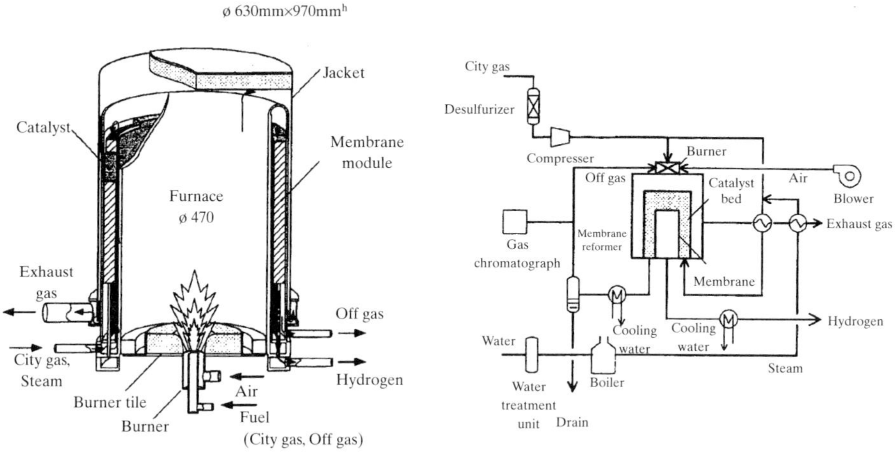

FIG. 5. Flow of the palladium membrane hydrogen system (Shirasaki et al. 2001)

of hydrogen. Figure 5 shows the structure of the system used for this process. However, this system is expensive, and there is little economic benefit owing to the intrinsic high cost of palladium. Thin membrane synthesis technology would overcome this barrier.

It is not clear how or when this system will become marketable, although the experimental feasibility has been verified. However, when a hydrogen production process which takes place below 500°C is realized on an industrial scale, many opportunities for using this heat source will be available, e.g., waste heat processes, gas turbine exhausts, and FBR. This chapter focuses on the possibility of a FBR for electric power generation and SER hydrogen production processes, as well as the conventional electrolysis process using a LWR, LWR–plutonium, and FBRs.

The following fundamental chemical–thermal balances are well known.
Heat:

$$CH_4 + H_2O \Rightarrow CO + 3H_2 - 206\,\text{kJ/mol}$$

$$CO + H_2O \Rightarrow H_2 + CO_2 + 41\,\text{kJ/mol}$$

$$CH_4 + 2H_2O \Rightarrow CO_2 + 4H_2 - 165\,\text{kJ/mol}$$

$$\text{Steam reforming reaction: } -206\,\text{kJ/mol}\ \ (CH_4 = 3H_2)$$

$$\text{Shift reaction: } +41\,\text{kJ/mol}\ \ (CO = H_2)$$

$$\text{Summing-up: } -165\,\text{kJ/mol}\ \ (CH_4 = 4H_2)$$

$$\text{Combustion of } H_2(\text{net})\text{: } +242\,\text{kJ/mol}\ \ (H_2) \text{ or } 968\,\text{kJ/mol}\,(4H_2)$$

$$\text{Combustion of } CH_4\text{: } +803\,\text{kJ/mol}\ \ (CH_4)$$

Assuming 40% energy conversion efficiency to FBR, the thermal heat of 1 GTOE-elec FBR and 12.167 GTOE methane generate 14.167 GTOE hydrogen. Theoretically, this method can be applied to coal and other hydrocarbon-based processes.

It should be noted that these hydrocarbon-based processes are not carbon-free. However, carbon sequestration technologies are applicable if needed.

Cost Assumptions of SERs or the Palladium Membrane Process

The purpose of this chapter is to assess the potential contribution of FBR-based hydrogen production processes. Since the material and heat balances of SERs are linear, imposing these process into the MARIA model is straightforward. The remaining problem is the cost. The assumptions used here are listed below.

1. The cost of producing hydrogen by electrolysis is assumed to be 4 Japanese yen per N-m^3, or \$129 per tons of oil equivalent (TOE) at 1990 prices.
2. Electric power generation in a FBR costs 10% more than in a LWR. If a FBR is used for hydrogen production, the power generation steam turbine unit is replaced by a SER or palladium membrane reactor. The maximum additional cost of a FBR–SER process for FBR power generation is assumed to be the same as in the electrolysis case above. The lowest additional cost is set at zero. In other words, the cost of a steam turbine FBR is equal to that of a SER. Needless to say, this lowest cost is no more than an assumption for these simulations.
3. The end-use cost of energy from hydrogen should also be included. MARIA incorporates industry, transportation, and other end-use sectors, and energy cost coefficients 10% higher than those of natural gas are assumed.

The above values represent the "reference costs" used in the rest of this chapter.

Simulation Results on the Contribution of Hydrogen Processes

I now assess the contribution of an FBR–SER process by MARIA with the Bern model based on the SRES-B2 scenario and 550 p.p.m.v. atmospheric carbon concentration, changing the additional cost of a hydrogen process to FBR-power generation, and including end-use energy cost coefficients. All other parameters are identical to those in the existing MARIA (Mori 2000a).

Figure 6 compares the hydrogen consumption paths in seven simulation cases with hydrogen process costs which are different from the reference value, and where the hydrogen end-use cost is 10% higher than that of natural gas. This figure suggests that there is no incentive to introduce FBR–SER under no carbon control policies when the hydrogen end-use cost is higher than that of natural gas. Even in the carbon concentration control case, the additional cost of the

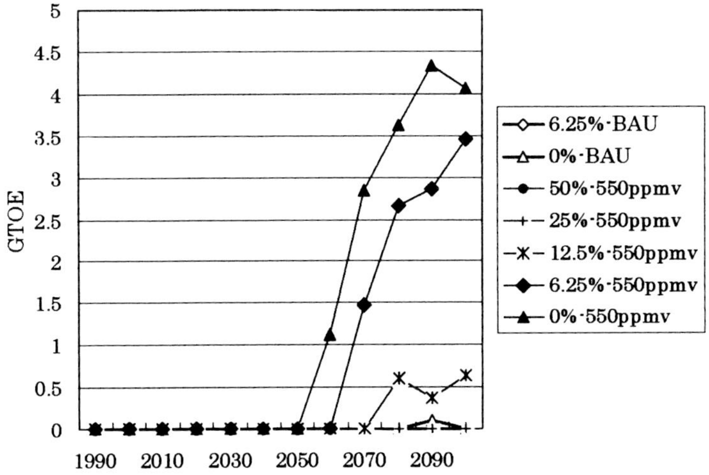

FIG. 6. Hydrogen consumption paths forecast by increasing hydrogen processing costs in cases of business-as-usual (*BAU*) and carbon control by 0%, 6.25%, 12.5%, 25%, and 50% of the reference-level cost, where hydrogen end-use costs are 10% higher than those of natural gas in all cases

hydrogen process to FBR power generation should be lower than the 12.5% cost of an electrolysis process. These findings with respect to the B2-marker are understandable since the direct use of natural gas and FBR power generation are more energy-efficient than using FBR–SER in no carbon control policy cases. In carbon control cases, the cost of hydrogen production is very important. When hydrogen process costs are low, the demand for FBR–SER power generation grows rapidly to 3.5–4.3 GTOE by the end of this century, which is more than 14%–17% of total final energy consumption, as shown below.

When the end-use cost coefficients of hydrogen decline to those of natural gas under the assumption of zero additional hydrogen processing costs to FBR power generation, the demands for FBR–SER hydrogen power generation increase deterministically, as shown in Fig. 7.

Figure 7 shows that the potential demand for hydrogen in carbon control policy cases will be around 6 GTOE, which is around 24% of the total final energy demand. Even in no carbon control policy cases, if there are no additional costs to FBR power generation or to natural gas end-use costs, FBR–SER hydrogen power generation will be used in the second half of this century. However, its value will diminish in 2100. These findings suggest that the potential of hydrogen production processes is high, but cost issues will limit their use.

Figure 8 compares the world final energy flow profiles in the B2-business-as-usual (B2-BAU) case with reference to hydrogen production and end-use costs, and with zero additional production and end-use costs.

Figure 9 shows the final demand profiles for hydrogen in carbon concentration control cases. As the energy costs of hydrogen decrease, its initial implementa-

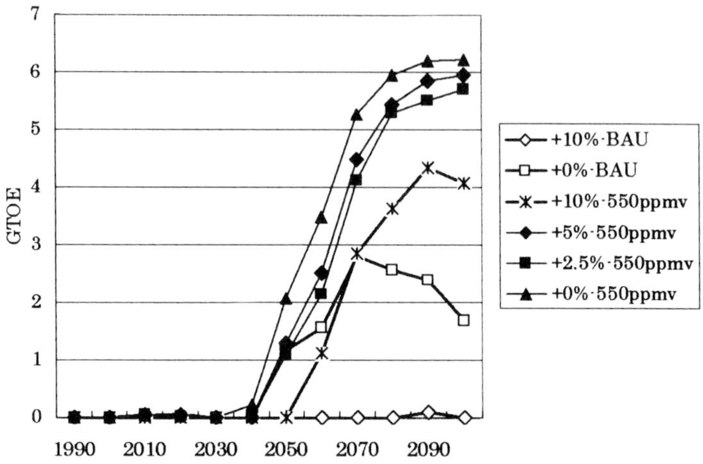

Fig. 7. Hydrogen consumption paths forecast by increasing hydrogen end-use cost coefficients to 0%, 2.5%, 5%, and 10% higher than those of natural gas, where the additional hydrogen processing cost for FBR power generation is set at zero in all cases

tion becomes earlier and its production increases. The detailed energy flows for BAU, and in cases of carbon control with reference hydrogen costs and those with no additional costs are compared in Fig. 10. This figure suggests that hydrogen input to the industry sector increases as hydrogen costs decrease. It can also be seen that the contribution of nuclear power increases toward the end of this century, as well as that of biomass.

Figure 11 summarizes the end-use hydrogen demands in cases of (A) no additional hydrogen processing costs for FBR power generation and reference end-use costs, and (B) no additional hydrogen processing and end-use costs. Hydrogen input increases uniformly in all sectors. Unlike the B2-BAU cases, hydrogen is mainly input to industry and other sectors, while biomass energy is mainly used in the transportation sector.

Interactions Between Nuclear and Biomass Power Sources

The previous section suggested that the use of nuclear power is essential to stabilize atmospheric carbon concentrations. However, although nuclear power is the most cost-effective option, it is not indispensable, as is shown in this section. Figure 12 shows the world primary energy flows in the B2-550 p.p.m.v. carbon control case, with no expansion of nuclear power after 2010. Biomass is widely used, and carbon sequestration has also risen to 8.8 billion carbon ton (GtC) in 2100, while it is 5.6 GtC in the B2-550 p.p.m.v. case.

Figure 13 shows that the loss of GDP where no nuclear expansion is assumed is more than 1.4%, while that in the B2-550 p.p.m.v. case is only 0.6%. The potential expansion of biomass power obviously mitigates the loss of GDP. Figure 14

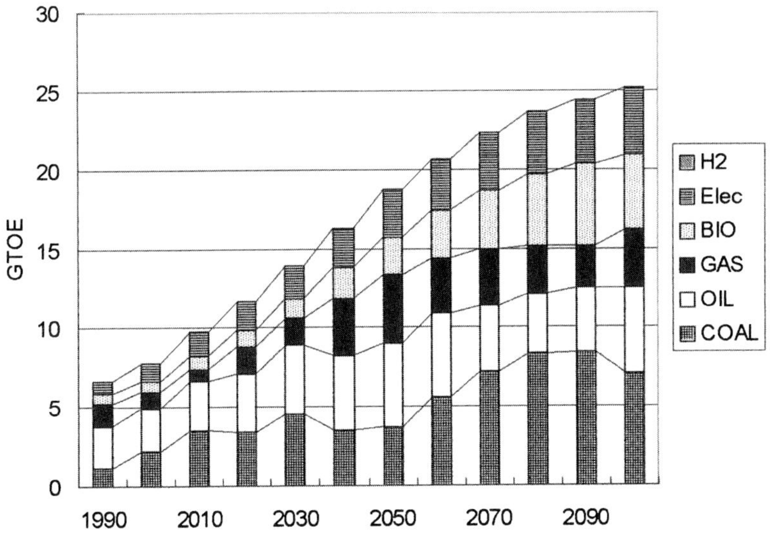

a

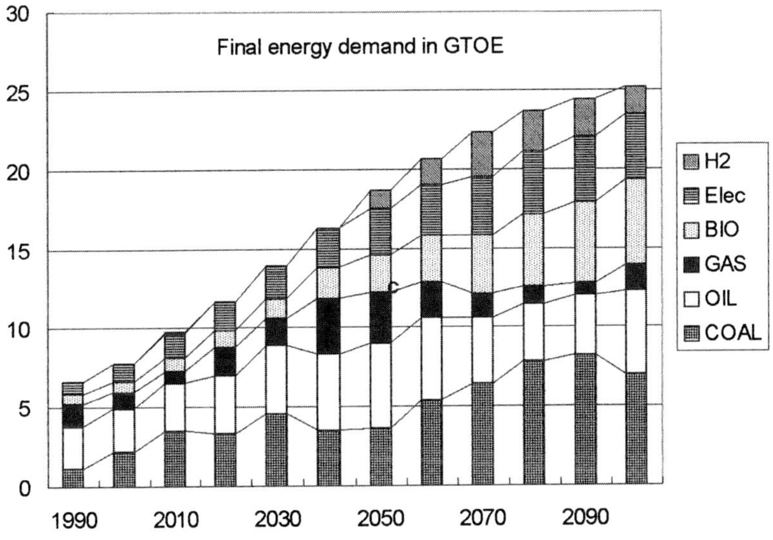

b

FIG. 8. **a** World final energy demand profiles in B2-marker with reference-level hydrogen production and end-use costs. **b** World final energy demand profiles in B2-marker with no additional hydrogen production and end-use costs

shows that a lower carbon concentration target increases the maximum loss of GDP, but this later decreases again. This suggests that the social structure can adapt to a lower-emissions economy. Figure 15 shows a similar situation when one compares the trajectory of B2U450 with that of B2U550.

This figure suggests that the impacts on whole economy of nuclear power saturation will be mitigated when the potential biomass supply increases. The

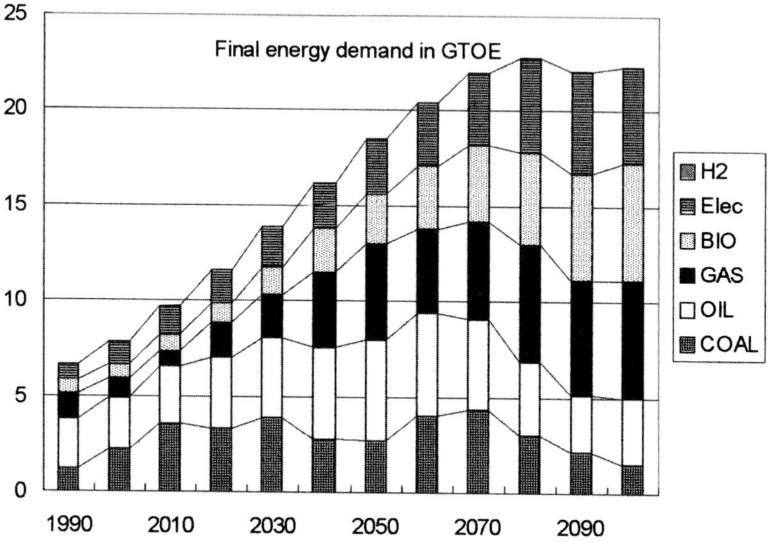

a

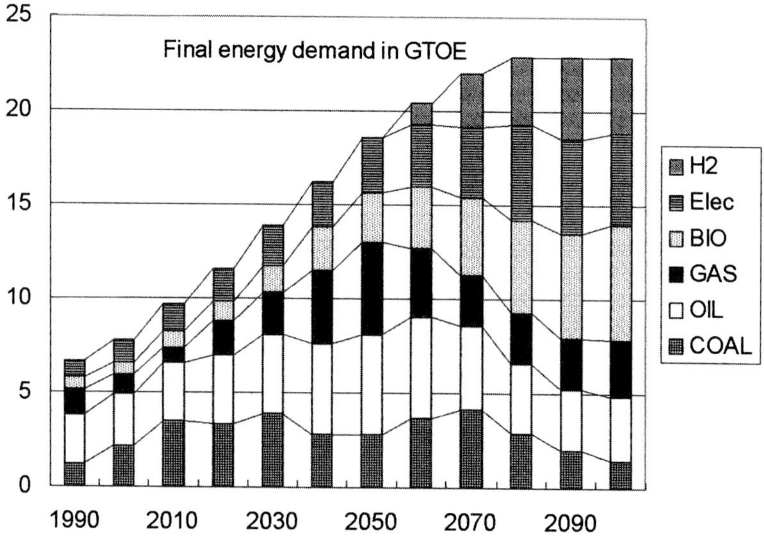

b

FIG. 9. **a** World final energy demand profiles in the B2-550 p.p.m.v. control case with reference-level hydrogen production and end-use costs. **b** World final energy demand profiles in the B2-550 p.p.m.v. control case with no additional hydrogen processing costs for FBR power generation, and reference-level end-use costs. **c** World final energy demand profiles in the B2-550 p.p.m.v. control case with no additional hydrogen processing and end-use costs

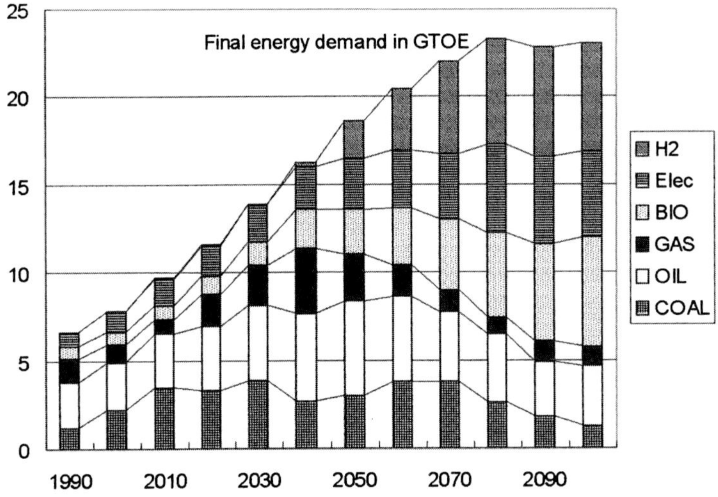

c

FIG. 9. *Continued*

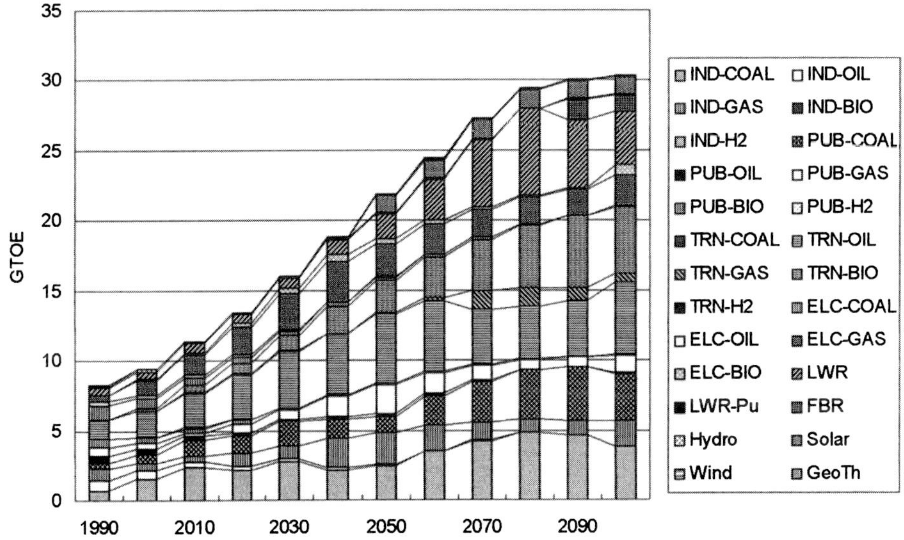

a

FIG. 10. **a** World primary energy flows in the B2-BAU case with reference-level hydrogen production and end-use costs. **b** World primary energy flows in the B2-550 p.p.m.v. control case with reference-level hydrogen processing costs for FBR power generation and end-use costs. **c** World primary energy flows in the B2-550 p.p.m.v. control case with no additional hydrogen processing and end-use costs

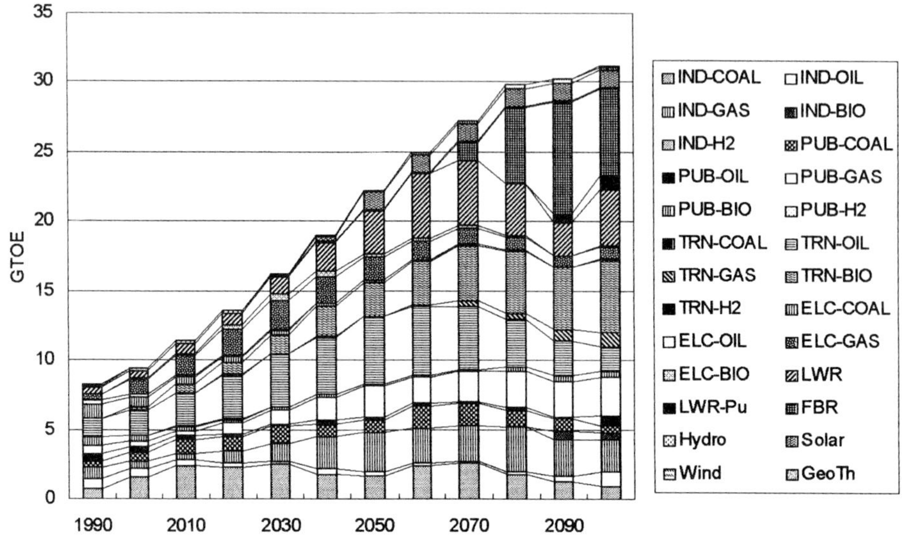

b

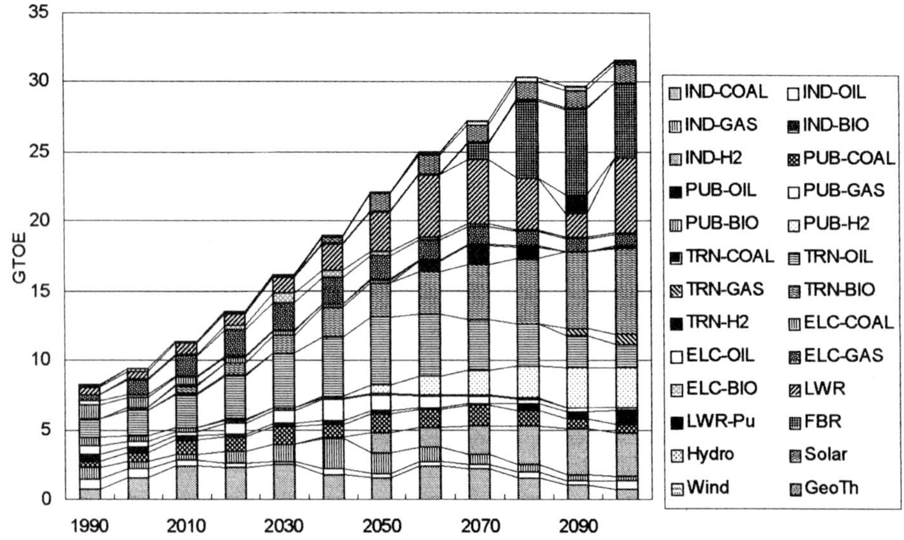

c

FIG. 10. *Continued*

model assumes that 10% of grassland and 5% of desert areas can additionally be converted to afforestation. Although the primary energy supply patterns do not change very much, the loss of GDP and the shadow prices of carbon emissions are apparently mitigated. Figures 13 and 14 compare GDP and shadow carbon emission prices in stabilization scenarios. The cases where atmospheric carbon concentrations are stabilized at less than 500 and 450 p.p.m.v. are also shown.

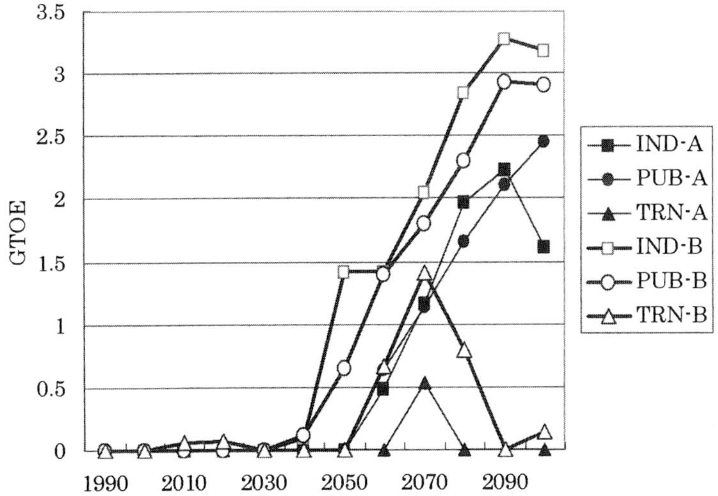

FIG. 11. Hydrogen end-use patterns forecast for industry (*IND*), transportation (*TRN*), and other sectors (*PUB*) in B2-550 p.p.m.v. control cases under (A) no additional hydrogen processing costs for FBR power generation and reference-level end-use costs, and (B) no additional hydrogen processing and end-use costs

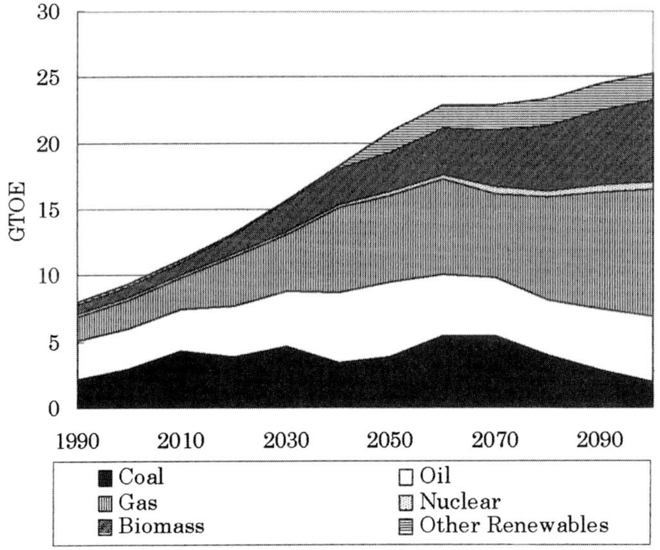

FIG. 12. World primary energy flows in the B2-550 p.p.m.v. control case with no nuclear power expansion

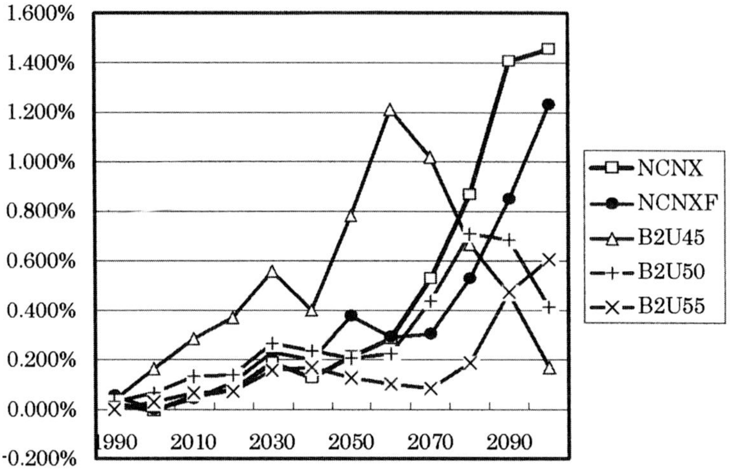

FIG. 13. Loss of world GDP when atmospheric carbon is stabilized. *B2U45*, 450 p.p.m.v; *B2U50*, 500 p.p.m.v; *B2U55*, 550 p.p.m.v; *NCNX*, 550 p.p.m.v. and no nuclear expansion; *NCNXF*, *NCNX* + increased potential biomass supply

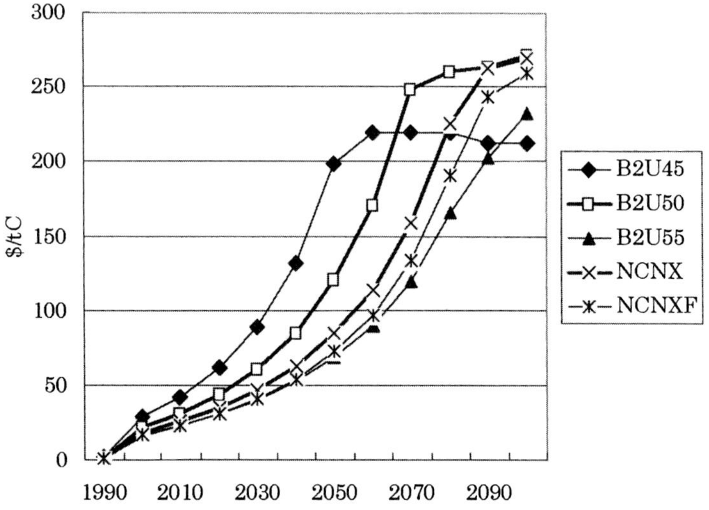

FIG. 14. Shadow prices of carbon emissions. *B2U45*, 450 p.p.m.v; *B2U50*, 500 p.p.m.v; *B2U55*, 550 p.p.m.v; *NCNX*, 550 p.p.m.v. and no nuclear expansion; *NCNXF*, *NCNX* + increased potential biomass supply

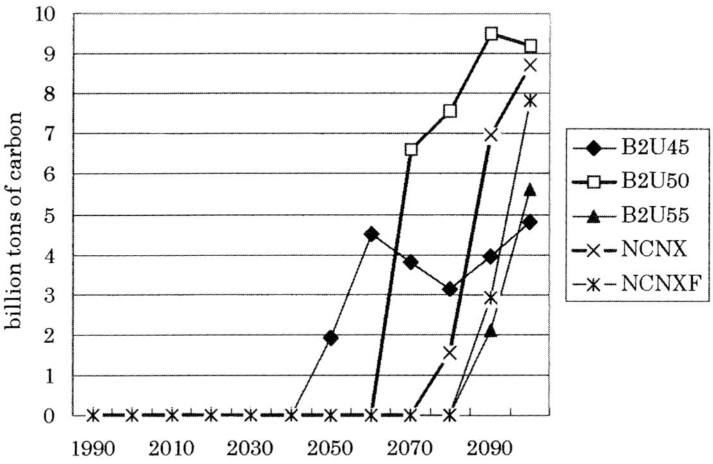

FIG. 15. Carbon sequestration. B2U45, 450 p.p.m.v; B2U50, 550 p.p.m.v; B2U55, 550 p.p.m.v; NCNX, 550 p.p.m.v. and no nuclear expansion; NCNXF, NCNX + increased potential biomass supply

Figure 15 shows how carbon sequestration technologies are implemented. As the conditions become more severe, earlier implementation of carbon sequestration is needed. The potential expansion of biomass supplies mitigates the need for these regrettable options.

Conclusion

This study assessed the potential contribution of new hydrogen production processes with FBR using MARIA with Bern-CC. Under a carbon concentration stabilization policy, FBR-based hydrogen can reach 6 GTOE in the second half of this century. However, the potential demand depends very much on supply costs. If the additional processing costs of hydrogen for FBR-electricity and the additional energy costs of natural gas can be reduced, FBR-based hydrogen can be used even in no carbon control cases.

However, options with a policy of no nuclear power expansion are also possible in carbon concentration control cases, where a lower primary energy demand with around 1% additional GDP loss are expected. Higher carbon sequestration implementation is also required in such cases.

Enhancement of the biomass energy supply potential would mitigate the burdens of carbon emission reduction which are clearly shown in GDP losses and the figures for the shadow prices of carbon emissions, even when the primary energy supply profiles are not very different.

However, low-temperature hydrocarbon-based steam-reforming processes are not only applicable to nuclear energy. New heat-cascading technologies should also be evaluated, as well as FBR-based processes. An assessment of these technologies should be central to future research.

Acknowledgments. The author is grateful to Dr. Masao Hori (Nuclear Systems Association, Japan) who organizes the nuclear–hydrogen committee in Japan, and who provided this research opportunity.

References

Hufton J, Weigel S, Waldron W, Nataraj S, Rao M, Sircar S (1999) Sorption-enhanced reaction process (SERP) for the production of hydrogen. Proceedings of the 1999 US DOE Hydrogen Program Review, NREL/CP-570-26938, Department of Energy, USA

Hufton J, Waldron W, Weigel S, Rao M, Nataraj S, Sircar S (2000) Sorption-enhanced reaction process (SERP) for the production of hydrogen. Proceedings of the 2000 Hydrogen Program Review, NREL/CP-570-28890, Department of Energy, USA

IPCC (Intergovernmental Panel of Climate Change) (1992), Climate Change 1992—The Supplementary Report to the IPCC Scientific Assessment, Cambridge University Press, UK

IPCC (Intergovernmental Panel of Climate Change) (1994) Radiative Forcing of Climate Change and An Evaluation of the IPCC IS92 Emissions Scenarios, Cambridge University Press, UK

IPCC (Intergovernmental Panel on Climate Change) (2000) Emissions scenarios. Nakicenovic N (ed) Cambridge University Press, UK

IPCC (Intergovernmental Panel on Climate Change) (2001a) Scientific assessment of climate change working group I to the third assessment report. Cambridge University Press, UK

IPCC (Intergovernmental Panel on Climate Change) (2001b) Third assessment report, working group 3. In press Joos et al. (1996) An efficient and accurate representation of complex oceanic and biospheric models of anthropogenic carbon uptake. Tellus 48B:389–417

Joos F, Bruno M, Fink R, Stocker TF, Siegenthaler U, LeQuéré C, Sarment JL (1996) An Efficient and accurate reprentation of complex oceanic and biospheric models of authropogenic carbon uptake. Tellus-B, Vol. 48, pp. 397–417

Mori S (2000a) The development of greenhouse gas emissions scenarios using an extension of the MARIA model for the assessment of resource and energy technologies. Technol Forecasting Soc Change 63:289–311

Mori S (2000b) Effects of carbon emission mitigation options under carbon concentration stabilization scenarios. Environ Econ Policy Stud 3:125–142

Mori S, Takahashi M (1999) An integrated assessment model for the evaluation of new energy technologies and food productivity. Int J Global Energy Issues 11(1–4):1–18

Shirasaki Y, Kobayashi K, Kuroda K (2001) New concept hydrogen production system based on the membrane reformer. Proceedings of the International Conference on Power Engineering, October 8–12, 2001, Xi'an, China

Weyant JP (ed) (1999) The cost of the Kyoto Protocol: a multi-model evaluation. Energy J, Special Issue

Wigley TML (1993) Balancing the carbon budget. Implications for projections of future carbon dioxide concentration changes. Tellus 45B:409–425

Assessments of Middle-Term Energy and Environmental Technology Options for Asian Regions Incorporating Resource and Quality Endowments by the ELSA Model

SHUNSUKE MORI* and TOMOHIRO FURUSE†

Summary. This chapter describes an energy/economy/environment model for Asian regions to assess the energy distribution structure and same simulations. Energy consumption in Asian countries is still increasing rapidly as the economic growth of this region continues the trend of the 1990s. Environmental pollution issues are also becoming more serious as the emission of pollutants from fossil fuel burning increases. In Asian countries, coal is still a major part of the primary energy supply, which causes serious air pollution. Technology options such as desulfurization facilities and fuel shifts to natural gas will mitigate the environmental issues. However, since the distribution of resources and the quality of the endowments of fossil fuels vary considerably among the regions, energy policies should take into account the interrelationships between domestic and international transportation issues.

In this chapter, we develop a middle-term energy/economy/environment model for the Asian regions, and disaggregate China and India into several regions to reflect the heterogeneity among the subregions of these countries. The model simulation results show that (1) coal will still be used extensively in most of the Asian regions under various environmental policies, (2) desulfurization systems will play a major role in meeting the energy supply conditions and mitigating sulfur emissions, (3) when low-cost and high-removal-rate desulfurization systems are not implemented, the demands for natural gas will become high and some countries will have to change their energy supply policies, and (4) a simultaneous carbon and sulfur emission control policy would provide a "win–win" situation for both Asian regions and Annex-I regions through clean development mechanism investments.

Key words. Asian regions, Energy and environment model, China, India, Energy transportation

* Department of Industrial Administration, Tokyo University of Science, 1-3 Kagurazaka, Shinjuku-ku, Tokyo 162-8601, Japan
† Norin Chukin Securities Co., Ltd., 1-7-2 Otemachi, Chiyoda-ku, Tokyo 100-0004, Japan

Introduction

Energy consumption in Asian countries is still increasing rapidly as the economic growth of this region continues the trend of the 1990s. Environmental pollution issues also appear more serious as the emission of pollutants from fossil fuel combustion increases. In Asian countries, coal supplies around 45% of the primary energy demand, and will still be a high priority resource in the near future because of its high availability and low supply costs. However, coal burning has two major problems. First, as is well known, coal burning causes serious air pollution and health effects since the sulfur content of coal is higher than that of other fuels. Although desulfurization facilities have already been developed, they are not widley used in Asian countries because of their high capital and operating costs (Li 2000) Second, like other fossil fuel resources, coal resources are unevenly distributed in Asian countries. They are mainly found in inland areas, while industry and urban areas have often developed near the coast. The transportation of coal between the production and consumption sites creates a high energy demand, especially in China and India. Furthermore, the quality of the coal, e.g., the contents of sulfur, sand, ash, calories, etc., also varies greatly. The so called clean-coal technologies have been developed to overcome such problems.

Oil demands in Asian countries are also increasing as the demand from the transportation sector grows. Because the oil production capacity in Asian regions cannot meet the increasing demand, it is projected that 90% of the oil supply of this region will depend on Middle-East countries in the near future.

Recently, natural gas resources have become the major energy source of many countries. This is because of the low sulfur content, the availability of pipeline transportation, and the high conversion efficiency when electric power is generated by combined gas turbines. However, uncertainties still remain in the existing estimations of the extent of natural gas resources. Other renewable energy resources, e.g., wind power, solar power, and biomass, will also contribute to a sustainable energy supply.

The mitigation of global warming has been the subject of international agreement since the Kyoto meeting in 1997 (COP-3). In spite of the fact that many developing Asian countries have not agreed with their reduction target for greenhouse gas (GHG) emissions, they do not totally oppose the flexibility measures in the Kyoto Protocol. The clean development mechanism (CDM) will promote international collaboration on GHG emission reduction.

The background given above suggests that an energy strategy is an issue of the highest priority, and that detailed regional characteristics should be considered when we assess resource and technology strategies for Asian countries. The purpose of this study was to develop a multiregional energy and environment model disaggregating the two major Asian countries, China and India, into subregions. Domestic energy transportation is also dealt with in detail in the model.

Background of Asian Energy Issues

The Asian region has shown rapid economic growth, especially in the 1980s. Asia–Newly Industrializing Economies countries had already initiated this economic growth in the mid-1970s, and were followed by the Association of Southeast Asian Nations (ASEAN) and China in the 1980s. India took off in the 1990s. The annual economic growth rate of the Asian region for 1985–1995 was 6.60%, while that of the whole world was 2.52% in the same period. Although the "financial crisis" in 1997 seriously affected this trend, Asian countries are still expected to recover their economic growth.

The demographic issues in Asian countries are still apparent. The population in the Asian region was 1.4 billion in 1950, and had increased to 3.4 billion in 1995. In 2050, it will reach 5.4 billion according to UN estimates.

Figure 1 shows the historical trends in annual primary energy demand growth rates. Energy consumption in the Asian region has increased at 5.9% per year from 1971 to 1994, which is around 2.7 times the world average growth rate. The total energy consumption in the Asian region was 1.5 billion tons of oil equivalent (GTOE) in 1994, which is 18.3% of the total world energy consumption. However, the per capita energy consumption in the Asian region still remains at 6.5% of that in North America. These high economic growth rates, together with high population growth rates and the current low per capita energy consumption level, suggest that the future energy demand in this region will increase enormously.

Figure 2 shows the primary energy supply mix of the world and of Asia in 1971 and 1994. The fraction of coal was around 56% in 1994. Although coal burning can cause serious environmental pollution, it will remain the main source of electric power generation in this region because of the abundant resources and low market price.

In China and India, energy transportation between production and consumption regions is particularly important because of their large land area.

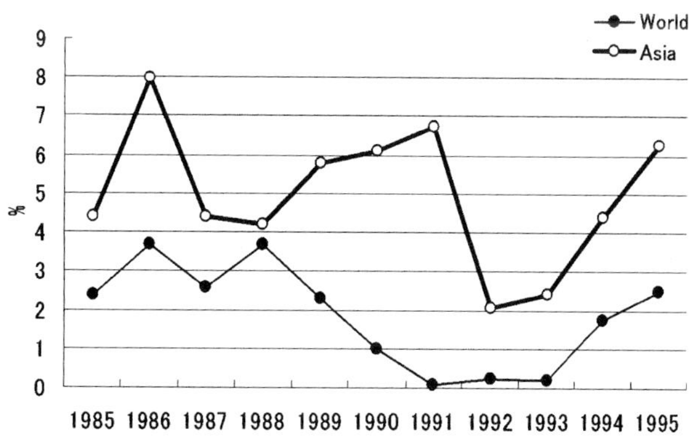

FIG. 1. Historical trends in annual primary energy demand growth rates in the world

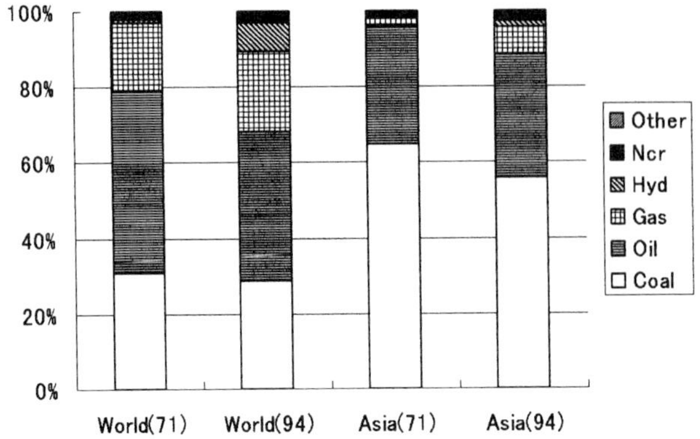

FIG. 2. Primary energy supply mix of Asia and the world in 1971 and 1994. *Ncr*, nuclear; *Hyd*, hydro power

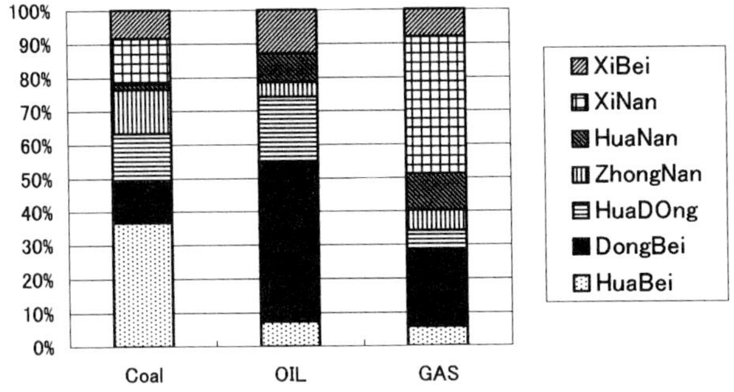

FIG. 3. Share of energy production by region in China

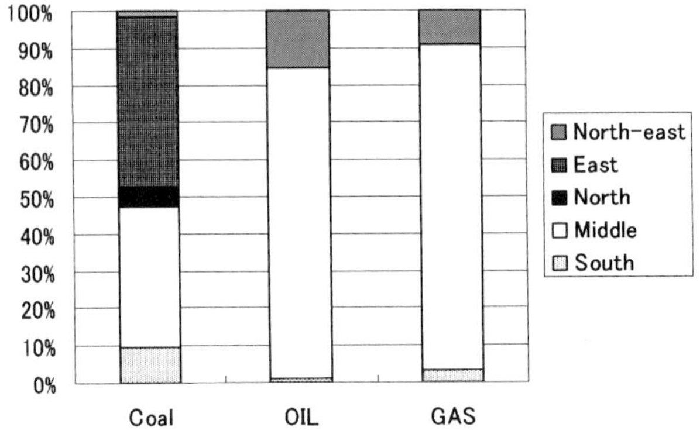

FIG. 4. Share of energy production by region in India

Figures 3 and 4 show the regional distributions of energy production in these two countries. These figures show that the distribution of resources varies considerably.

Based on this observation, one can conclude that the following three issues are significant for future energy strategies: (1) to meet the future increase in energy demands; (2) to utilize the existing energy resources efficiently, and to incorporate technological choices; (3) to consider both environmental and energy transportation issues.

Framework of the Model

Classification of Regions

The aim of this study was to assess energy technologies, including energy transportation infrastructures, to deal with the geographical distribution of fossil energy resources and quality. We focused on the Asian area shown in Fig. 5, and derided it into the 15 subregions listed in Table 1. The largest city in the subregion was used to represent its energy demand node in the energy supply network model. We then developed a multiregional energy–economy model, including domestic and international energy transportation, to treat the geographic resource distribution in detail. The contribution of desulfurization options is assessed in the model.

We used Indonesia, Russia, and the Middle-East countries as the oil and natural gas production countries. Indonesia, Australia, and South Africa were assumed to be the coal-supply countries for Asia.

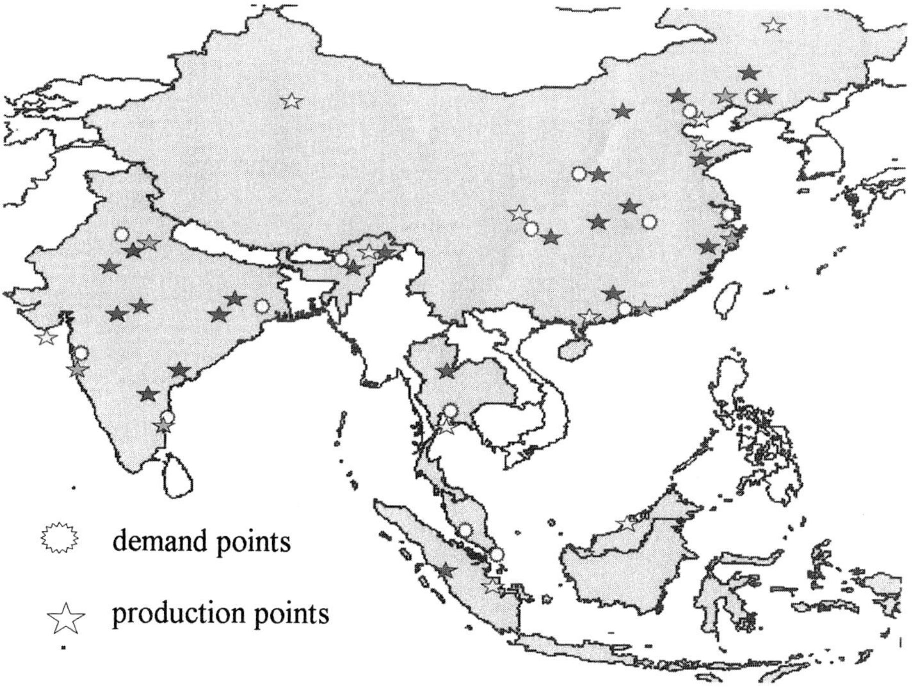

FIG. 5. The Asian region, and the energy demand and production points of the model

TABLE 1. The Energy-demand subregions of the model

Country	Subregions	Demand node cities
China	HuaBei	Beijing
	DongBei	Shengyang
	HuaDong	Shanghai
	ZhongNan	Wuhan
	HuaNan	Guangzhou
	XiNan	Chongping
	XiBei	Xian
India	South India	Madras
	Middle India	Bombay
	North India	Delhi
	East India	Calcutta
	Northeast India	Gauhati
Thailand		Bangkok
Malaysia		Kuala Lumpur
Singapore		Singapore

Structure of the Model

The structure of the model is shown in Fig. 6. As in existing energy/economics models, this model incorporates production functions which consist of capital stock, labor, and energy inputs by region. The output is distributed among investment, consumption, and energy cost.

The energy supply block involves production, pretreatment of coal, transportation, and conversion. This model focuses on SO_2 and CO_2 as air pollutants caused by combustion processes. Sulfur dioxide can be removed by desulfurization facilities.

The objective function is the maximization of the discount sum of consumption, as in GLOBAL 2100 (Manne and Richels 1992).

Model Formulations

The indicators are defined as follows:

— i, region (1, HuaBei; 2, DongBei; 3, HuaDong; 4, ZhongNan; 5, HuaNan; 6, XiNan; 7, XiBei; 8, South India; 9, Middle India; 10, North India; 11, East India; 12, Northeast India; 13, Thailand; 14, Malaysia; 15, Singapore; 16, Indonesia; 17, Australia; 18, South Africa; 19, Russia; 20, the Middle East)
— k_f, fossil fuels (1, coal; 2, oil; 3, natural gas)
— k_r, carbon-free energy sources (1, hydraulic power; 2, nuclear power)
— t, time period

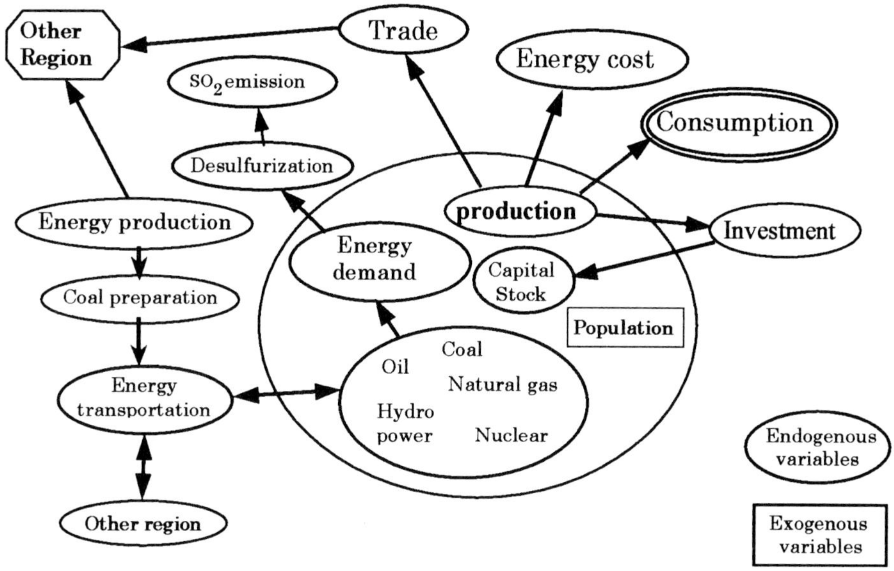

FIG. 6. Structure of the model (one region)

Energy Resources

The assumptions about energy resources are shown in Table 2 (MITI 1998). Table 3 summarizes the initial resource extraction costs (Inagaki et al. 1996). It is assumed that the extraction costs rise as cumulative production increases (Rogner 1997).

Energy Production and Demand

The relationships among energy production, energy export, import, and supply are expressed by simplified linear equations.

$$\mathrm{FEP}_{i,k_f,t} = \sum_{j=1}^{15} \mathrm{FEMex}_{i,j,k_f,t} \qquad (1)$$

where

$\mathrm{FEP}_{i,k_f,t}$ is k_f fossil energy production in region i in period t (million ton),
$\mathrm{FEMex}_{i,j,k_f,t}$ is k_f energy transportation from production region i to demand region j in period t (million ton).

Energy Transportation and Supply

Losses from energy transportation are one of the major key features in assessing energy strategies, especially in China and India, since there are other pos-

TABLE 2. Fossil energy resources and potential hydraulic and nuclear power supplies

	Coal (million ton)	Oil (million ton)	Natural gas (million ton)	Hydropower (MTOE)	Nuclear power (MTOE)
HuaBei	183296.0	472.6	174.9	3.0	0.0
DongBei	8878.4	1585.1	586.5	3.0	0.0
HuaDong	18616.0	597.3	221.0	7.3	145.98
ZhongNan	9164.8	68.7	20.6	9.8	0.0
HuaNan	1432.0	65.5	9.7	6.0	236.19
XiNan	30644.8	82.0	56.0	115.8	0.0
XiBei	34368.0	459.5	522.0	20.7	0.0
South India	13127.2	62.8	21.6	18.5	0.9
Middle India	48895.0	584.3	354.6	3.5	0.7
North India	1061.8	0.0	2.9	6.6	0.5
East India	142265.6	0.0	0.0	3.4	0.0
Northeast India	889.7	147.9	113.7	19.6	0.0
Thailand	1422.0	51.6	171.2	0.7	0.0
Malaysia	0.0	588.0	1375.4	5.25	0.0
Singapore	0.0	0.0	0.0	0.0	0.0
Indonesia	6412.0	756.5	1302.3	—	—
Australia	23346.0	0.0	0.0	—	—
South Africa	24243.0	0.0	0.0	—	—
Russia	0.0	0.0	9996.9	—	—
Middle East	0.0	22546.7	8061.5	—	—

MTOE, million tons of oil equivalent

TABLE 3. Resource extraction costs in US dollars per ton

	Coal	Oil	Natural gas
China	39.29	55.78	95.20
India	26.91	55.78	85.40
Thailand	19.64	55.78	85.40
Malaysia	—	55.78	85.40
Indonesia	39.29	55.78	85.40
Australia	39.29	55.78	—
South Africa	39.29	55.78	—
Russia	—	55.78	77.00
Middle East	—	55.78	95.20

sible options. For instance, (1) raw coal transportation by railways (existing), (2) washed and cleaned coal transportation, (3) natural gas pipeline transportation, (4) electric power transportation by high-voltage direct current transmission, (5) decentralized energy supply systems, etc.

In our model, it is assumed that electricity generated by nuclear and hydraulic power is transported, and that the energy losses from transportation are approximated by linear equations.

$$\mathrm{FEMin}_{i,j,k_f,t} = \mathrm{FEMex}_{i,j,k_f,t} - \mathrm{FEMls}_{i,j,k_f,t} \tag{2}$$

$$\begin{aligned}&\mathrm{FEMls}_{i,j,k_f,t}\\&\quad=(\mathrm{RLS}_{k_f}\times\mathrm{RID}_{i,j}+\mathrm{WLS}_{k_f}\times\mathrm{WID}_{i,j}+\mathrm{PLS}_{k_f}\times\mathrm{PID}_{i,j})\\&\qquad\times\mathrm{FEMex}_{i,j,k_f,t}\end{aligned}\tag{3}$$

$$\mathrm{FE\_S}_{j,k_f,t}=\sum_{i=1}^{20}\mathrm{FEMin}_{i,j,k_f,t}\times\mathrm{Hp}_{i,k_f}\tag{4}$$

where

$\mathrm{FEMin}_{i,j,k_f,t}$ is k_f fossil energy imports from energy production region i to demand region j in period t (million ton),
$\mathrm{FEMls}_{i,j,k_f,t}$ is the energy loss and consumption of k_f fossil energy transportation from region i to region j in period t (million tons of oil equivalent, MTOE),
RLS_{k_f} is the rate of energy loss of k_f fossil energy by railway transportation per 1000 km (%),
$\mathrm{RID}_{i,j}$ is the distance of railway energy transportation between region i and region j (1000 km),
WLS_{k_f} is the rate of energy loss of k_f fossil energy by ship transportation per 1000 km (%),
$\mathrm{WID}_{i,j}$ is the distance of ship energy transportation between region i and region j (1000 km),
PLS_{k_f} is the rate of energy loss of k_f fossil energy by pipeline transportation per 1000 km (%),
$\mathrm{PID}_{i,j}$ is the distance of pipeline transportation between region i and region j (1000 km),
$\mathrm{FE\_S}_{j,k_f,t}$ is the total supply of k_f fossil energy for region j in period t (MTOE),
Hp_{i,k_f} is the energy conversion coefficient by production region i and for k_f fossil energy (kcal/kg).

Energy Demand

Figure 7 shows an outline of the energy supply–demand flow in the model. This model incorporates five primary energy sources, two secondary energy products, and three final energy demand sectors. The energy demands for transportation and the other sectors are given exogenously.

Since there are so many power generation options, their assessment is the key issue in building up technology strategies. Eight fired power generation technologies are incorporated in our model. There are four coal-based technologies, i.e., a conventional coal-fired plant, an integrated gasification combined cycle with a low efficiency (IGCC-low), and integrated gasification combined cycle with a high efficiency (IGCC-high), and a solid oxide fuel cell (SOFC), one oil-based power plant (OLF), and three gas-based technologies, i.e., a conventional gas-fired plant (GSF), small gas turbines (SGT), and a gas combined cycle (GCC). In the equations below, these are represented by the letter m.

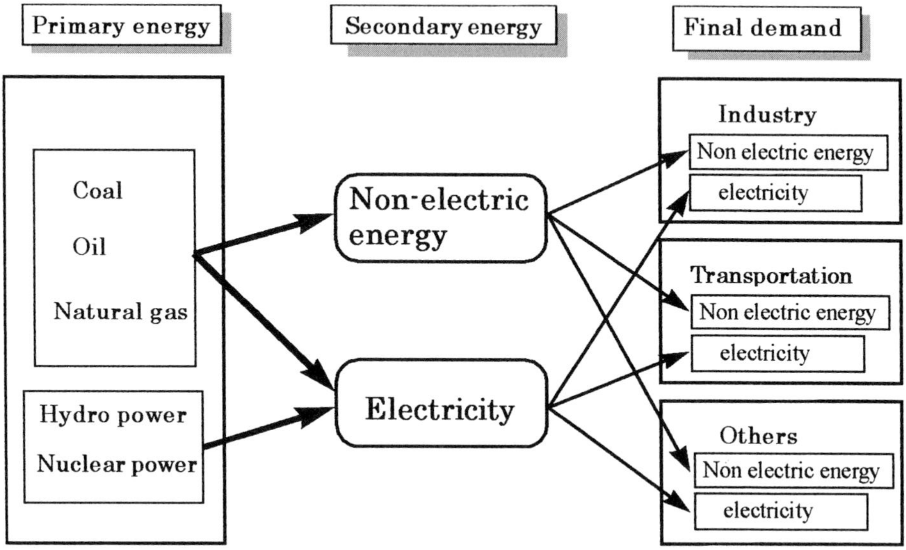

FIG. 7. Energy flow structure

$$FE\_s_{j,k_f,t} = FE\_NE_{j,k_f,t} + \sum_m FE\_E^m{}_{j,k_f,t} \tag{5}$$

$$FEEP_{j,k_f,t} = \sum_m PP^m{}_{k_f,t} \times FE\_E^m{}_{j,k_f,t} \tag{6}$$

$$EC\_NE_{j,t} = \sum_{k_f=1}^{3} FE\_E_{j,k_f,t} \tag{7}$$

$$EC\_E_{j,t} = \sum_{k_f=1}^{3} FEEP_{j,k_f,t} + \sum_{k_r=1}^{2} RE\_S_{j,k_r,t} \tag{8}$$

$$EC\_NE_{jt} = EC\_NEI_{jt} + EC\_NEP_{jt} + EC\_NET_{jt} \tag{9}$$

$$EC\_E_{jt} = EC\_EI_{jt} + EC\_EP_{jt} + EC\_ET_{jt} \tag{10}$$

where

$FE\_NE_{j,k_f,t}$ is the k_f fossil energy supply for the nonelectric energy demand of region j in period t (MTOE),

$FE\_E^m{}_{j,k_f,t}$ is the k_f fossil energy supply for electric power generation technology m of region j in period t (MTOE),

$FEEP_{j,k_f,t}$ is the electric power generated by k_f fossil energy for region j in period t (MTOE),

$PP^m{}_{k_f,t}$ is the energy conversion efficiency of power generation technology m for k_f fossil energy in period t (%),

$EC\_NE_{j,t}$ is the total nonelectric energy demand of region j (MTOE),
$EC\_E_{j,t}$ is the total electric power demand of region j (MTOE),
$RE\_S_{j,k_f,t}$ is the electric power generated by nuclear and hydraulic power in region j in period t (MTOE),
$EC\_NEI_{j,t}$ is the nonelectric energy demand of the industry sector of region j in period t (MTOE),
$EC\_NET_{j,t}$ is the nonelectric energy demand of transportation sector of region j in period t (MTOE),
$EC\_NEP_{j,t}$ is the nonelectric energy demand of other sectors of region j in period t (MTOE),
$EC\_EI_{j,t}$ is the electric power demand of industry sector of the region j at the period t (MTOE),
EC_EE_{jt} is the electric power demand of the transportation sector of region j in period t (MTOE),
EC_EP_{jt} is the electric power demand of other sectors of region j in period t (MTOE).

Energy Transportation

We impose upper-limit constraints on energy transportation among regions. The transportation capacities are assumed to be expandable in response to an increase in demand. The equations relating to railways are shown below.

$$RC_{j,t} \geq \sum_{i=1}^{20} \sum_{k_f=1}^{3} FEMex_{i,j,k_f,t} \times RID_{i,j} \tag{11}$$

$$WC_{j,t} \geq \sum_{i=1}^{20} \sum_{k_f=1}^{3} FEMex_{i,j,k_f,t} \times WID_{i,j} \tag{12}$$

$$PC_{j,t} \geq \sum_{i=1}^{20} \sum_{k_f=1}^{3} FEMex_{i,j,k_f,t} \times PID_{i,j} \tag{13}$$

$$RC_{j,t+1} = \delta \times RC_{j,t} + NRC_{j,t} \tag{14}$$

where

$RC_{j,t}$ is the transportation capacity of the railway for demand region j in period t (ton km),
$WC_{j,t}$ is the transportation capacity of shipping for demand region j in period t (ton km),
$PC_{j,t}$ is the transportation capacity of the pipeline for demand region j in period t (ton km),
δ is the depreciation rate per decade (%),
$NRC_{j,t}$ is the newly implemented transportation capacity of the railway for demand region j in period t (ton km).

Sulfur Dioxide Emission

The emissions of pollutants in a region are calculated by the total consumption of oil and coal multiplied by the coefficients for their sulfur and carbon contents. These values were extracted from the literature (NISTEP 1991), as shown in Table 4.

$$SO_{j,t} = \sum_{i=1}^{20} \sum_{k_f=1}^{3} FEMin_{i,j,k_f,t} \times PSO_{i,k_f} \times Up_{k_f} \tag{15}$$

$$S_{j,t} = SO_{j,t} - RSO_{j,t} \tag{16}$$

where

$SO_{j,t}$ is the atmospheric sulfur emissions of region j in period t (metric ton),
PSO_{i,k_f} is the sulfur content (%),
Up_{k_f} is the emission coefficients,
$S_{j,t}$ is the sulfur emissions from fuel burning of region j in period t (metric ton),
$RSO_{j,t}$ is the sulfur removed by desulfurization facilities in region j in t (metric ton).

We adopted two-stage desulfurization processes: (1) the coal preparation process (50% desulfurization, 25% energy loss, and 40% weight reduction), (2) simple desulfurization equipment in a postcombustion process (70% desulfurization and 70% capacity utilization rate), and (3) high-removal-rate desulfur-

TABLE 4. Sulfur content in weight percent

Source	Coal	Oil
HuaBei	0.58	0.40
DongBei	1.26	0.40
HuaDong	1.62	0.40
ZhongNan	1.04	0.40
HuaNan	1.04	0.40
XiNan	3.12	0.40
XiBei	1.76	0.40
South India	0.60	—
Middle India	0.60	0.50
North India	0.60	—
East India	0.64	—
Northeast India	4.10	0.50
Thailand	2.40	0.50
Malaysia	—	0.50
Singapore	—	—
Indonesia	0.50	0.50
Australia	0.55	—
South Africa	0.50	—
Middle East	—	3.00

TABLE 5. Cost assumptions for desulfurization in $ per ton of sulfur

	Simple removal system (70%)	High-removal system (99%)
Coal	423	999
Oil	846	2000

ization equipment (99% desulfurization and 70% capacity utilization rate). Desulfurization systems are also applicable to oil refineries. The sulfur removal costs assumed are shown in Table 5. We also imposed hypothetical constraints on the sulfur emission limits according to region.

$$S_{j,t} \leq \overline{\mathrm{SO}_j} \tag{17}$$

Economic Activity Block

An economic activity block aggregates regional industries into one sector in a similar way to the ETA–MACRO and GLOBAL 2100 (Manne and Richels 1992) models. For simplicity, the Cobb–Douglas production function with four inputs, i.e., capital stock, labor, electric energy, and nonelectric energy, was employed for each region. To distinguish explicitly between "modern industrialization" and traditional activities, the Putty–Clay formulation was employed. The following equations represent the economic activities in region j in period t.

$$\mathrm{YN}_{j,t} = (1+\mu_{j,t}) \times A_j \times \mathrm{IS}_{j,t}^{\alpha} \times \mathrm{LN}_{j,t}^{\beta} \times \mathrm{NEN}_{j,t}^{\gamma} \times \mathrm{EEN}_{j,t}^{1-\alpha-\beta-\gamma} \tag{18}$$

$$Y_{j,t} = (1-d) \times Y_{j,t-1} + \mathrm{YN}_{j,t} \tag{19}$$

$$L_{j,t} = (1-d) \times Y_{j,t-1} + \mathrm{LN}_{j,t} \tag{20}$$

$$K_{j,t} = (1-d) \times K_{j,t-1} + \mathrm{IS}_{j,t} \tag{21}$$

$$\mathrm{EC\_NEI}_{j,t} = (1-d) \times \mathrm{EC\_NEI}_{j,t-1} + \mathrm{NEN}_{j,t} \tag{22}$$

$$\mathrm{EC\_EI}_{j,t} = (1-d) \times \mathrm{EC\_EI}_{j,t-1} + \mathrm{EEN}_{j,t} \tag{23}$$

$$C_{j,t} = Y_{j,t} - \mathrm{IS}_{j,t} - \mathrm{E\_Cost}_{j,t} \tag{24}$$

where $Y_{j,t}$ is the gross output, $\mathrm{A_j}$, μ_t, α, β, γ are the parameters of the production function, $K_{j,t}$ is the capital stock, $L_{j,t}$ is the total labor input, $\mathrm{LN}_{j,t}$ is the labor input for new investment, $\mathrm{EE}_{j,t}$ is the total electric power input, $\mathrm{NE}_{j,t}$ is the is total nonelectric power input, $\mathrm{EEN}_{j,t}$ is the electric power input for new investment, $\mathrm{NEN}_{j,t}$ is the nonelectric power input for new investment, $C_{j,t}$ is the consumption in region j, $\mathrm{IS}_{j,t}$ is the investment for region j, and $\mathrm{E\_Cost}_{j,t}$ is the energy cost of region j.

Energy costs consist of extraction costs, power generation costs, liquefaction costs for the transportation sector, transportation costs, trade transaction costs, and desulfurization costs. These parameters were extracted from the literature (NISTEP 1991; Suzugaki, et al. 1996).

$$\mathrm{E\_Cost}_{j,t} = \mathrm{EPC}_{j,t} + \mathrm{EEC}_{j,t} + \mathrm{HC}_{j,t} + \mathrm{MC}_{j,t} + \mathrm{IMC}_{j,t} + \mathrm{RSC}_{j,t} \tag{25}$$

where $EPC_{j,t}$ is the extraction cost, $EEC_{j,t}$ is the power generation cost, $HC_{j,t}$ is the liquefaction cost, $MC_{j,t}$ is the transportation cost, $IMC_{j,t}$ is the net import cost, and $RSC_{j,t}$ is the desulfurization cost.

In this model, we adopted a simple objective function to represent the consumption maximization and equity distribution among regions, i.e.,

$$\max \sum_t (1-r)^t \sum_j L_{j,t} \ln(C_{j,t}/L_{j,t}) \tag{26}$$

The parameters of the production functions were estimated based on the literature (World Bank 1998; China 1999). These are summarized in Table 6.

Table 7 shows the assumptions which were made based on reference power generation plants in Japan (Sugiyama 1999a). However, a recent study reports

TABLE 6. Parameters of production functions

Location	A	α	β	γ
HuaBei	3.05927	0.419	0.508	0.03204
DongBei	2.72534	0.395	0.534	0.02845
HuaDong	3.08931	0.343	0.588	0.01969
ZhongNan	5.66465	0.283	0.650	0.02451
HuaNan	2.92704	0.375	0.561	0.02094
XiNan	6.99616	0.295	0.636	0.02679
XiBei	4.85964	0.421	0.512	0.02476
South India	1.99709	0.327	0.629	0.02883
Middle India	1.57530	0.356	0.593	0.02863
North India	2.00289	0.411	0.543	0.02879
East India	2.49888	0.431	0.525	0.02884
Northeast India	2.32319	0.379	0.577	0.02883
Thailand	5.33968	0.579	0.402	0.01187
Malaysia	4.86781	0.572	0.394	0.02251
Singapore	3.47396	0.522	0.465	0.00951

TABLE 7. Reference power generation costs

	Construction cost (million \$/Gw)	Capacity factor (%)	Annualization rate (%)	Efficiency (%)
Coal-fired plant	1500	70	17	28
IGCC-low	1700	70	17	37
IGCC-high	2000	70	17	38
SOFC	2800	70	17	42
Oil-fired plant	1250	70	17	35
Gas-fired plant	1200	70	17	37
Small GT	900	70	17	33
GCC	2000	70	17	46
Hydraulic power	2000	45	13	—
Nuclear power	2000	75	17	—

* Efficiency is assumed to increase by around 0.5% per decade. ** Hydraulic and nuclear power inputs are provided exogeneously. IGCC, integrated gasification combined cycle; SOFC, solid oxide fuel cell; GT, gas turbine; GCC, gas combined cycle

much lower costs for China (Imanaka and Yamaji 2001). In the present study, the costs of fuel-burning power generation systems are adjusted by multiplying the appropriate coefficient, i.e., 35% for China and 50% for other regions.

Simulations

Simulation Scenario 1. Contribution of Desulfurization

The model simulates four decades, 1995–2035. Our model consists of 19000 variables and 16000 constraints, some of which are nonlinear. The model runs on GAMS software. One simulation takes around 30h on an IBM-compatible PC with Pentium III 1.2-GHz.

In this section, we compare the following three simulation scenarios:

— Case 1, business as usual (BAU);
— Case 2, sulfur emissions are controlled to no more than 110% of the 1995 value by implementing desulfurization;
— Case 3, sulfur emissions are controlled to no more than 110% of the 1995 value without implementing high-removal-rate desulfurization.

Case 3 was designed to assess the contribution of high-removal-rate desulfurization equipment in comparison with cases 1 and 2.

Simulation Results

Figure 8 shows the results of the simulation on GDP in Case 1 (BAU). Economic growth will continue in the first half this century. Table 7 summarizes the economic growth rates for 1995–2000, 2000–2010, 2010–2020, and so on. Li (2000) forecasts that the annual economic growth rates for China will be 8.3% in 1997–2000, 7.1% in 2000–2010, 6.1% in 2010–2020, and 5.1% in 2020–2030. We can see that the results given in Table 8 are basically compatible with these figures. However, detailed projections of economic growth are not currently available for other regions. Harmonization with other projections with BAU will be needed to assess the options by sensitivity analysis. Figures 9 and 10 show the SO_2 and CO_2 emissions, respectively, in the BAU case. There are some other studies on sulfur emissions in Asian regions (NISTEP 1991; Sugiyama 1997, 1999b; Li 2000). For instance, with respect to sulfur pollution in China, Li (2000) estimated that 26.3 million tons of SO_2 was emitted in 1996 assuming that 80% of the sulfur in fuels is emitted, while Sugiyama (1999b) estimated 32.6 million tons of SO_2 in 1995. Li (2000) also estimated that SO_2 emissions in 2030 will be 87.3 million tons (reference case), which is almost identical to Sugiyama's projection (Sugiyama 1999), and 82.1 million tons if high-efficiency power generation in used, while SO_2 emissions in 1995 and 2030 are 29.56 and 92.07 million tons, respectively, in the energy linkage structure of Asia (ELSA) model. Even in the historical 1995 case, SO_2 emission estimates in China did not harmonize. There are a few possible reasons for this. Li (2000) assumes that the sulfur content

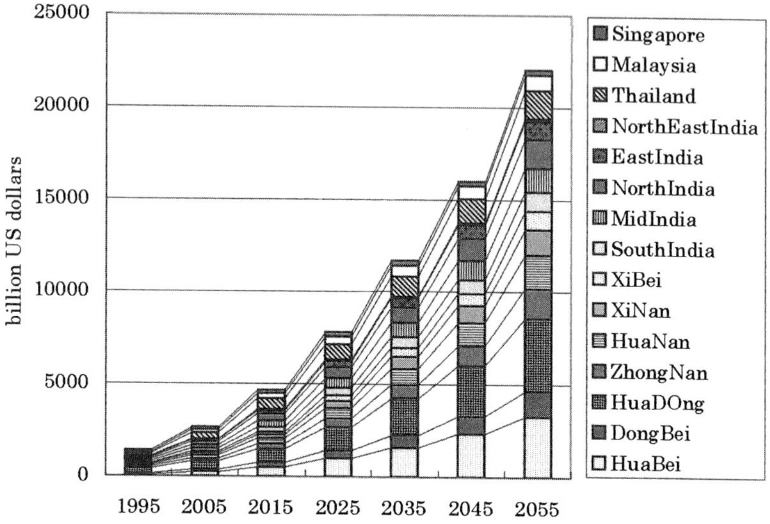

FIG. 8. Simulation results for GDP (billion $US) by region in Case 1 (BAU)

TABLE 8. Summary of annual GDP growth rates by region (%)

	1995–2000	2000–2010	2010–2020	2020–2030	2030–2040	2040–2050
HuaBei	12.95	8.57	7.47	5.73	4.31	3.66
DongBei	6.14	5.96	6.34	5.35	4.18	3.89
HuaDong	6.02	5.47	5.58	4.92	3.78	3.48
ZhongNan	6.30	5.68	5.78	5.16	3.96	3.87
HuaNan	7.51	6.20	6.03	5.22	4.25	3.93
XiNan	8.14	6.37	5.98	5.24	4.07	3.93
XiBei	9.57	7.68	7.26	5.82	3.88	3.83
China total	7.58	6.38	6.25	5.29	4.05	3.72
South India	7.33	5.67	4.73	3.92	3.11	2.78
Middle India	7.34	5.73	4.88	4.00	3.30	2.67
North India	9.88	6.82	5.27	4.32	3.70	3.01
East India	9.63	6.90	5.46	4.54	3.73	3.36
Northeast India	7.80	6.00	4.86	4.12	3.40	3.19
India total	8.38	6.23	5.06	4.18	3.46	2.94
Thailand	9.05	6.12	4.32	3.11	2.22	1.61
Malaysia	9.53	6.67	4.87	3.49	2.46	1.78
Singapore	6.72	4.26	2.47	1.40	0.50	0.04

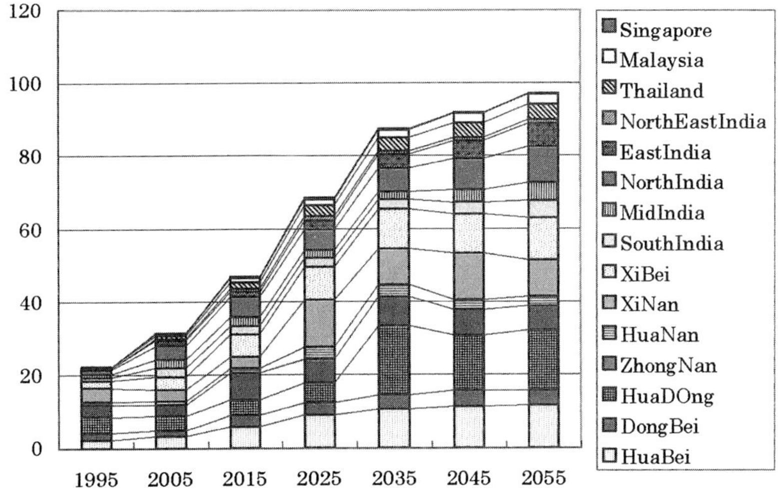

FIG. 9. Simulation results for SO_2 emissions (million tons of S) by region in Case 1 (BAU)

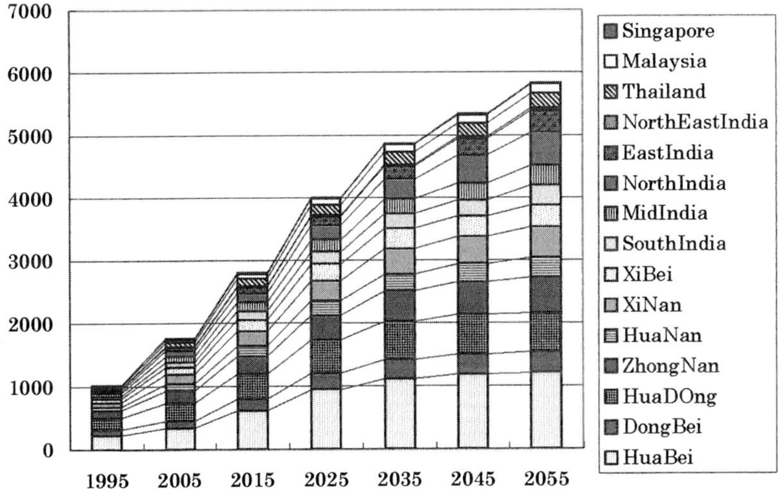

FIG. 10. Simulation results for CO_2 emissions (million tons of C) by region in Case 1 (BAU)

of coal is uniformly 1.15%, while the assumptions in ELSA range from 0.58% to 3.12%, as shown in Table 4. This may be the reason why SO_2 emissions in ELSA are around 10% higher than those given by Li. In this way, one can say that the outputs from ELSA are basically compatible with Li's study. Li and the ELSA model both assume that 80% of the sulfur combusted in fuels is emitted into the atmosphere, and the rest remains in the burner. However, Sugiyama does not mention this point.

Figures 11 and 12 show coal and gas supplies to China and India, respectively. As is often pointed out, China will have to import fossil energy sources even in the case of coal. This is because the imported high-quality coal will be cheaper than the domestic low quality coal, especially in the coastal industrialized zones of China. Figure 11 suggests that Russia and Malaysia will be the major suppliers of natural gas to China as the demand for natural gas increases. Similarly, the Middle East will be the major natural gas supplier to India, and South Africa will provide coal for India.

Figure 13 shows the fuel-based power generation patterns of China and India. One can see that coal based technologies are used extensively. IGCC is mainly used in China, while SOFC plays a major role in India. These differences are caused by the efficiency, the cost of equipment, and the value of electricity.

In Case 2, when all regions control their sulfur emissions so that they do not exceed 110% of the 1995 level, high-removal-rate desulfurization equipment is implemented for both coal and oil. The GDP and the primary energy structure, except for the natural-gas supply, do not change very much. Figure 14 compares the natural-gas supply from the Middle East region to Asian countries in Cases 1 and 2. One can see that the natural-gas demand for India increases rapidly, and that the demands of other regions are crowded out. Since the total supply of fossil fuels is restricted, the natural-gas supply from the Middle East in 2055 in Case 2 is lower than that in Case 1. The natural-gas demand from China therefore decreases as the imports from the Middle East are lowered.

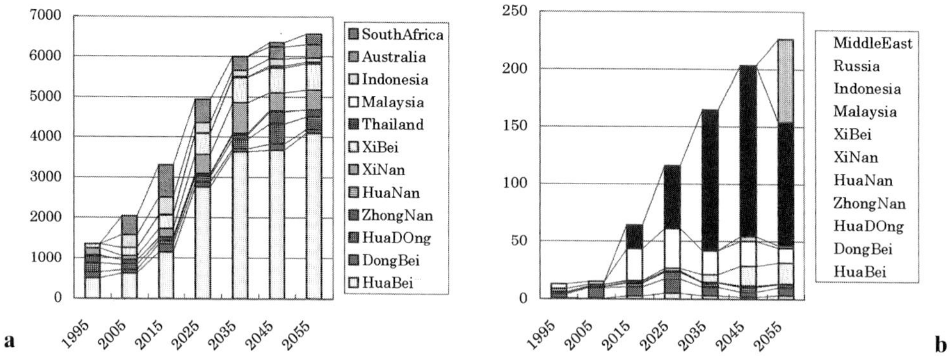

FIG. 11. Coal and gas supplies to China (MTOE). **a** Coal. **b** Natural gas

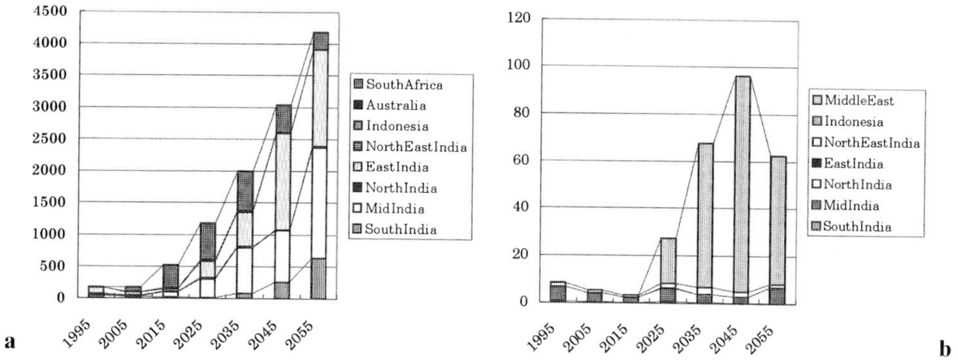

FIG. 12. Coal and gas supplies to India (MTOE). **a** Coal. **b** Natural gas

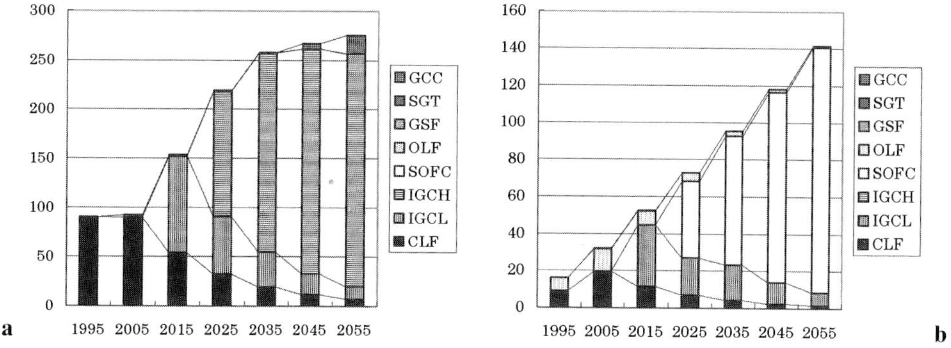

FIG. 13. Power generation technology profiles in China and India (MTOE). **a** China. **b** India. *GCC*, gas combined cycle; *SGT*, small gas turbines; *GSF*, gas fired power plant; *OLF*, oil fired power plant; *SOFC*, solid oxide fuel cells; *IGCH*, IGCC-high; *IGCL*, IGCC-low; *CIF*, coal fired plant

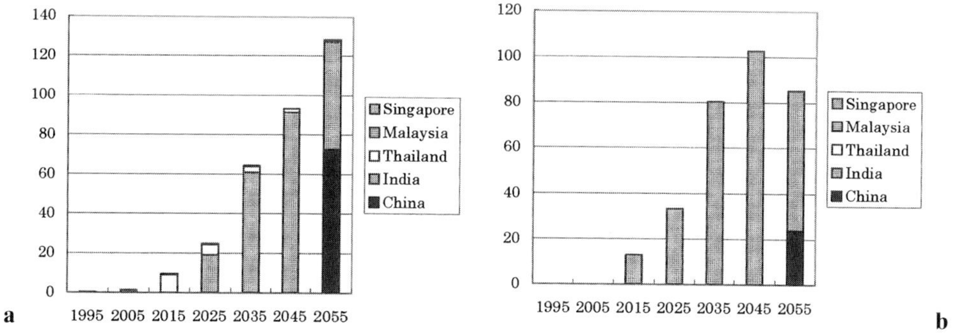

FIG. 14. Natural gas supply (million ton) from the Middle East region to Asian countries in Cases 1 and 2. **a** Case 1. **b** Case 2

TABLE 9. Marginal costs of sulfur emissions reduction in Case 2 in US$ per ton of sulfur

	1995	2005	2015	2025	2035	2045	2055
China avg.	0.0	150.0	268.0	855.5	1115.3	931.2	985.1
India avg.	0.0	619.1	657.5	876.5	1156.3	1486.9	1857.2
Thailand	0.0	700.4	1847.0	1646.8	1645.4	1450.8	1618.6
Malaysia	0.0	1950.3	1804.4	1488.5	1117.5	1770.9	2046.2
Singapore	0.0	1852.5	1674.3	1350.2	1609.5	1099.4	1943.3

TABLE 10. Summary of annual GDP growth rates by region in Case 3 (%)

	1995–2000	2000–2010	2010–2020	2020–2030	2030–2040	2040–2050
China total	7.59	6.37	6.20	5.19	3.90	3.81
India total	8.29	6.20	5.07	4.15	3.37	2.84
Thailand	8.92	5.98	4.12	2.85	1.89	1.26
Malaysia	9.32	6.43	4.49	3.06	1.99	1.30
Singapore	6.63	4.20	2.34	1.24	0.44	0.05

Table 9 shows the marginal prices of sulfur emissions reductions. These figures suggest how desulfurization options are adopted by different regions. When high-removal desulfurization equipment is needed for an oil refinery, the marginal price comes to $2000 per ton of sulfur. Because of the terminal effects of the dynamic optimization procedure, the marginal cost exceeds this in the case of Malaysia in 2055.

Case 3 represents an extreme scenario in order to evaluate the contribution of high-removal-rate desulfurization equipment. Table 10 shows the change in GDP growth rates in Asian countries. By comparing Table 10 with Table 8, one can see that the economic growth rates of China and India do not change much, while those of other countries, i.e., Thailand, Malaysia, and Singapore, are seriously affected. This has already been suggested by Table 8.

In this case, the natural-gas supply to China changes drastically. Figure 15 shows that the demand for natural gas in China declines rapidly. In 2025, the natural-gas demand in China in Case 3 is around two thirds of that in Case 2, and it decreases to less than one third in 2055. This counterintuitive observation can be explained by the "crowding-out" of the natural-gas supply from Russia. Figure 16 compares the natural-gas supply profiles from Russia in Cases 2 and 3. Although the total supply from Russia is almost identical in Cases 2 and 3, the profiles of the importers are completely different. Thailand, Malaysia, and Singapore need natural gas from Russia, and therefore the share for China declines.

Table 11 shows the marginal costs of sulfur emission reductions in Case 3. Needless to say, such high values are not realistic. However, this table does indicate how valuable efficient sulfur removal systems could be in Asian regions.

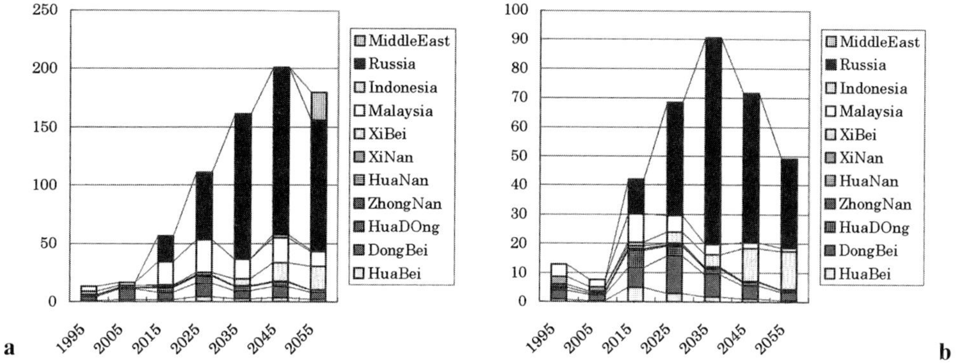

FIG. 15. Natural-gas demands by China in Cases 2 and 3 (MTOE). **a** Case 2. **b** Case 3

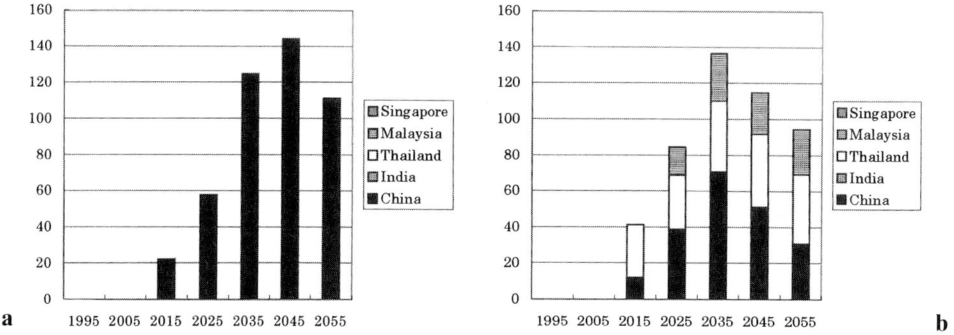

FIG. 16. Natural-gas supply from Russia to Asian countries in Cases 2 and 3 (million ton). **a** Case 2. **b** Case 3

TABLE 11. Marginal costs of sulfur emission reductions in Case 3 in US$ per ton of sulfur

	1995	2005	2015	2025	2035	2045	2055
China avg.	0.0	149.7	333.5	2105.0	11198.8	23709.2	46377.8
India avg.	0.0	614.4	657.7	650.8	5579.3	12922.3	46961.6
Thailand	0.0	699.9	32373.5	48703.0	103631.0	190878.0	348925.0
Malaysia	0.0	6218.5	33297.0	106302.0	164277.0	252875.0	444643.0
Singapore	0.0	34768.3	5116.2	258667.0	282399.0	215170.0	318009.0

Simulation Scenario 2. Ancillary Benefits of CDM

Since the clean development mechanism (CDM) proposed at COP-3 was accepted at COP-7, Annex I countries have been able to explore opportunities to reduce carbon emissions in non-Annex I countries at low marginal costs. On the other hand, the developing countries are mainly interested in the mitigation of domestic regional environmental pollution, and not so much in global warming issues, because the former problem is more urgent.

Since some carbon emission reduction options can reduce sulfur emissions at the same time, these would provide a "win–win" outcome, or "ancillary benefits," as pointed out in the IPCC Third Assessment Report (IPCC 2001). In order to see the effects of carbon emission reduction policies, we considered two cases.

- Case 4, Asian countries reduce carbon emissions by 10% of BAU (Case 1) levels without sulfur emission reductions.
- Case 5, Asian countries reduce carbon emissions by 10% of BAU (Case 1) levels and reduce sulfur emissions so that they do not exceed 110% of the 1995 level.

High-removal sulfur emission options are available in both cases. Here, we are interested in the change in marginal costs of carbon and sulfur emission reductions. Table 12 compares the marginal carbon emission reduction costs in Cases 4 and 5. It can be seen that the marginal costs in Case 5 are often lower than those in Case 4.

Table 13 shows the marginal costs of sulfur emission reductions in Case 5. Comparing Table 13 with Table 9, it is dear that a win–win situation exists in China and India. However, other regions which are starting to become industrialized show a negative outcome. If high sulfur removal rate options are taken away in Case 5, the marginal costs become high, as is shown in Table 14, where a win-win situation exists in all countries.

TABLE 12. Comparison of the marginal costs of carbon emission reductions in US$ per ton of carbon in Cases 4 and 5

	1995	2005	2015	2025	2035	2045	2055
Case 4							
China	0	4.35	62.86	92.03	179.44	323.11	726.14
India	0	0.00	47.50	61.52	57.55	79.76	98.18
Thailand	0	11.55	90.74	80.26	107.80	132.55	161.68
Malaysia	0	6.15	94.90	90.41	129.99	162.36	194.16
Singapore	0	0.00	472.25	340.06	504.86	549.98	656.83
Case 5							
China	0	3.47	61.32	90.67	173.30	315.19	685.28
India	0	0.00	41.69	56.22	59.26	77.51	91.86
Thailand	0	21.00	63.74	80.14	107.60	133.97	160.40
Malaysia	0	14.76	96.90	89.07	123.22	160.38	185.80
Singapore	0	36.90	249.33	279.06	493.81	522.09	700.80

TABLE 13. Marginal costs of sulfur emission reductions in Case 5 in US$ per ton of sulfur

	1995	2005	2015	2025	2035	2045	2055
China avg.	0.0	151.6	277.6	555.7	809.9	963.8	1269.4
India avg.	0.0	619.3	658.3	655.2	869.6	929.4	1355.4
Thailand	0.0	349.5	1800.7	1763.1	1752.4	1778.3	2053.6
Malaysia	0.0	655.6	1811.2	1774.9	1759.8	1783.7	2056.6
Singapore	0.0	1809.3	1673.9	1664.1	1734.3	1570.1	2865.2

TABLE 14. Marginal costs of sulfur emission reductions in Case 5 without high-sulfur-removal-rate equipment in US$ per ton of sulfur

	1995	2005	2015	2025	2035	2045	2055
China avg.	0.0	212.2	343.9	911.3	6164.7	22308.2	43666.4
India avg.	0.0	613.2	658.4	652.5	2415.9	11097.3	39525.2
Thailand	0.0	349.7	25973.6	42539.9	92484.8	175235.0	305916.0
Malaysia	0.0	732.0	31887.0	91108.1	170227.0	228479.0	397491.0
Singapore	0.0	41768.9	850.3	265832.0	324166.0	348591.0	296125.0

Conclusion

We have developed a model to assess energy and environmental technologies for Asian regions, taking into account the geographical distribution of resources and the variation in quality. Since the dynamics of the economies in this region are so active and uncertain, it may be too early to recommend middle-term energy strategies. There are many other issues which were not considered in our research. For instance, the technology options for energy-intensive industries are not dealt with explicitly. "Clean coal" technologies other than desulfurization, including coal liquefaction and gasification for integrated gas combined cycles (IGCC) and fuel cells (IGFC), must be taken into account for as long as coal continues to be used as a major energy source, and environmental control policies must be implemented. Greenhouse gas emission control policies should also be assessed, and must include local environmental pollution mitigation measures.

Nonetheless, our study has provided some significant findings on energy strategies.

1. Coal will still play a major role in the primary energy supply portfolio even if sulfur control is enforced. The cost of desulfurization facilities is much lower than the "fuel switch" to gas options, since the supply of natural gas will not meet energy demands and environmental measures. This is because of its geographical distribution, and the technical and financial constraints which work against the expansion of an infrastructure for gas use.
2. Clean coal technologies will significantly mitigate the costs of sulfur emission control policies under the increasing demand for energy.
3. Coal production within the Asian regions will not be enough to meet the demand, not because of the resource endowment, but for transportation reasons.

4. The Asian regions will have to depend on such non-Asian countries as Australia, Russia, and South Africa for their fossil fuel imports. This finding suggests the importance of establishing a marine fuel transportation infrastructure, as well as gas pipelines across continents.

5. The clean development mechanism (CDM) may provide win–win situations for both Annex I and non-Annex I countries, so that they will be able to reduce the marginal costs of both carbon emission and sulfur emission reductions. This was seen for both China and India. Even in the countries which are just beginning to become industrialized, this situation appears when efficient sulfur removal options are not available.

Acknowledgments. We express our thanks to Mr. Katano who initiated the model for India.

References

China (1999) China Energy Statistics, National Bureau of China

Imanaka K, Yamaji K (2001) Power system modeling with coal transportation for Shandong, China. 17th Energy System/Economy/Environment Conference, Japan Society of Energy and Resources

IPCC (Intergovernmental Panel on Climate Change) (2001) Climate change 2001 mitigation. Contribution of Working Group III to the 3rd Assessment Report of the IPCC. In: Metz B, Davidson O, Swart R, Pan J (eds) Cambridge University Press, Cambridge, Chap 9

Li D (2000) An econometric analysis of the economy, energy and environment of China in 2030. 16th Energy System/Economy/Environment Conference

Manne A, Richels RG (1992) Buying greenhouse insurance. MIT Press, Cambridge

MITI (1998), Energy Statistics. Ministry of International Trade and Industry

NISTEP (National Institute of Science and Technology Policy) (1991) Analysis of the structure of energy consumption and the dynamics of emissions of atmospheric species related to global environmental change. NISTEP Report No. 21, NISTEP

Rogner H-H (1997) An assessment of world hydrocarbon resources. Annu Rev Energy Environ 22:217–262

Sugiyama T (1997) SOX emissions of east Asia: a historical comparative analysis of SOX emissions of east Asian countries and its implication on the Chinese emission outlook. Socioeconomic Research Center Report No.Y97005, Central Research Institute of Electric Power Industry

Sugiyama T (1999a) China provincial energy and emission model (CPE model). Socioeconomic Research Center Report No.Y98009, Central Research Institute of Electric Power Industry

Sugiyama T (1999b) Atmospheric pollutant emission model for chinese states CRIEPI Report, Central Research Institute of Electric Power Industry

Suzugaki T, Kikuchi S, Fujii Y, Yamaji K (1996) Optimal allocation on energy transportation infrastructure in Asia–Eurasia region. 12th Energy System/Economy/Environment Conference, Japan Society of Energy and Resources

World Bank (1998) International statistics of economics. World Bank

Analysis of the Optimal Configuration of the Energy Transportation Infrastructure in Asia and Eurasia

Yasumasa Fujii and Taketo Hayashi

Summary. It has become important to decide what energy-related infrastructures, such as transcontinental natural gas pipelines and international electricity grids, should be constructed in Asia/Eurasia, and how the energy demands of this area should be satisfied securely, economically, and environmentally benignly over the next several decades. The purpose of this study was to investigate the possible future configuration of the energy- and CO_2-related infrastructure in Asia/Eurasia, which is a neighboring region to Japan.

Using a linear programming technique, we have been developing a large-scale energy-related infrastructure model which intertemporally minimizes the sum of the discounted total energy systems costs until the year 2050. The model explicitly involves intraregional transportation networks for fuels, electricity, and recovered CO_2 among about 90 nodes in Asia/Eurasia. The model illustrates concrete geographical distributions of the demand and supply of various primary energy types, CO_2 recovery and disposal, and the transportation flows of the fuels, electricity, and recovered CO_2 among the nodes. The nodes are connected by plausible land and/or ocean transportation routes. Coal freight trains, oil pipelines, natural gas pipelines, power transmission lines, and CO_2 pipelines are the specific measures considered for land transportation. We assume ocean transportation routes for coal, oil, and natural gas between each pair of coastal nodes in the model. Coal bulk carriers, oil tankers, and liquefied natural gas (LNG) tankers are the specific ocean transportation measures. The specific capacity of each transportation route is determined as the result of minimizing the total energy system cost.

Although a great deal of uncertainty remains, the preliminary results indicate that the development of gas production and transportation infrastructures appears to be a robust energy supply option for Asian countries, and that the economic validity of the development of region-wide electricity grids among Asian countries is not necessarily obvious. The results suggest that transporting coal and

---

Department of Electrical Engineering, University of Tokyo, 7-3-1 Hongo, Bunkyo-ku, Tokyo 113-8656, Japan

natural gas by rail or pipeline and generating electricity close to the energy-consuming cities is generally more economical than generating electricity at the mine-mouth or well-head and transmitting electricity by power transmission lines.

Key words. Energy transportation infrastructure, Asia, CO_2 recovery and disposal, Energy system model, Linear programming

Introduction

While the energy demands in China, Southeast Asia, and East Asia are projected to grow substantially over the coming decades, there has been much concern about the rapid increases in anthropogenic CO_2 emissions from fossil fuel burning, as well as increases in atmospheric CO_2 concentration, which is believed to influence the problems that cause global warming.

Coal is an abundant and widely distributed fossil fuel in Asia and Eurasia, and is expected to continue to be a major energy resource. Although the price of coal per unit of calorific value has been relatively low in this region, this growing demand for coal will not be met without the extensive development of transportation infrastructures such as railroads and bulk carriers. In the case of crude oil, this resource is not as plentiful as coal, and is unevenly distributed. Oil supplies for Asia and Eurasia continue to be increasingly dependent upon the Middle East, and such over-dependency of oil procurement on a single geopolitical region may potentially aggravate the energy securities of these countries. Natural gas is a clean and high-quality fuel. Its combustion generates less CO_2 than any other fossil fuel on a per-calorie basis. From the viewpoint of environmental protection, natural gas is the best substitute for oil and coal. However, enormous capital investment in a transportation infrastructure (e.g., liquefied natural gas tankers, and liquefaction and regasification plants, as well as extensive pipelines) in Asia and Eurasia will be required in order to increase the percentage of natural gas in the total primary energy supply for this region.

In such circumstances, the development and exploitation of energy resources in Asia and Eurasia, i.e., East Siberia and the Russian Far East, have attracted considerable attention. It has become increasingly important to decide how the primary energy requirements for this region should be securely and economically provided, as well as what energy infrastructures (e.g., transcontinental natural gas pipelines and long-distance power transmission lines) should be constructed, with particular attention to CO_2 emission abatement from fossil fuel use.

In response to these questions, the purpose of this study was to obtain insights into the optimal future configuration and operation of an energy infrastructure in Asia and Eurasia, and also the potential roles of CO_2 recovery and sequestration technologies. For this purpose, we used a linear-programming technique to develop a large-scale energy system model, which intertemporally minimizes the sum of the discounted total energy system costs until the year 2050 (Horikawa et al. 2000; Fujii et al. 2000; Fujii and Yamaji 1998). The model explicitly involves intraregional transportation networks of fuels and recovered CO_2 among 84 representative network nodes for Asia and Eurasia, as well as energy conversion facilities, including those for hydrogen production and methanol synthesis. The model tries to illustrate concrete geographical distributions of the demand and supply of various fuels by node, CO_2 recovery and sequestration by node, and the transportation flows of the fuels and recovered CO_2 among the nodes.

In the following sections, an outline of the current version of the energy model and a summary of its computational results are presented.

Energy Infrastructure Model for Asia and Eurasia

Geographical Coverage of the Energy Model

The geographical coverage of the model is the whole world. As shown in Fig. 1, the Asian region has 84 representative nodes of large cities and production sites, and the rest of the world is disaggregated into five regions. The model has 58 spatially distributed nodes, which represent energy-consuming areas in Afghanistan, Australia, Bangladesh, Bhutan, Brunei, Cambodia, China, India, Indonesia, Iran, Iraq, Japan, Kazakhstan, Kyrgyz, Laos, Malaysia, Mongolia, Myanmar, Nepal,

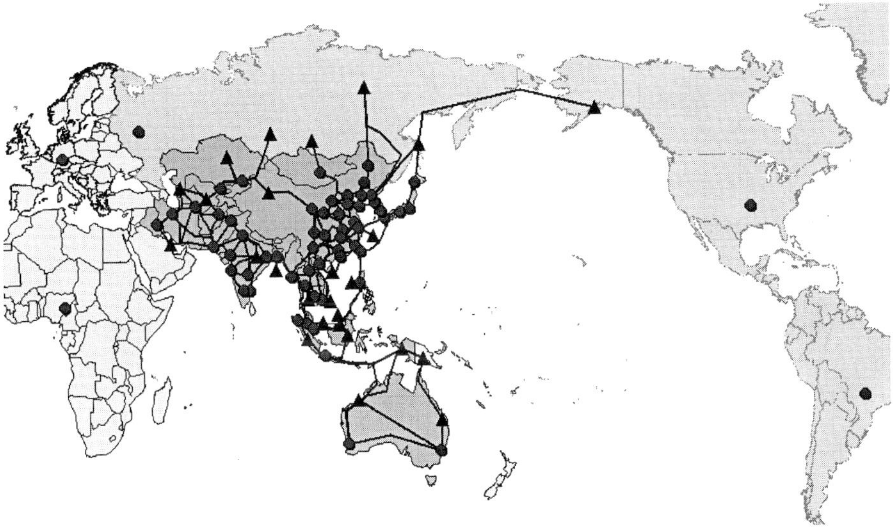

FIG. 1. Representative nodes and land transportation routes considered in the model. The Asian regional energy system is modeled as an energy network with 84 representative nodes, and the rest of the world is disaggregated into five regions. The nodes are connected by plausible land and/or ocean transportation routes, but the ocean transportation routes are not indicated in the figure. *Circles* indicate the city nodes which have both energy production and consumption, and *triangles* indicate energy production nodes which have no indigenous energy demands

North Korea, Pakistan, Philippines, Singapore, South Korea, Tadzhikistan, Thailand, Turkmenistan, Uzbekistan, and Vietnam. The model also assumes 26 energy production nodes, which include those in East Siberia and the Russian Far East.

As shown in Fig. 1, the model assumes an energy transportation infrastructure network of 84 nodes. The nodes are connected by plausible land and/or oceanic transportation routes. Coal freight trains, oil pipelines, natural gas pipelines, long-distance power transmission lines, and CO_2 pipelines are considered as the specific means of land transportation. We assume oceanic transportation routes for coal, oil, and natural gas between each pair of coastal nodes in the model, but these oceanic routes are not shown in Fig. 1. Coal bulk carriers, oil tankers, and LNG tankers are considered as the specific means of oceanic transportation. In the current version of the model, oceanic transportation of CO_2 and internode transportation of hydrogen and methanol are also taken into account. The specific capacity of each transportation route is determined as the result of the minimization of the total energy system costs.

Outline of the System Structure of the Energy Model

Figure 2 shows the assumed possible energy flow at each node in this model. Fossil fuel gasification, methane and methanol syntheses, hydrogen production,

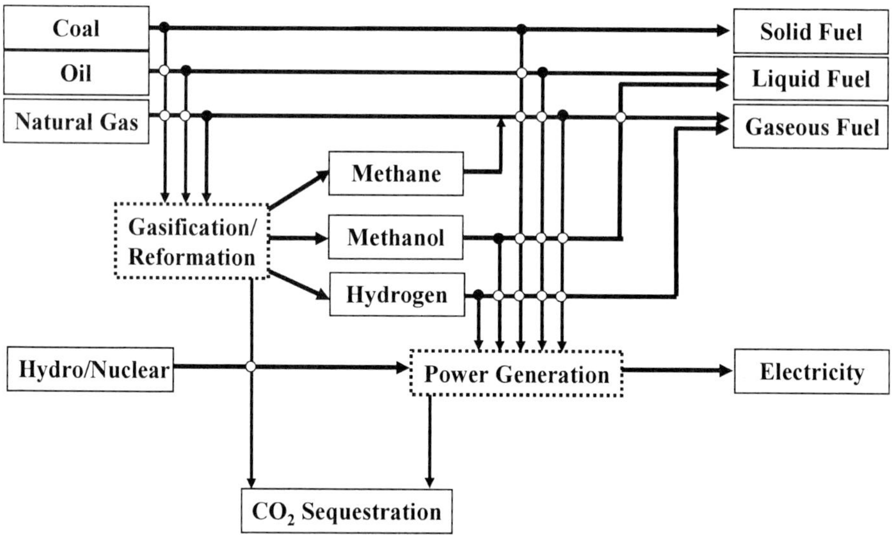

FIG. 2. Assumed possible energy flow at each node. The model deals with five primary energy resources, i.e., coal, oil, natural gas, hydroelectric power, and nuclear power. Various energy conversion processes, such as methanol synthesis and hydrogen production through gasification/reformation, are taken into account, in addition to power generation. Final consumption is divided into four sectors, i.e., solid fuel, liquid fuel, gaseous fuel, and electricity

and electric power generation are considered as the technological options for energy conversion. Each node has the possibility of having any one of the energy conversion facilities. An elaborate integration of these conversion plants, together with CO_2 recovery facilities, results in a large supply of low-carbon-intensive fuels with low additional CO_2 emissions from their conversion processes. Such an integrated energy system can be expected to contribute to remarkable reductions in CO_2 emissions from end-use sectors.

With respect to the electricity generation sector, the model explicitly takes into account the daily load duration curves expressed simply by four time-periods (instantaneous peak, peak, intermediate, and off-peak), as seen in Fig. 3, in order to determine how each type of power plant will be operated in accordance with the diurnal variation in electricity demands. This is because the capacity factors of electric power plants are presumed to have a large influence on their economic characteristics. In the model, the future contributions of nuclear and hydraulic power plants are exogenously determined prior to the cost minimization in this study.

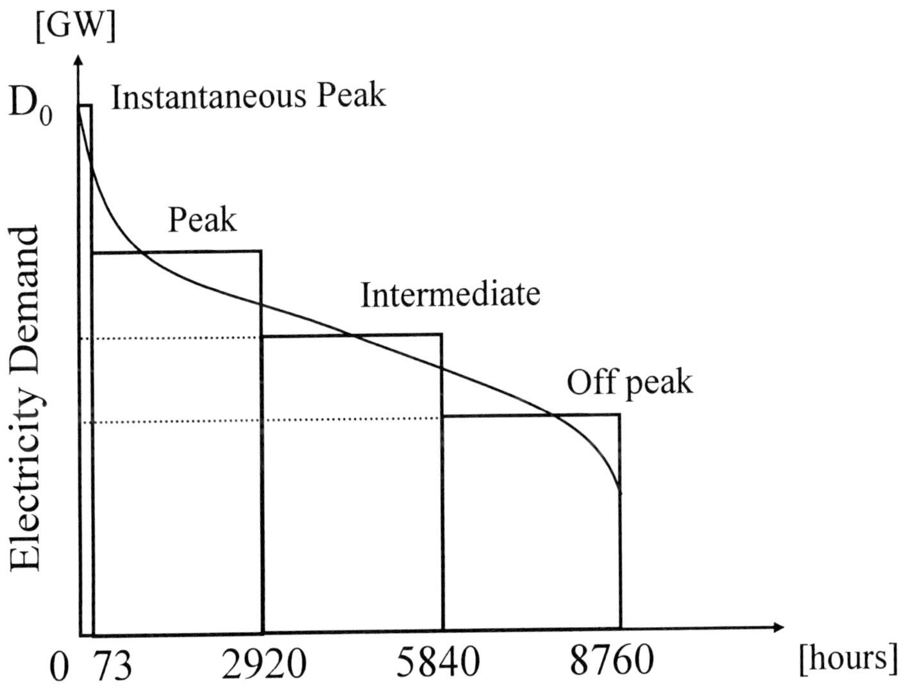

FIG. 3. Assumed load-duration curve. To determine how each type of power plant will be operated in accordance with diurnal variations in electricity demand, the model takes into account the daily load-duration curves for four time-periods, i.e., instantaneous peak, peak, intermediate, and off-peak. The *horizontal axis* is the annual load-duration time, and the maximum duration time is 8760 h

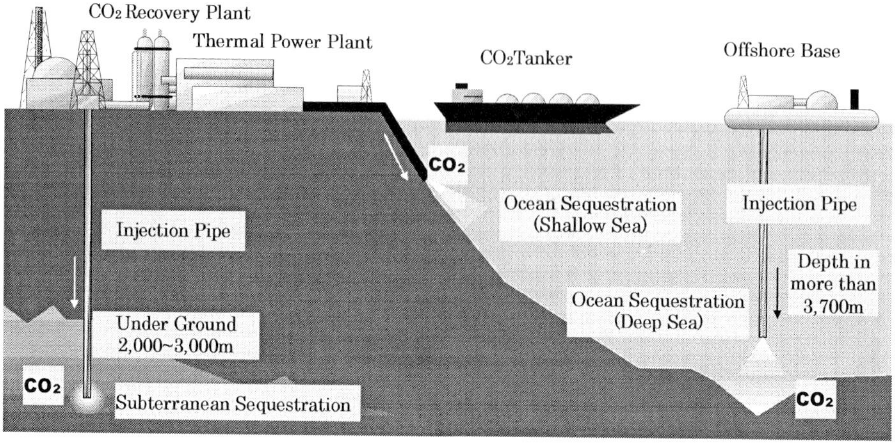

FIG. 4. CO_2 recovery and sequestration system. The model takes into account CO_2 recovery from various energy conversion processes, e.g., thermal power plants, and methods of both subterranean and ocean sequestration

One of the notable features of the model is that it can explicitly analyze the roles of processes of CO_2 recovery and sequestration in the energy system. Figure 4 illustrates the CO_2 recovery and sequestration processes. As specific measures for CO_2 recovery, the model takes into account both chemical absorption from the flue gas of thermal power plants, and physical absorption from the output gases of fossil fuel reforming processes. There are two major methods of CO_2 sequestration: ocean sequestration and subterranean sequestration. Subterranean sequestration is classified into three types: (1) injection of CO_2 into oil wells for enhanced oil recovery (EOR) operations; (2) storage of CO_2 in depleted natural gas wells; (3) sequestration of CO_2 in aquifers. This model takes account of the first two sequestration methods, and can assess their future potentials by node.

In the case of ocean sequestration, unquestionably the storage capacity of the ocean is sufficiently large, but it is very difficult to estimate specific costs for the secure deposition of CO_2 in the ocean. This is because many uncertainties exist, e.g., changes in the pH of seawater, clathrate formation on the seabed, and the resultant ecological impacts. Notwithstanding these serious uncertainties about ocean sequestration, we introduced it into the model to get an insight into the economic feasibility of ocean sequestration as one of the technological options.

It is assumed that the recovered CO_2 is not only disposed of, but is also recycled as a chemical feedstock for methanol synthesis. This option can build up a type of carbon cycle within the energy system, but the amount of CO_2 thus recycled is limited by the regional capacity for hydrogen provision.

Mathematical Formulation of the Model

This model is mathematically formulated as a multi-period intertemporal linear optimization problem with linear inequality and equality constraints. The constraints represent the supply and demand balances of each type of energy by node, energy, and CO_2 balances in energy conversion processes, and the state equations for several intertemporal dynamics such as the depletions of fossil fuel resources and subterranean CO_2 reservoir capacities, the capital vintage of various facilities in the energy system, and so forth. The objective function of the problem is defined as the sum of the discounted total energy system costs distributed over time, which include fuel production costs, and the fixed costs for plant, comprising capital and maintenance costs, energy transportation costs, CO_2 recovery and sequestration costs, and carbon taxes, all of which are leveled out for all plant. The supply-cost curves of fossil fuels by node are expressed as stepwise linear functions with respect to the amounts of their cumulative productions. More mathematical descriptions can be found in Horikawa et al. (2000), Fujii et al. (2000), and Fujii and Yamaji (1998).

The model seeks the optimal regional development paths for the energy-related infrastructure for the years from 2000 to 2050, at intervals of 10 years, using a linear-programming technique. Therefore, the model does not take into account any nonlinear effects such as economies of scale with respect to unit construction costs for various facilities, especially pipelines. Furthermore, for simplicity, all the variables in the model are treated as continuous real numbers, although some, such as those expressing the number of tankers, should be treated as discrete integer numbers in the real world.

Assumed Data

Reference Energy Demand Scenarios

The final consumption sector of the energy infrastructure model is disaggregated into the following four types of secondary energy carriers: (1) gaseous fuel, (2) liquid fuel, (3) solid fuel, and (4) electricity. In the case of electricity consumption, as mentioned above, the model explicitly takes into account the daily load duration curves, expressed simply by four time periods: instantaneous peak, peak, intermediate, and off-peak.

In the model, the future energy consumption is given exogenously as reference scenarios by type, node, and year. The reference scenarios of energy consumption by type of secondary energy and by subregion are illustrated in Fig. 5. These calculations were based upon the method proposed by Murota and Ito (1996), where per capita income is presumed to be the main parameter determining the future trends of per capita final energy consumption by type of secondary energy. According to this method, as per capita income increases, grid energies, namely gaseous fuel and electricity, are also assumed gradually to increase their share of total energy consumption. The energy supplies of relatively low-income countries are assumed to be dependent upon solid fuel. Table 1 gives the assumed annual

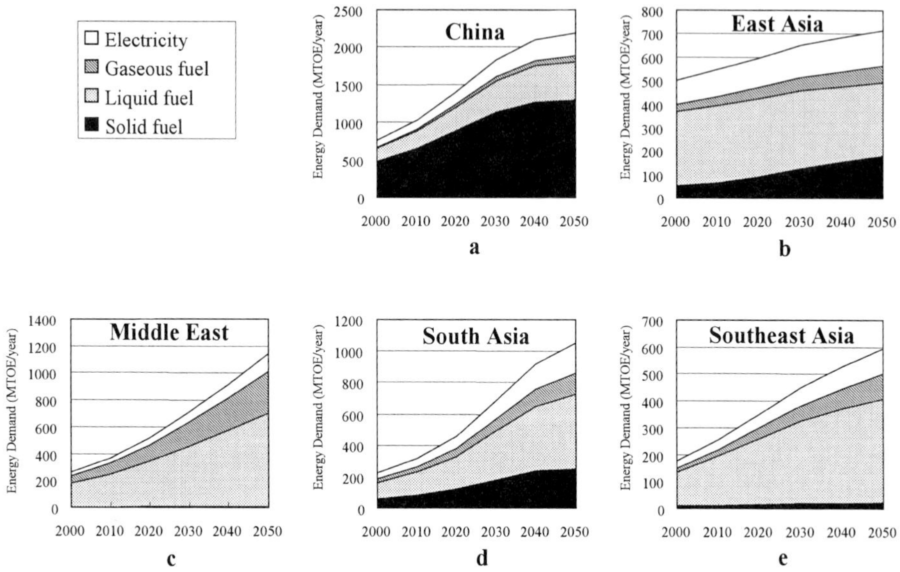

FIG. 5. Reference energy demand scenarios by subregion. The final consumption sector of the model is disaggregated into four types of secondary energy carrier: (1) gaseous fuel, (2) liquid fuel, (3) solid fuel, and (4) electricity. Huge demands for solid fuel are expected in China in the first half of this century. **a** Demand scenario for China. **b** Demand scenario for East Asia. **c** Demand scenario for the Middle East. **d** Demand scenario for South Asia. **e** Demand scenario for Southeast Asia

TABLE 1. Assumed annual average GDP growth rates from 2000 to 2050

Country	%/year	Country	%/year
Japan	0.5	Thailand	3.5
South Korea	2.2	Myanmar	4.1
North Korea	5.2	Malaysia	3.3
China	4.7–6.0[a]	Singapore	1.2
Hong Kong	1.1	Indonesia	5.2
Taiwan	2.4	Philippines	5.3
Mongolia	4.4	Australia	1.6
Vietnam	5.5	Bangladesh	4.5
Laos	5.9	India	5.9
Cambodia	6.1		

[a] In the model, China consists of 19 different city nodes, and the growth rates of their economies are assumed to range from 4.7 to 6.0%/year

average GDP growth rates from 2000 to 2050 for the major countries in the model. The amounts of energy consumption by node were mainly calculated by using data on the geographical distribution of population.

Resource Amounts and Production Costs

The amounts of coal, crude oil, and natural gas resources in this study were mainly derived from JPDA (1991), Masters et al. (1994), World Petroleum Assessment (2000), and the Coal Energy Center (1999). Geographical distributions of fossil fuel resources were partly estimated on the basis of proven reserves. Figure 6 shows our assumptions of the amounts of fossil fuel resources by subregion. The fuel resources are divided into four grades depending on their unit production costs, with a grade 1 resource being the most economic. With the wide variety of economic and geological conditions, the production costs of fossil fuels can be estimated only with considerable uncertainty. Since each node must have various economic grades of resources, the production-cost curve of each node is expressed as a step-wise linear function with respect to their amounts of cumulative production. Figure 7 shows the assumed production costs of fossil fuels as functions of their respective resource amounts.

Infrastructure Construction Costs

In this section, we show the unit construction costs of each element of the infrastructure. In the model, these cost parameters were mainly derived from the estimates made in various publications (JPDA 1991; CEDIGAZ 1995; Yamaji and Fujii 1995; Suzugaki et al. 1999; IEA 1996, 1997, 1998; IPCC 1995; MITI 1998; Komiyama et al. 1989). Table 2 gives the assumed characteristics of the energy

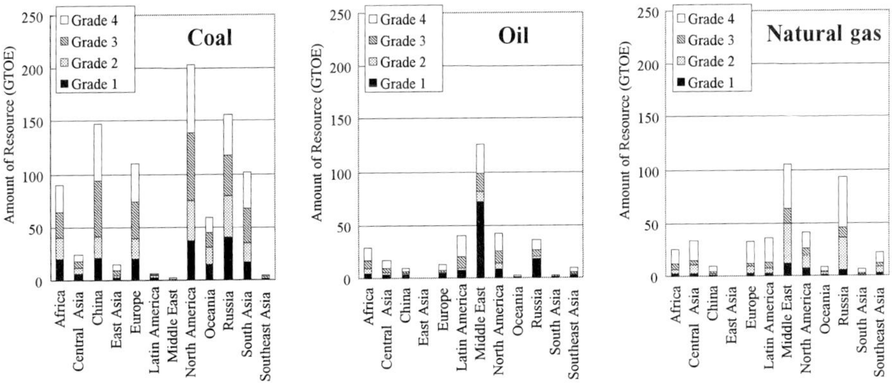

a b c

FIG. 6. Amounts of fossil fuel resources by subregion. The fuel resources are divided into grades 1–4 according to their unit production costs. Grade 1 is the most economic. **a** Coal resources. **b** Oil resources. **c** Natural gas resources

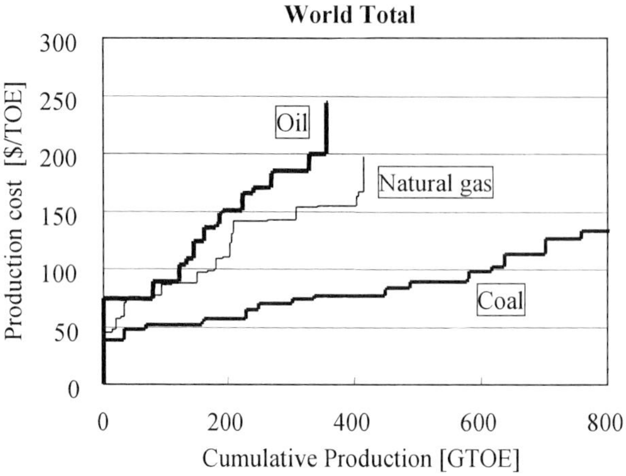

FIG. 7. Assumed production costs of coal, oil, and natural gas. Each node has the four grades of fuel resources. The production costs of each grade of the resources may differ in the model according to the node. The aggregated cost curve for the production of each fuel is expressed as a step-wise linear function with respect to their amount of cumulative production

TABLE 2. Assumed characteristics of energy transportation facilities

	Unit	Unit cost	Transportation loss (%)
Coal freight train	US$/(TOE/year)	45.4 *L*	2.2 *L*
Oil pipeline	US$/(TOE/year)	6.2 *L*	2.3 *L*
Natural gas pipeline	US$/(TOE/year)	22 *L*	2.3 *L*
Methanol pipeline	US$/(TOE/year)	12.6 *L*	2.3 *L*
Hydrogen pipeline	US$/(TOE/year)	35.2 *L*	3.5 *L*
Bulk coal carrier	US$/(TOE/year)	0.94 *L* + 0.78	—
Oil tanker	US$/(TOE/year)	0.61 *L* + 0.5	—
Liquefied natural gas tanker	US$/(TOE/year)	6.07 *L* + 97.6	0.2 *L*
Methanol tanker	US$/(TOE/year)	1.23 *L* + 1.02	—
Liquefied hydrogen tanker	US$/(TOE/year)	13.6 *L* + 213.8	0.2 *L*
DC power transmission line	US$/kW	89.7 *L* + 23.8	10 *L*

DC, direct current; *L*, transportation distance; unit 1000 km

transportation facilities. Their fixed and variable costs are expressed as linear functions with respect to transportation distance *L* [unit 1000 km]. The costs of the ships were deduced from the unit construction cost of ships and their typical operational pattern for loading and unloading. Pipeline costs were assumed to vary between 100% and 200% of the values in Table 2 depending on their specific geographical route conditions. For example, an offshore pipeline is assumed to cost twice as much as a land pipeline. For simplicity, the energy losses associated with ocean transportation of coal and oil are neglected. The lifetime of the pipelines is assumed to be 60 years, and that of the ships and the liquefaction and re-gasification plants to be 30 years.

Table 3 gives the assumed characteristics of thermal power plants. Slight improvements in the thermal conversion efficiency of the power plants are assumed over the simulation period, and the variations are listed in Table 3.

TABLE 3. Assumed characteristics of fossil-fuel-fired power plants

	Construction cost (US$/kW)	Efficiency (%)
Coal-fired	1300	27–39
Oil-fired	750	29–43
Natural-gas-fired	850	34–49
Methanol-fired	1650	33–49
Hydrogen-fueled	1850	32–47
IGCC	2000	31–46

IGCC, integrated coal gasification combined cycle

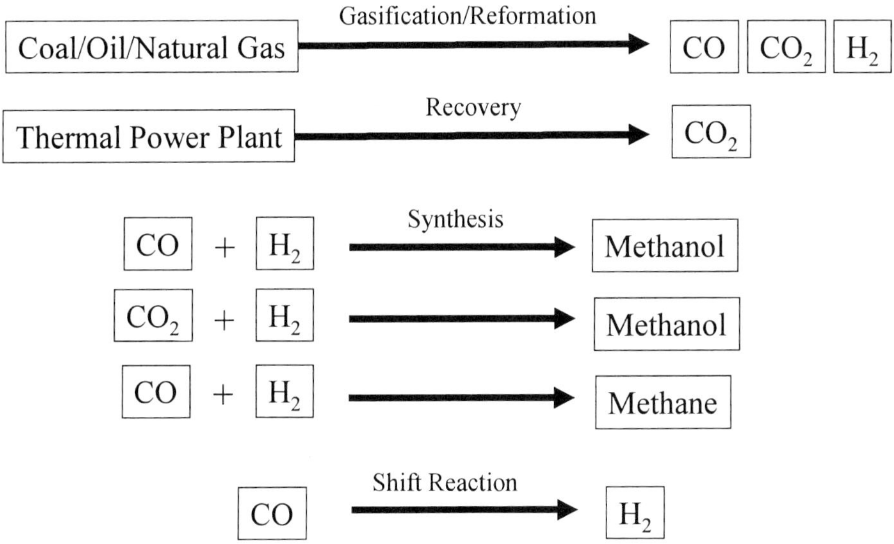

FIG. 8. Energy conversion processes assumed in the model. In addition to power generation, the model takes account of various types of energy conversion processes which involve CO, H_2, and CO_2. The specific conversion processes are coal gasification, oil gasification, natural gas reformation, CO_2 recovery, and two different processes of methanol and methane synthesis

In order to reduce the size of the infrastructure model, the future contributions of nuclear and hydraulic power plants are given exogenously as scenarios by node and year. In this study, we assume that the contribution of these non-fossil power plants will be rather modest, and expect that their annual generation levels will be kept constant at their present levels over the simulation period. The lifetime of electric power generation plants is assumed to be 30 years.

In addition to thermal electric power plant, various types of energy conversion plant, such as a coal gasification plant and a methanol synthesis plant, were introduced as technological options in the model. Figure 8 shows the energy conver-

TABLE 4. CO_2 recovery transportation and sequestration costs

	Unit	Fixed cost	Variable cost
CO_2 pipeline	US$/(tC/year)	54 L + 11.3	1.3 L
CO_2 recovery	US$/(tC/year)	30	0
CO_2 liquefaction	US$/(tC/year)	32	0
Depleted gas-well injection	US$/tC	0	45
Ocean sequestration	US$/tC	0	100–124

L, transportation distance; unit 1000 km

sion (fuel reforming and synthesis) flows. We assumed two types of synthesis method for methanol. The lifetime of these energy conversion plants is assumed to be 30 years.

CO_2 Recovery, Sequestration, and Recycling

As previously mentioned, one of the notable features of the model is that it can explicitly analyze the roles of processes of CO_2 recovery and sequestration in the energy system. The related costs are shown in Table 4. We assumed that the oil production cost by EOR is 70 US$ per ton of oil equivalent (TOE), and that CO_2 can be disposed of at a rate of 0.6 tons of carbon per ton of oil equivalent (t-C/TOE) of recovered oil. With respect to depleted gas-well injection, the storage capacities are derived from the simple assumption that one CO_2 molecule can replace one CH_4 molecule. For ocean sequestration, the recovered CO_2 is assumed to be liquefied, and then to be transported to offshore sequestration sites by tanker. We assumed three ocean sequestration costs depending on the transportation distance from a port with a shipment of liquefied CO_2 to the nearest offshore sequestration site.

Simulation Results from the Model

This section presents some of the simulation results from this energy infrastructure model. The study assumed three policy cases, a business-as-usual case (BAU), an investment constraint case (INC), and a carbon tax case (CTX). The BAU case does not anticipate either investment constraints or CO_2 abatement policies over the specified time-horizon. In the INC case, we assume that the amount of investment in the energy transportation infrastructure is limited to under 0.5%–1.0% of GDP for specific countries in Asia. In the CTX case, we simply assume the introduction of certain rates of carbon tax ranging from 100 US$/t-C to 500 US$/t-C with a central value of 300 US$/t-C.

BAU Case

Figure 9 shows the calculated flows of coal in 2030 and 2050. Extensive railroad transportation of coal can be seen among the nodes of Chinese cities. Owing to

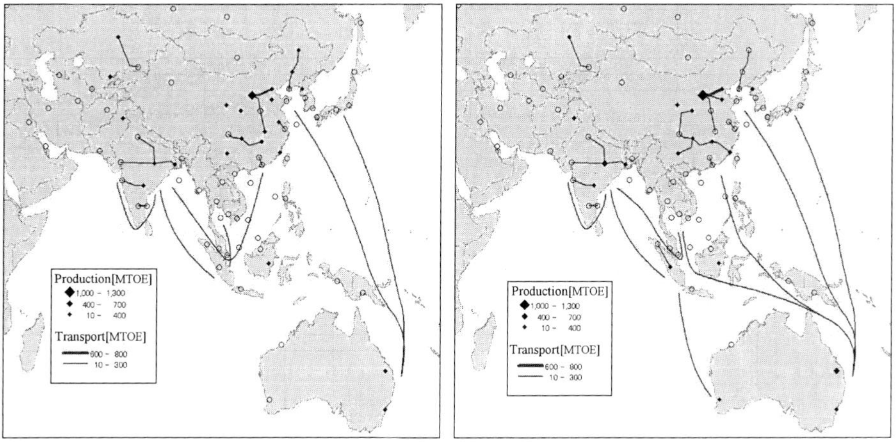

a **b**

FIG. 9. Coal production and transportation in the business as usual (BAU) case. *Black rhombuses* on the maps indicate the locations of coal production sites, and the size of the rhombus corresponds to the annual coal production level at that node. *Solid lines* on the maps indicate coal transportation routes, and the width of the line corresponds to the annual coal transportation level on that route. Transportation directions are not indicated explicitly. **a** Coal production and transportation pattern in the year 2030. **b** Coal production and transportation pattern in the year 2050

the relatively high transportation costs of coal freight trains, the coal requirements for most of the regions, such as the industrialized coastal region of China, North Korea, South Korea, and Japan, are provided by Australia by means of bulk coal carriers. Because we assume relatively low production costs for coal in India, the coal demands in Southeast Asia are partly satisfied by coal from India.

Figure 10 shows the calculated flows of oil in 2030 and 2050. The oil requirements in 2020 and 2050 are provided almost exclusively by the Middle East. It is interesting to note that some oil production can be found around the nodes of the Tarim Basin (Urumchi) and the Caspian Sea (Fig. 10).

Figure 11 shows the calculated flows of natural gas in 2030 and 2050. As the figure shows, supply sources of natural gas are geopolitically diversified over the region. For instance, China is provided with natural gas not only by Southeast Asia, but also by the regions of the former Soviet Union. In the case of Japan, LNG tankers are predicted to be the main means of transportation, but there is also a long-distance gas pipeline between the Russian Far East of Sakhalin and the northern part of Japan.

Figure 12 shows the time-profiles of total regional fossil fuel production by type, and Fig. 13 shows the time-profiles of total regional electricity generation by type. It should be noted that the contributions of hydroelectric and nuclear power stations are determined exogenously under the scenario represented in this figure. From these figures, coal is expected to be the dominant primary energy

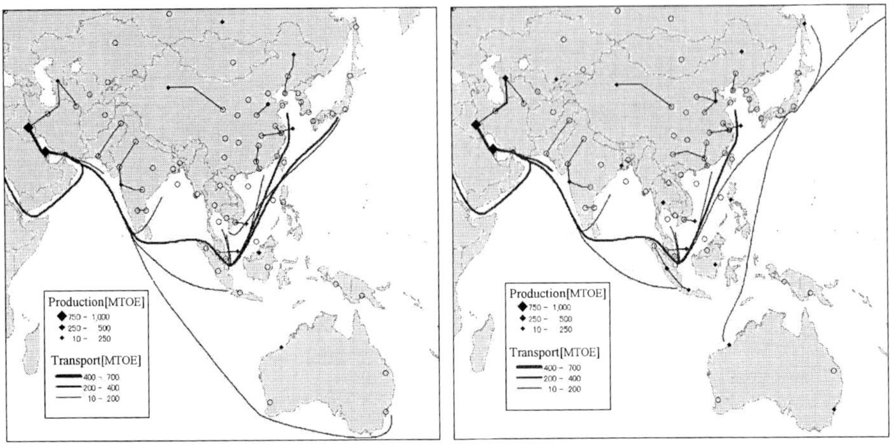

a b

FIG. 10. Oil production and transportation in the BAU case. *Black rhombuses* on the maps indicate the locations of oil production sites, and the size of the rhombus corresponds to the annual oil production level at that node. *Solid lines* on the maps indicate oil transportation routes, and the width of the line corresponds to the annual oil transportation level on that routes. Transportation directions are not indicated explicitly. **a** Oil production and transportation pattern in the year 2030. **b** Oil production and transportation pattern in the year 2050

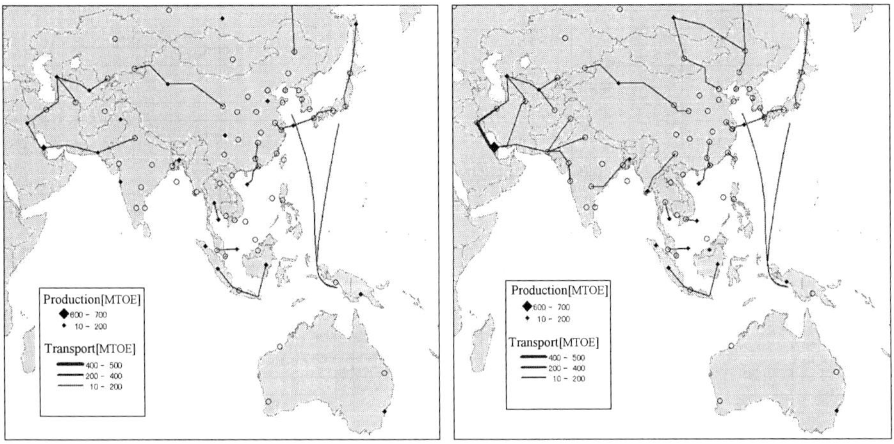

a b

FIG. 11. Natural gas production and transportation in the BAU case. *Black rhombuses* on the maps indicate the locations of the natural gas production sites, and the size of the rhombus corresponds to the annual production level of natural gas at that node. *Solid lines* on the maps indicate the transportation routes of the fuel, and the width of the line corresponds to the annual transportation level of the fuel on that route. Transportation directions are not indicated explicitly. **a** Natural gas production and transportation pattern in the year 2030. **b** Natural gas production and transportation pattern in the year 2050

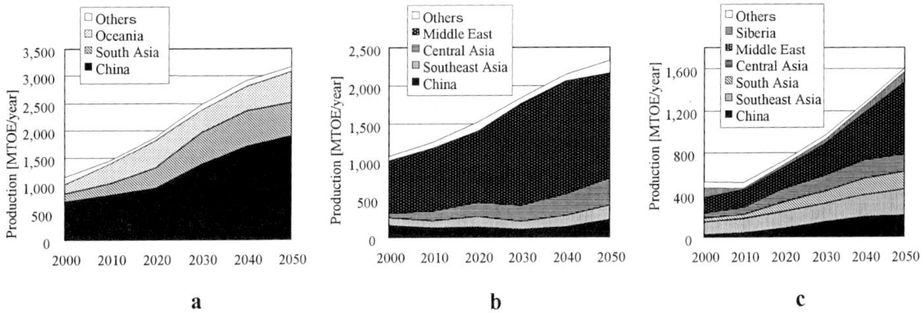

FIG. 12. Time-profiles of regional total fossil fuel production by type and subregion in the BAU case. The subregions are China, Southeast Asia, South Asia, Central Asia, Middle East, Oceania, Siberia, and others. **a** Coal production. **b** Oil production. **c** Natural gas production

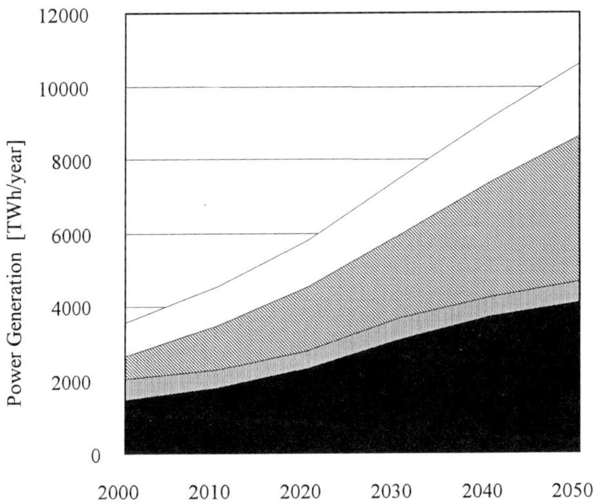

FIG. 13. Electricity generation by energy source in the BAU case. Coal (*black*) is expected to be the dominant primary energy source for electric power generation in Asia and Eurasia, and natural gas (*hatched shading*) is the second most important. The contributions of hydroelectric and nuclear power stations (*white*) are exogenously determined under the scenario represented in this figure. The share of oil-fired power generation (*dotted shading*) is expected to be rather small over the next five decades

source in the BAU case, especially for electric power generation in Asia and Eurasia. However, even in the BAU case, natural gas is estimated to become the second most important fuel for power generation.

Figure 14 shows the computational results for electric power generation mixes by node. This figure suggests that transporting coal and natural gas by rail or pipeline and generating electricity close to the energy-consuming cities is gener-

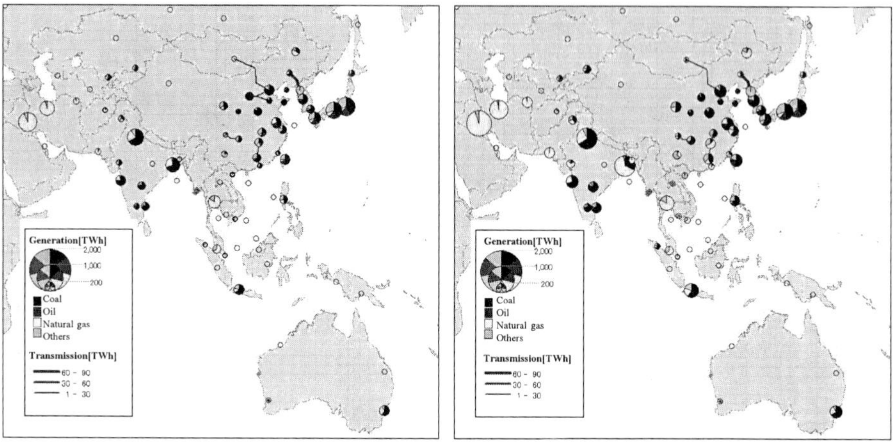

a b

FIG. 14. Electricity generation and transmission in the BAU case. The pie charts on the maps indicate electricity generation mixes by energy source, and the size of the chart corresponds to the annual electricity generation level at that node. *Solid lines* on the maps indicate power transmission routes, and the width of the line corresponds to the annual transmission level of that routes. Transmission directions are not indicated explicitly. **a** Electricity generation and transmissions pattern in the year 2030. **b** Electricity generation and transmissions pattern in the year 2050

ally more economical than generating electricity at a mine-mouth or well-head and transmitting electricity by power transmission lines.

Figure 15 shows the time-profiles of CO_2 emissions in Asia and the share of cumulative emissions from 2000 to 2050 by subregion. The figure indicates a three-fold increase in CO_2 emissions in Asia as a whole within the next five decades. The sub-region of China is estimated to be by far the largest source, followed by the subregion of South Asia including India. The amounts of CO_2 emissions in Japan are expected to be almost constant over the simulation period. These figures do emphasize the importance of initiating concrete actions for the abatement of CO_2 emissions in this region.

INC Case

One of the results of the BAU case is that the amount of energy transported expands rapidly in accordance with the rapid growth in energy demands by the Asian countries. However, this outcome requires a large amount of investment money, and it is not easy for developing countries to procure enough money to construct the extensive infrastructure needed. Therefore, in the INC case, we sought the optimal configuration for an energy transportation infrastructure in Asia, while considering specific upper limits for investment money in some major developing countries, i.e., China, India, and Pakistan. The estimated investment

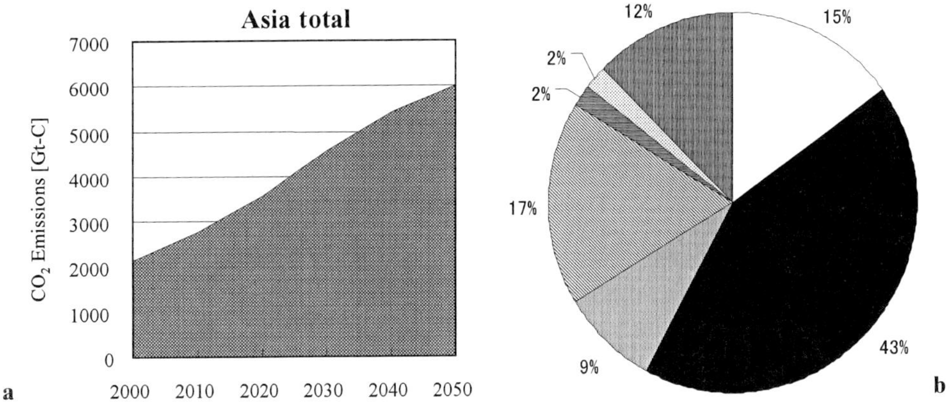

FIG. 15. CO_2 emissions in Asia in the BAU case. **a** Time-profiles of CO_2 emissions in Asia from 2000 to 2050. **b** Share of cumulative CO_2 emissions from 2000 to 2050 by subregion: East Asia (*white*), China (*black*), Southeast Asia (*dense dotted shading*), South Asia (*oblique hatching*), Central Asia (*horizontal hatching*), Oceania (*sparse dotted shading*), and the Middle East (*vertical hatching*)

FIG. 16. Investment in energy transportation infrastructure in China, and in India and Pakistan in the BAU case. The values are percentage ratios to their respective GDP. *Black squares*, values for China; *white circle*, aggregated values for India and Pakistan

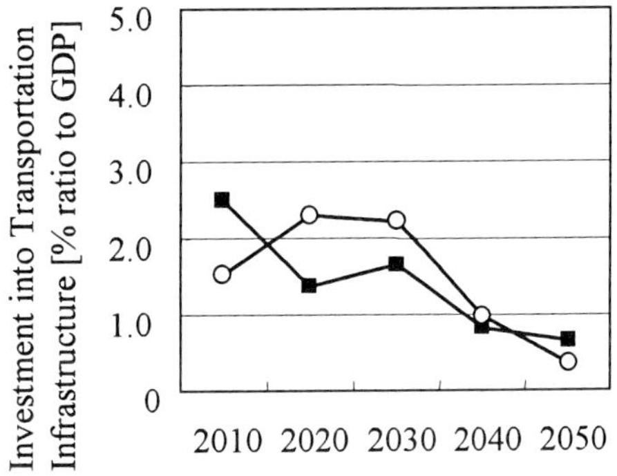

in the energy transportation infrastructure in these countries is shown in Fig. 16 for the BAU case. The infrastructures which were taken into account are land transportation infrastructures (oil pipelines, natural gas pipelines, power transmission lines, coal freight trains, etc.), ocean transportation infrastructures (oil tankers, LNG tankers, bulk coal carries, etc.), and other related equipment (liquefaction plants, AC–DC converters, etc.).

The amounts of infrastructure investment by type in China and in India and Pakistan together are shown in Fig. 17. As the constraints on investment become stricter in China and in India and Pakistan, the investment in coal freight trains decreases significantly, and the rate of investment in oil pipelines increases in consequence.

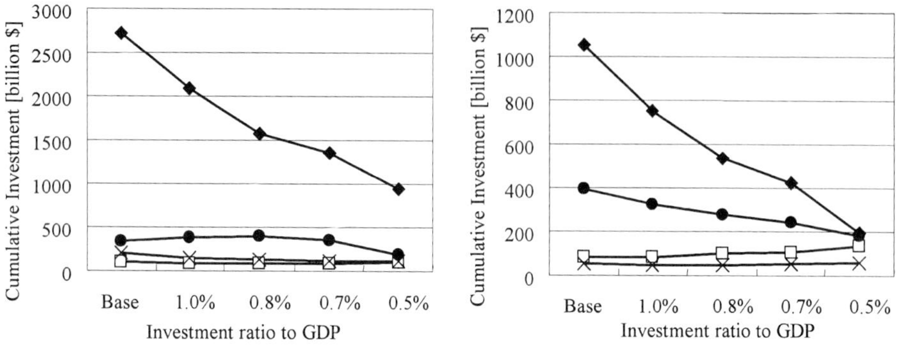

FIG. 17. Cumulative investment from 2000 to 2050 under different investment constraints. The vertical axes are the cumulative investment by transportation type: coal freight train (*black diamonds*), oil pipeline (*white squares*), natural gas pipelines (*black circles*), and other (*crosses*). The horizontal axes are the upper limits of investment expressed in terms of percentage ratios to GDP. **a** Cumulative investment by transportation type from 2000 to 2050 in China. **b** Cumulative investment by transportation type from 2000 to 2050 in India and Pakistan

The investment in natural gas pipelines in China decreases when 0.5% of GDP becomes the upper limit of annual amount of the investment in the energy transportation infrastructure. Although investment in natural gas pipelines has gradually been decreasing, investment in oil pipelines is increasing in line with the increase in oil consumption.

To illustrate the influence of investment constraints on the optimal energy flow in Asia, we show the simulation results for the case where an upper limit on the amount of annual investment of 0.7% of GDP was arbitrarily assumed. The state of coal production and transportation in this scenario in 2050 is shown in Fig. 18. When compared with the BAU case, the ocean transportation of coal to China varies significantly. Domestic coal production in China and India does not catch up owing to the lack of a transportation infrastructure (i.e., railroad systems), and it is anticipated that the energy systems of these countries will be dependent upon overseas coal.

The state of oil production and transportation is shown in Fig. 19. The demand for oil in the electric power sectors is estimated to increase in both China and India. The simulation results indicate that although India will expand its oil imports mainly from the Middle East, China will do so from Alaska as well as the Middle East.

The state of natural gas production and transportation is shown in Fig. 20. Compared with the BAU case, the development of Chinese natural gas pipelines is greatly delayed in the INC case.

FIG. 18. Coal production and transportation in 2050 in the INC case. *Black rhombuses* on the map indicate the locations of coal production sites, and the size of the rhombus corresponds to the annual coal production level at that node. *Solid lines* on the map indicate coal transportation routes, and the width of the line corresponds to the annual coal transportation level on that route. Transportation directions are not explicitly indicated

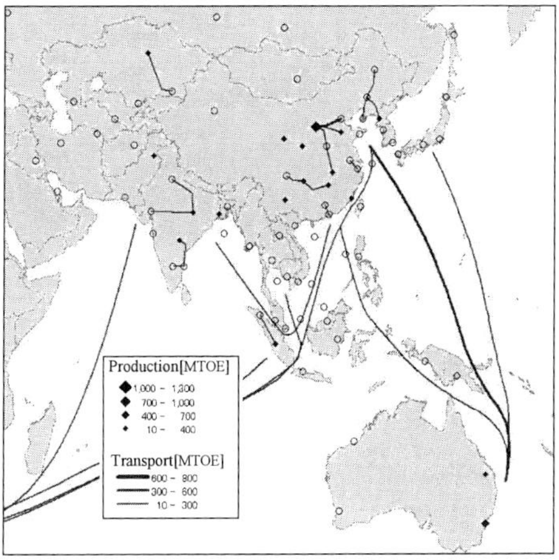

FIG. 19. Oil production and transportation in 2050 in the INC case. *Black rhombuses* on the map indicate the locations of oil production sites, and the size of the rhombus corresponds to the annual oil production level at that nodes. *Solid lines* on the map indicate oil transportation routes, and the width of the line corresponds to the annual oil transportation level on that route. Transportation directions are not explicitly indicated

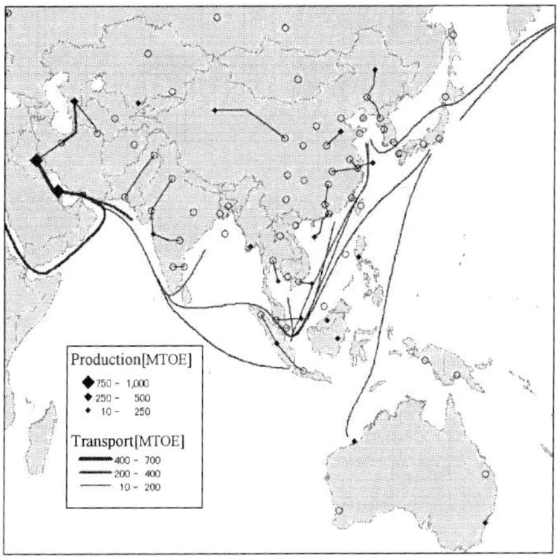

FIG. 20. Natural gas production and transportation in 2050 in the INC case. *Black rhombuses* on the map indicate the locations of the production sites of natural gas, and the size of the rhombus corresponds to the annual production level of natural gas at that node. *Solid lines* on the map indicate the transportation routes of the fuel, and the width of the line corresponds to the annual transportation level of the fuel on that route. Transportation directions are not explicitly indicated

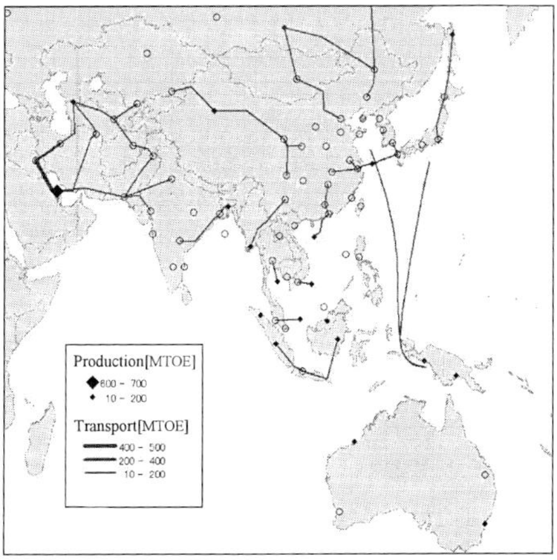

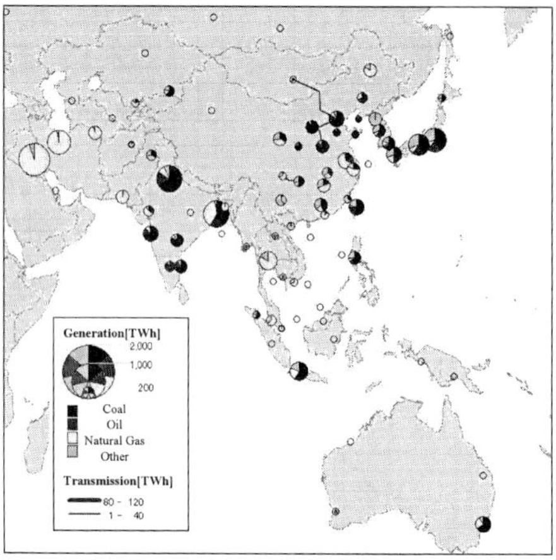

FIG. 21. Electricity generation and transmission in 2050 in the INC case. Pie charts on the map indicate electricity generation mixes by energy source. The size of the chart corresponds to the annual electricity generation level at that node. *Solid lines* on the map indicate power transmission routes, and the width of the line corresponds to the annual transmission level on that route. Transmission directions are not explicitly indicated

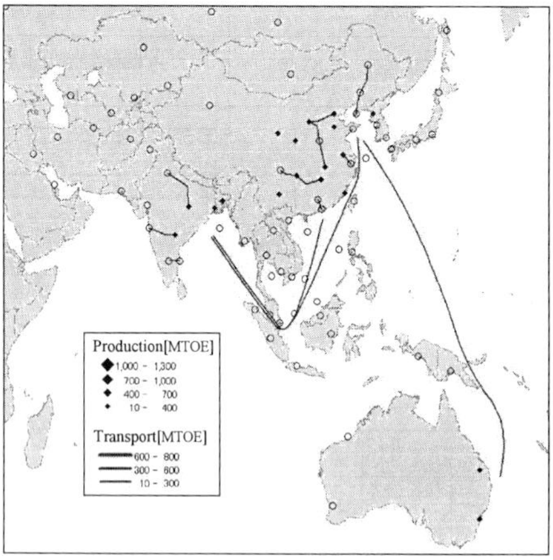

FIG. 22. Coal production and transportation in 2050 in the CTX case. *Black rhombuses* on the map indicate the locations of coal production sites, and the size of the rhombus corresponds to the annual coal production level at that node. *Solid lines* on the map indicate coal transportation routes, and the width of the line corresponds to the annual coal transportation level on that route. Transportation directions are not explicitly indicated

The computational results of electricity generation and transmission in 2050 is shown in Fig. 21. If the investment constraints in China were to become very strict, electricity generation by coal-fired power plants would not grow after the year 2020. Electricity generation by oil-fired power plants increases noticeably compared with the BAU case, and that by gas-fired power plants in China increases a little.

In the case of India and Pakistan, electricity generation by coal-fired power plants does not increase at all, and decreases considerably compared with the BAU case.

CTX Case

The computational results for energy production and transportation in 2050 are shown in Figs. 22–25. As compared with the results in the BAU case, coal

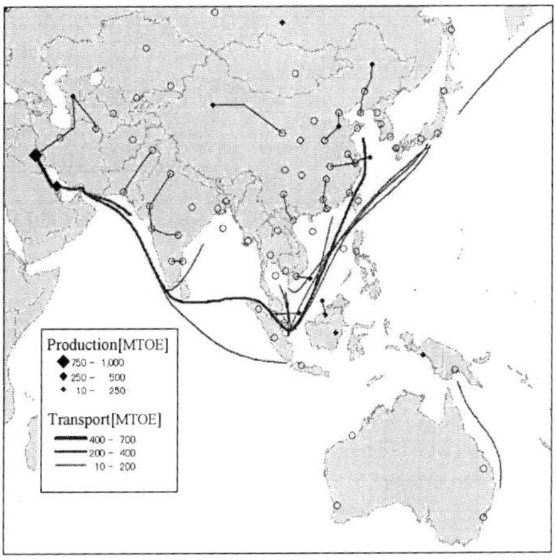

FIG. 23. Oil production and transportation in 2050 in the CTX case. *Black rhombuses* on the map indicate the locations of oil production sites, and the size of the rhombus corresponds to the annual oil production level at that node. *Solid lines* on the map indicate oil transportation routes, and the width of the line corresponds to the annual oil transportation level on that route. Transportation directions are not explicitly indicated

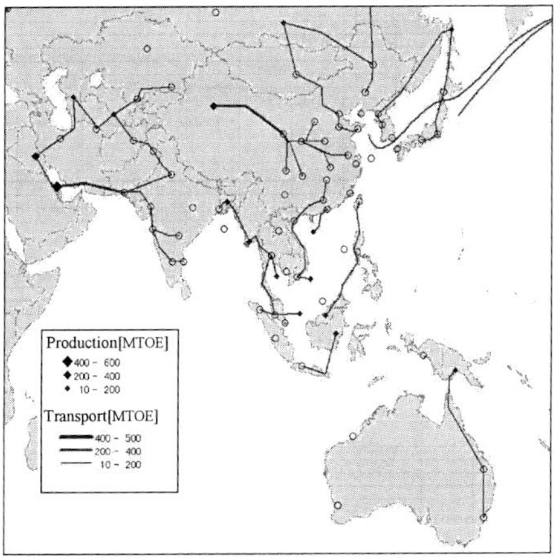

FIG. 24. Natural gas production and transportation in 2050 in the CTX case. *Black rhombuses* on the map indicate the locations of the production sites of natural gas, and the size of the rhombus corresponds to the annual production level of natural gas at that node. *Solid lines* on the map indicate the transportation routes of the fuel, and the width of the line corresponds to the annual transportation level of the fuel on that route. Transportation directions are not explicitly indicated

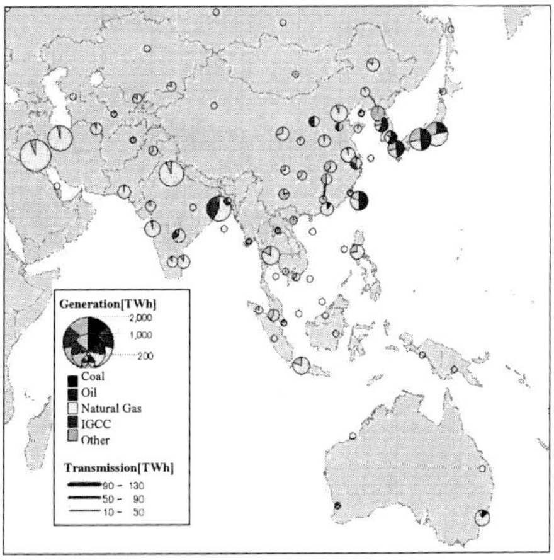

FIG. 25. Electricity generation and transmission in 2050 in the CTX case. Pie charts on the map indicate electricity generation mixes by energy source. The size of the chart corresponds to the annual electricity generation level at that respective node. *Solid lines* on the map indicate power transmission routes, and the width of the line corresponds to the annual transmission level on that route. Transmission directions are not explicitly indicated

production is substantially reduced in the CTX case, and coal is replaced by natural gas. The major reduction in coal consumption occurs in the electricity generation sectors in China and India. The carbon tax has practically no effect on the topology of the oil supply infrastructure. Natural gas will form a larger share of the total fossil energy supplies, and its contribution to the electricity generation sector will become enormous over the simulation period. A more extensive pipeline network for natural gas transportation is expected in the western part of India, and the eastern part of China is linked to gas wells in the western territories, as well as those in East Siberia, with long-distance natural gas pipelines of much larger capacities. The model takes account of the fact that hydrogen will serve as a substitute for natural gas in the gaseous fuel demands, but estimates that the amount of hydrogen consumed in Asia will remain insignificant, even in the CTX case. It should be noted that the emerging of hydrogen for fuel-cell vehicles was not taken into account in this model.

Figure 26 shows the reduction in CO_2 emissions with different carbon tax rates from 100 to 500 $/t-C during the period between 2000 and 2050. The higher the rate of carbon tax, the lower the amount of CO_2 emitted into the atmosphere. The relation between the rates of carbon tax and the rates of CO_2 emissions reduction is also shown in Fig. 26. The decrease in the gross CO_2 emissions is achieved by changes in the electricity generation mix. Figure 27 shows the total abatement costs for CO_2 emissions reductions. The abatement costs are defined as the difference between the total energy system costs in the CTX case and those in the BAU case. The abatement cost rises exponentially as the amount of CO_2 emissions reduction increases.

Figure 28 shows the calculated flows of recovered CO_2 between 2000 and 2050 with a carbon tax of 300 US$/t-C. The simulation results indicate that ocean sequestration of CO_2 is expected to play an important role in areas such as Japan and South Korea where there is very little capacity for CO_2 subterranean sequestration. On the other hand, in China and India, CO_2 pipeline transportation and

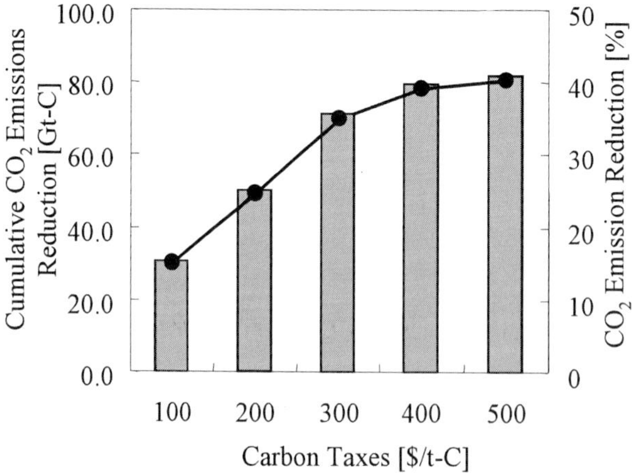

FIG. 26. Cumulative CO_2 emissions reduction with different carbon taxes. The bar graph indicates the cumulative amount of CO_2 emissions reductions, and the line graph indicates the emission reduction rates compared with the emission levels in the BAU case. The scales of the line graph and the bar graph are indicated along the right and left vertical axes, respectively

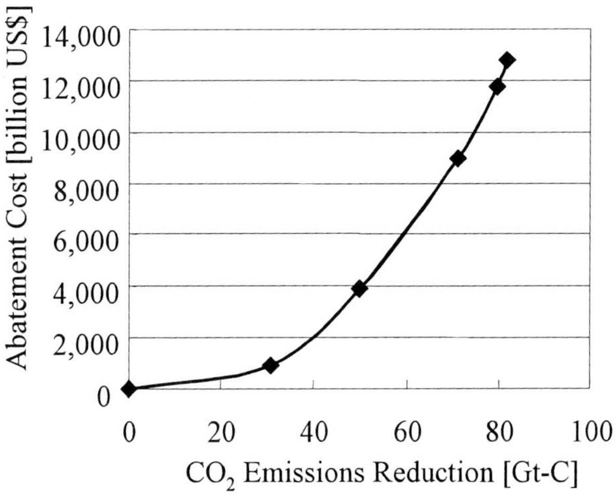

FIG. 27. Cumulative CO_2 emissions reduction from 2000 to 2050 and the total abatement cost. The abatement costs are defined as the difference between the total energy system costs in the CTX case and those in the BAU case

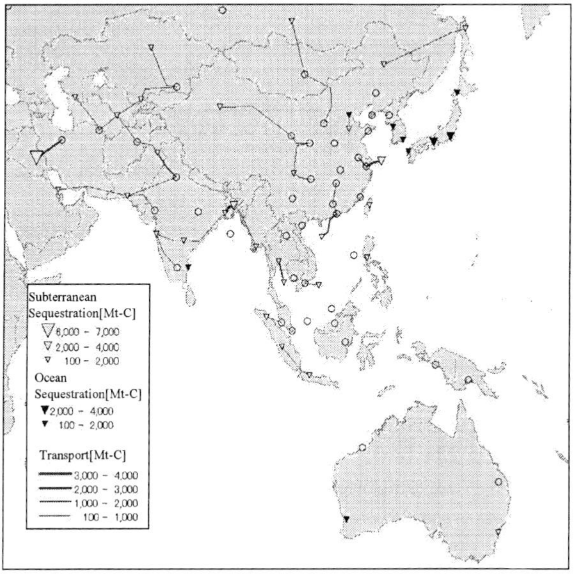

FIG. 28. Cumulative amounts of CO_2 sequestration and transportation with a carbon tax of 300 US$/t-C. *White triangles* on the map indicate the locations of subterranean CO_2 sequestration sites, and the *black triangles* at the coastal nodes on the map indicate the seaports for CO_2 shipment for ocean sequestration. The size of the triangle corresponds to the cumulative amount of CO_2 sequestration at that node. *Solid lines* on the map indicate CO_2 transportation routes, and the width of the line corresponds to the cumulative amount of CO_2 transportation on that route. Transportation directions are not explicitly indicated

subterranean sequestration are expected. Note that the positions of the dark triangles at the coastal nodes in this figure indicate the ports for CO_2 shipments, but not sequestration sites. The ocean sequestration sites are not explicitly shown in Fig. 28.

Concluding Remarks

The purpose of this study was to obtain insights into the possible future configuration and operation of energy- and CO_2-related infrastructures in Asia and Eurasia, where energy demands are increasing rapidly. This chapter has presented an outline of the energy system model built in the study, and shown some of the results obtained. Bearing in mind the considerable uncertainties of various assumptions made in the model, the results of the simulation are tentatively summarized below.

1. In the BAU case, coal will be the dominant primary energy source, especially for power generation, in most Asian countries, and natural gas will become the second most important primary energy source. Most of the oil requirements of the Asian and Eurasian regions will continue to be provided almost exclusively by the Middle East.
2. The results indicate that transporting fuels by rail or pipeline and generating electricity close to the energy-consuming cities is generally more economical than generating electricity at the mine-mouth or well-head and transmitting electricity by power transmission lines.
3. The development of gas production and transportation infrastructures appears to be a robust energy supply option for Asian countries. An increased reliance on natural gas would provide Asian countries with more geographically diversified energy supply structures, thus improving the security of their energy procurement.
4. The necessity of developing region-wide electricity grids among Asian countries is not obvious. However, in some cases, we find a few intercity routes for power transmission lines in the optimal solution of the model.
5. Investment constraints on the energy transportation infrastructure in some Asian countries may decrease the use of their domestic coal, and raise their level of dependence on oil and natural gas, as well as on coal imported from overseas countries.
6. In the CTX case, the model estimated that an extensive network of natural gas pipelines would be developed in China and East Asia. Neither investment constraints nor carbon taxes seemed to have a significant influence on the optimal configuration of region-wide electricity grids.
7. It seems unlikely that the electric power systems in Japan will be linked to those of neighboring countries, mainly because there are few economic advantages in doing so. When it comes to regional energy grids, Japan may have to give priority to international natural gas pipelines.

This study is continuing, and the following research topics are to be incorporated into future studies:

— further improvements in the accuracy of data on fossil fuel resources and production costs;
— sensitivity analyses of future energy demand scenarios;
— an extension of the energy system model described in this paper to include various nonfossil fuels;
— consideration of some nonlinear effects of the infrastructure, such as economies of scale.

References

CEDIGAZ (1995) Planned gas pipeline around the world. CEDIGAZ, Paris

Coal Energy Center (1999) World outlook for coal mines (in Japanese). Coal Energy Center, Tokyo

Fujii Y, Yamaji K (1998) Assessment of technological options in the global energy system for limiting the atmospheric CO_2 concentration. Environ Econ Policy Stud 1: 113–139

Fujii Y, Horikawa Y, Yamaji K (2000) Assessment of the future potentials of CO_2 sequestration technologies through the use of an energy infrastructure model for Asia/Eurasia. 5th International Conference on Greenhouse Gas Control Technologies. August 13–16, 2000, Cairns Convention Center, Australia, p 935–940

Horikawa Y, Nonaka S, Satomi N, Fujii Y, Yamaji K (2000) Optimal planning of energy transportation infrastructure in Asia (in Japanese). Proceedings of the 16th Energy System/Economics/Environment Conference of the Japan Society of Energy and Resources, Tokyo, p 241–246

IEA (1996) Asia gas study. International Energy Agency, Paris

IEA (1997) Energy technologies for the 21st century. International Energy Agency, Paris

IEA (1998) World energy outlook. International Energy Agency, Paris

IPCC (1995) Energy supply mitigation options. In: Climate Change 1994, Intergovernmental Panel on Climate Change, IPCC Second Assessment Report (SAR), Working Group II. Cambridge University Press, Cambridge, chap 19

JPDA (1991) Studies on resources of oil and natural gas. JPDA working group for resource assessment (in Japanese). Japan Petroleum Development Association, Tokyo

Komiyama H, Kamikojima K, Kataoka T, Tsuchiya Y, Nozaki T, Matsumura Y, Matsuhashi R (1989) Evaluation of IES (Integrated Energy Systems) from the aspect of chemical engineering (in Japanese). In: Proceedings of the 6th Energy System/Economics Conference of the Japan Society of Energy and Resources, p 43–46, Tokyo

Masters CD, Attanasi ED, Root DH (1994) World petroleum assessment and analysis. In: Proceedings of the 14th World Petroleum Congress, Stavanger, Norway. Wiley

MITI (1998) Energy '98 (in Japanese). Ministry of International Trade and Industry of Japan, Tokyo

Murota Y, Ito K (1996) Global warming and developing countries. Energy Policy 24:1061–1077

Suzugaki T, Kikuchi S, Fujii Y, Yamaji K (1999) Optimal configuration of energy transportation infrastructure in Asia/Eurasia (in Japanese). In: Proceedings of the 15th Energy System/Economics/Environment Conference of the Japan Society of Energy and Resources, Tokyo, p 349–354

World Petroleum Assessment (2000) Description and results. US Geological Survey, USGS Digital Data Series DDS-60, Multi Disc Set Version 1.0

Yamaji K, Fujii Y (1995) Global energy strategies (in Japanese). Denryoku-shinpou-sha, Tokyo

Local and Global Environmental Concerns Related to India's Energy Requirements

Ritu Mathur* and Tetsuo Tezuka†

Summary. This chapter highlights some of the major issues in India's energy use, and seeks to identify areas that must be addressed in order to move toward a sustainable energy and environmental scenario. First, we sketch the likely energy trajectory for India for the next 40 years using a bottom-up energy optimization model. The impacts of the probable energy mix in the future are examined in terms of their global environmental effects and the pressure which is likely to be exerted on the economy from increased energy imports. Second, power generation is identified as being one of the foremost areas of concern. A modeling approach is used to address various specific technical and policy issues related to the supply of coal to power plants, since coal is expected to continue to play a dominant role, at least in the next few decades, based on current expectations regarding the penetration of alternative fuels and technologies.

Here, we advocate a multidimensional energy policy as the panacea for India's energy problems in the next few decades. We also recommend specific options for optimizing the coal supply and utilization in the power sector. It is argued that significant benefits may accrue to the economy merely by reallocating existing coal-supply links to power plants. Moreover, the existing beneficiation technology for indigenous thermal coal was found to be unsuitable. This is crucial in the light of a recent environmental directive requiring the use of coal which produces less than 34% ash in many power plants. In India, this might result in an increase in coal imports for power generation. Adopting a differential freight structure for transporting washed coal is proposed as a policy option that could enhance the attractiveness of indigenous coal beneficiation, and thus protect the domestic coal market. Technologically, a switch to an integrated gasification combined cycle (IGCC) and supercritical coal-based technologies was found to be more attractive than the existing coal beneficiation technology.

---

*Tata Energy Research Institute, Darbari Seth Block, Habitat Place, Lodi Road, New Delhi 110003, India
†Graduate School of Energy Science, Kyoto University, Yoshidahonmachi, Sakyo-ku, Kyoto 606-8501, Japan

We concluded that it would be valuable to push technologies such as IGCC by using mechanisms such as the clean development mechanism (CDM) in a cooperative effort with developed countries to move toward long-term sustainability. At the same time, the need to develop coal beneficiation technologies which are suitable for Indian coals is stressed so that existing power plants are ensured of a supply of coal of the requisite quality.

Keywords. Indian energy sector, Coal utilization, Technological co-operation

Introduction

In the next few decades, it is estimated that the energy demands of developing countries will exceed those of the developed world, leading to a formidable challenge to provide "adequate" energy resources to fuel these demands. On the other hand, another sword of Damocles hanging precariously over our heads is the environmental threat from local and global pollutants resulting from the extraction, conversion, and use of energy. In this context, it is pertinent not only to estimate future levels of energy requirements in large and rapidly developing economies such as India, but also to identify and successfully address the major energy–environment concerns associated with the future energy mix.

Although numerous estimates of energy demand and supply have been made for India, there are few projections that include both traditional and commercial energy sources. Moreover, the estimates have been made over different time frames and with varying assumptions of demands and fuel supplies. The Hydrocarbon Vision projects the requirements for hydrocarbons in the Indian economy until 2025. The Ministry of Coal (1997) provided the coal requirements and production targets until 2011–2012, while the MNES (Ministry of Non-Conventional Energy Sources) (1999) provided an estimated future potential of various renewable energy forms. The Five-Year Plans made by the Government of India give targets for enhanced production capacity for each of the major energy resources based on estimated increases in demand. However, more often than not these targets have failed to materialize, especially in the case of the power sector. We postulate that this is because of over-ambitious targets and a lack of integrated and ad hoc energy–environment–economy policies. The Indian energy sector is faced with numerous constraints and conflicting objectives, and the probable energy mix over the next few decades is not easy to visualize in the absence of an integrated modeling framework. Apart from the usual characteristics of a developing country such as a large population, extensive use of non-commercial traditional fuels, and the lack of properly documented information on energy consumption, pricing, and supply, the Indian energy economy is beset with numerous paradoxes that make the task of estimating future energy needs rather formidable. Prominent among these is the coexistence of modernization and backwardness, the geographical vastness and the consequent diversity in climatic conditions as well as in energy use patterns, the coexistence of an organized and an unorganized sector, and the co-existence of abject poverty and great affluence. Therefore, a bottom-up energy sector model was used first to visualize the requirements and mix of energy across various end-uses, bearing in mind the fuel and technological constraints which will probably be imposed in the future, and using expert judgment in conjunction with documented information as available. Since the framework of the energy sector model was also used to examine future environmental impacts, a deliberately cautious approach was adopted toward the extent of the penetration of efficient fuels and technologies. This was in order to avoid erring towards a scenario which fails to give a correct forecast of energy and environmental threats.

The energy sector model indicated that energy requirements are estimated to increase to 6.4 times their present level, while CO_2 emissions would increase 8.8 times from 1996–1997 to 2036–2037 under a business-as-usual (BAU) scenario. This indicates that the Indian economy is nowhere near being able to stabilize its energy requirements and emission levels, let alone move toward a reduction of environmental emissions from the current levels.

Although the Indian Government has not signed the Kyoto Protocol, the country is conscious of the need to protect the environment, at both local and global levels, and to decrease energy use through modernization. Various policies at local and national levels have already been initiated in a bid to reduce environmental damage due to developmental activity and modernization. Moreover, planners and researchers are already involved in identifying, developing, and adopting cleaner technologies and options. Collaborative research with the more advanced countries, aimed at identifying options and measures that lead to sustainable development, can further help the economy to progress along the development curve while incorporating the "learning benefits" of the developed world.

This chapter is based on results of the TERI–JST Collaborative Program that was conducted under the auspices of the Core Research for Evolutional Science and Technology (CREST) Program of the Japan Science and Technology Corporation (JST).

Overview of India's Energy Sector Issues

India is a huge developing country with over one billion people and a land area of 3288000 km^2 (WDR 1999–2000). Since independence, there have been considerable changes in the structure of its economy and in the pattern of energy use. The liberalization of the Indian economy since 1991 has further resulted in enhancing the process of industrial development. Moreover, the country has experienced rapid urbanization, with the urban population increasing from 23% in 1980 to 28% in 1998. The increase in per capita commercial energy consumption from 352 kgoe (kilograms of oil equivalent) in 1980 to 476 kgoe in 1996 (WDR 1999–2000) reinforces the positive correlation between economic development and an improvement in the average standard of living. With the expand-

TABLE 1. End-use demands considered in the model

Sector	End uses
Agriculture	Land preparation, irrigation
Commerce	Lighting, cooking, space conditioning
Industry	Caustic soda, soda ash, aluminum, iron and steel, cement, cotton textiles, paper, fertilizer, sugar, other industries
Residential	Lighting, cooking, heating, cooling, water-heating, electrical appliances
Transport	Passenger and freight movements

ing population of the country, it is intimidating to visualize the colossal energy requirements in the future as modernization continues to propel the per capita requirements of the country toward the world average of 1684 kgoe in 1996 (WDR 1999–2000).

India was estimated to emit about 997 MT (million tonnes) of CO_2 in 1996 (WDR 1999–2000), while the per capita emissions were 1.1 tonnes. Even with constant per capita emissions over the next 40 years, CO_2 emissions would increase to about 1500 MT. In fact, recent figures for per capita emissions indicate a continuously increasing trend from 0.5 tonnes in 1980 to 1.1 tonnes in 1996. It is therefore imperative to focus attention on developing sustainable levels and patterns of energy use in the future in order to protect the local as well as the global environment.

Methodological Framework

Description of the Energy Sector Model

At the outset, this study used a bottom-up model developed for the Indian energy sector to examine the likely fuel-use pattern for the country over a 40-year modeling period extending from 1996–1997 to 2036–2037. Energy demands were disaggregated across various end-uses among the five major energy consuming sectors, i.e., agriculture, commerce, industry, residential, and transport, as shown in Table 1.

Energy demands were estimated and provided exogenously for each of these end-uses based on assumptions regarding the main driving forces, i.e., population and GDP. Corresponding supply options for fuels (domestic extraction, indigenous production, and technological processes, and end-use equipment) were provided to fulfil each of the demands. The IEA–MARKAL framework (Fig. 1) was then used to choose and allocate the various competing fuels and technologies based on least-cost criteria for the energy system as a whole.

Energy Sector Model: Assumptions

The energy sector model for India was set up with 1996–1997 as the base year, and extended over eight time-periods of 5 years each (coinciding with the 5-year

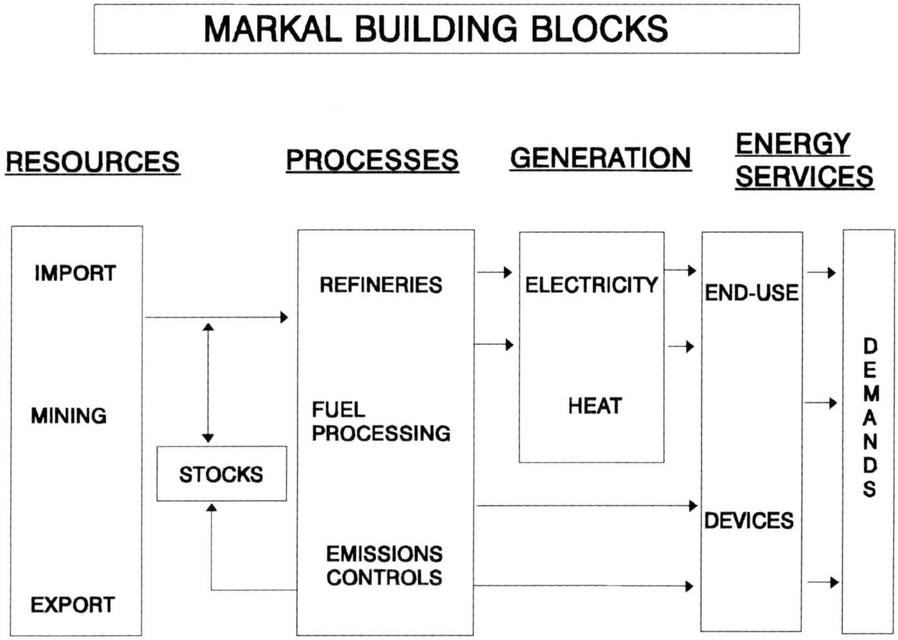

FIG. 1. Basic structure of the MARKAL framework

plans of the Government of India) until 2036–2037. End-use energy demands were forecast over the modeling period on the basis of the projected population and GDP. Thus, these were the drivers of the energy sector model for India.

Population projections until 2050 were used, as estimated by the Population Foundation of India. Further, the total population in the base year was divided into urban and rural using the urbanization index as in the Census Report (1991). The urbanization index was assumed to increase linearly from 26.9% in 1996–1997 to 45% by 2036–2037. The average size of urban and rural households was 4.5 and 4.9, respectively, in 1994–1995 (Planning Commission Report). It was assumed that these would decrease linearly to 4 and 4.5, respectively, by 2036–2037. Urban households were further divided into three income categories, i.e., low, medium, and high, in order to account for varying behavior and energy-use patterns across different standards of living. These categories correspond to the 1994–1995 income levels in the range Rs 75 000–150 000 per annum, Rs 150 000–300 000 per annum, and above Rs 300 000 per annum. Urban households were distributed by income category for 1994–1995 based on the actual proportions indicated by the NCAER survey results of 1996. Over the 40-year modeling period, it was assumed that living standards would improve, and the proportion in the high-income category would increase from 30% in 1994–1995 to 40% in 2036–2037, while the proportion in the low-income group would decrease from 37% in 1994–1995 to 30% by 2036–2037. Although the model assumed an overall GDP growth rate of 6%, the GDP was disaggregated into agriculture, industry, and services based on the relationships between sectoral and overall GDP using previous data. Consequently, the agriculture, industry and services sectors exhibit an annual average growth rate of 2.8%, 6.7%, and 6.5%, respectively. The economic growth rate of 6% is slightly lower than that recommended by the Planning Commission for the next decade (7%–8% growth rate was forecast for the last few years, but this was not achieved), but this is considered to be a more realistic and achievable figure. A combination of end-use demand estimation methods, process models, and econometric techniques were adopted to forecast end-use demands in each of the sectors.

On the supply side, there was provision in the model to use various fuels and technologies to meet different demands. Constraints on the availability of fuels (commercial and traditional) and technologies (existing and futuristic) were imposed based on current trends in resource availability, production, imports, and exports, as well as current policies toward the development of the infrastructure and production capacities. These constraints were largely based on documented information regarding the production of fuels and plans for capacity enhancement, supplemented by informed judgement provided by experts in the field.

It was assumed that the production of domestic coking coal would increase from 40.5 MT[1] in 1996–1997 to 60 MT in 2036–2037, while the production of noncoking coal would increase from 245.5 MT in 1996–1997 to 800 MT in 2036–2037. Although India has huge coal resources, its production is inadequate in terms of the quality of coal required. Consequently, about 10 MT of coking coal is imported every year for use in the iron and steel industry. Recently, the import of noncoking coal has started because of the environmental regulations requiring power plants located in sensitive areas and metropolitan cities to use coal with an ash content lower than 34%.

Crude oil production was 32.9 MT in 1996–1997 (TEDDY 2000–2001; TERI) while similar levels of crude oil were also imported. The production of crude oil has remained more or less stagnant over the past decade and this trend is expected to continue, thereby imposing great pressure on imports of oil and oil products. In the model, it was assumed that oil refinery capacity would increase in line with current plans from 62.87 MT in 1996–1997 to 112 MT by 2000–2001 and 179 MT by 2006–2007.

Although the production of natural gas was 23.3 bcm (billion cubic meters) in 1996–1997, indigenous availability is expected to be reduced to 16 bcm by 2011–2012[2] in line with the projections of the Sub-Group on Utilization of Natural Gas. Accordingly, higher levels of gas use in the future would need to be met through greater imports of gas.

[1] TEDDY 2000/01, TERI. 2001
[2] Hydrocarbon Vision 2025. Government of India

Renewable energy technologies such as biogas plants, solar energy, wind energy, small hydroelectric power plants, and biomass energy are believed to have great potential in India, and accordingly are included in the model.[3]

In the model, it was assumed that traditional fuels would have a declining role in future because of expectations of modernization and improved lifestyles, as well as a move toward the availability of sustainable firewood. Accordingly, it was assumed that the supply of fuelwood would decline from the estimated level of 169 MT in 1996–1997 to 51.2 MT by 2020.[4] It was assumed that the supply of dung and crop residue would remain at a constant level of 100 MT and 63.5 MT, respectively, over the modeling time-frame, based on the premise that its use would be restricted by behavioral changes associated with improved lifestyles rather than by limitations on their availability.

Based on an understanding of the developmental capacity of the Indian power sector, it is assumed that additions based on hydroelectric and nuclear power, renewables, and gas would continue to be rather limited within the current technopolitical environment. Most of the additional power generation capacity is therefore expected to continue to be based on existing sub-critical coal technology. All prices used in the model (of fuels, processes, and end-use devices) are at 1996–1997 levels.

The BAU scenario thus developed is considered to be the probable energy scenario for India. The model is based on an objective of cost minimization in the entire energy sector, and thereby provides the optimal energy mix that satisfies all given demands at the least possible cost to the economy.

Six alternative scenarios were developed by modifying the constraints on the future availability of various fuels and technologies. These scenarios were established in order to examine the extent of monetary savings to the economy and reductions of CO_2 emissions (if any) as compared with BAU, or with the most likely scenario for the country.

The alternative scenarios developed for comparison with BAU were:

1. greater penetration of efficient technologies (Alt 1);
2. improved performance of end-use equipment (Alt 2);
3. demand-side policy (Alt 3);
4. supply-side scenario (Alt 4);
5. integrated energy planning scenario (Alt 5);
6. emissions abatement scenario (Alt 6).

In the course of designing these alternative scenarios and validating baseline trends, discussions with experts in all relevant sectors indicated that many technological changes were unlikely to materialize in the absence of policies and directives which encouraged their implementation and follow-up, especially in

---

[3] MNES 2000. Government of India

[4] Based on 93 million ha of forest area in 2020, and a sustainable fuelwood yield of 55 tons/km²

the domestic, commercial, and transport sectors. Moreover, changes in these sectors are likely to be spread across numerous end-uses. The Indian power sector was generally viewed as one with substantial scope for diversity, although it was unanimously felt that coal would remain the most prominent source of energy in the next few decades.

In the light of the continuing dominance of coal in Indian industry, a coal-based power generation sector model (COLINK) was specifically developed using generalized algebraic modeling systems (GAMS). The COal-LINKage model is a LP optimization model structured under the general framework of the transportation problem, and this was used to examine various options by which the coal-predominant energy structure of the country could be improved in the future.

The COLINK model is static in nature and was set up for three time-periods which represent the last years of the country's 5-year plans. While the year 1996–1997 represents the base year, the scenario analyses were carried out for 2001–2002 and 2006–2007. The scenarios were designed to allow alternative linkages for coal movements, include the possibility of introducing washeries, charging differential freight rates for transportation of clean coal, and examining the scope for alternative generation technologies vis-à-vis the washing of coal. An analysis of differences in system costs, reductions in emission levels, and variations in coal production and supply patterns in the alternative cases as compared with the base case was used to examine the role of various influencing factors.

Description of the COLINK Model

In its most simplistic form, the COLINK model (Fig. 2) is set up as a general transportation problem, where coalfields and ports represent the supply nodes, while coal-based power plants represent the demand nodes.

Coal-based power generation at each of the plant nodes (based on planned additions to capacity in each year) served as the driving force of the model. We therefore assume that the planned additions to capacity are such that they correspond to demand requirements in the future. The model then makes choices about the source and route of coal from various coalfields and washeries to meet the power generation requirements in each time-period so that the total cost of the system is minimized. Accordingly, the quantity of coal selected for washing was also determined by the model based on the relative cost-effectiveness of the washery under various scenarios.

Balance equations as well as process-control equations in the model ensure that the quantity of coal mined from the coalfields is equal to that transported to the washeries and plants, that the quantities output from each washery relate to the input quantities, and that washery yields, the calorific value of input coal, and grade-wise production limits for each of the coalfields limit the off-take and supplies to the washeries and power plants.

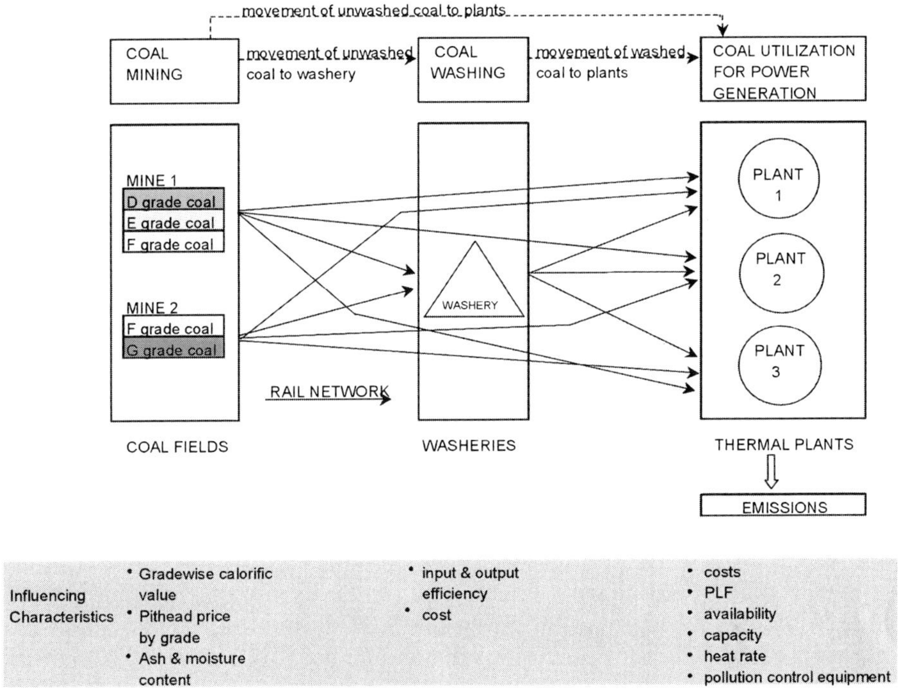

FIG. 2. Structure of the COLINK model

Description of the Scenarios in the COLINK Model

The BAU and alternative scenarios simulated with the COLINK model are described below.

BAU Scenario. The base case was developed and validated for 1996–1997 using documented data of coal supplies and power generation and based on the existing set of coal linkages. The BAU scenario envisages an extension of the base-case assumptions for the years 2001–2002 and 2006–2007. The coal supply was assumed to increase in line with the production plans of the supply agencies, and the existing coal linkages were expected to continue. It was also assumed that there would be no beneficiation of coal by washing in the BAU scenario.

Free Linkage Scenario. The authors hypothesize that the existing linkages for coal supplies to power plants may not be the most economic for the entire system. Discussions with railway authorities indicated that track capacity is not a limiting factor, and that larger quantities of coal could be moved between the demand and supply nodes even in the immediate short-term without any additional capacity. Against this background, a free linkage (FL) scenario was set up for the immediate short-term (2001–2002) to examine whether overall system costs could be reduced if alternative linkages were allowed. Unlike the BAU scenario, the free linkage scenario allows the movement of coal between all coalfields and power plants. The model then has an option to choose a set of linkages that may result in lower overall system costs than in the BAU scenario.

Scenario for a Washery with Single and Multiple Ash Reduction Targets. In the light of the notification of September 1997, a washery scenario was set up to examine whether the use of washed coal was an economically preferable option. Accordingly, the model was provided with the option of introducing washeries to wash coal to a 34% ash level in the single-target scenario, and to 34%, 30%, and 25% ash levels in the multiple-target scenario. The single- and multiple-target washery scenarios were evaluated for the year 2006–2007, and it was assumed that a maximum of 4 Mt washery capacity could exist at each of the coal supply nodes by this year.

Dual-Price Scenario. Given that there has not been much progress in India with regard to the setting up of washeries for thermal coal, a scenario was developed to examine whether the beneficiation of thermal coals could be promoted by adopting a differential freight structure for transporting washed and unwashed coal. The dual-price scenario attempts a sensitivity analysis of thermal coal beneficiation-with progressively higher levels of discount on freight charges for washed coal. This scenario was examined for the year 2006–2007.

Technological Improvement Scenario. The technological improvement scenario was developed to examine the competitiveness of alternative power generation technologies vis-à-vis the option of coal washing. Along with the option of introducing coal washeries, this scenario allows the introduction of more efficient generation technologies such as the integrated coal gasification combined cycle (IGCC) and supercritical technologies instead of existing subcritical technologies. This scenario indicates the preferred technology for investment in terms of the least-cost option when moving towards cleaner coal-based power generation.

Findings

This section first examines the outcomes of the energy sector model in terms of the probable energy mix and the associated concerns of the energy sector over the next 40 years. Then, specific results from the COLINK model are discussed to highlight policy options that could be used to optimize coal utilization in the power sector.

Results of the Energy Sector Model

Emerging Trends in the Energy Mix

The optimal level and pattern of energy use, as provided by the BAU scenario of the energy sector model, indicated that the overall useful energy re-

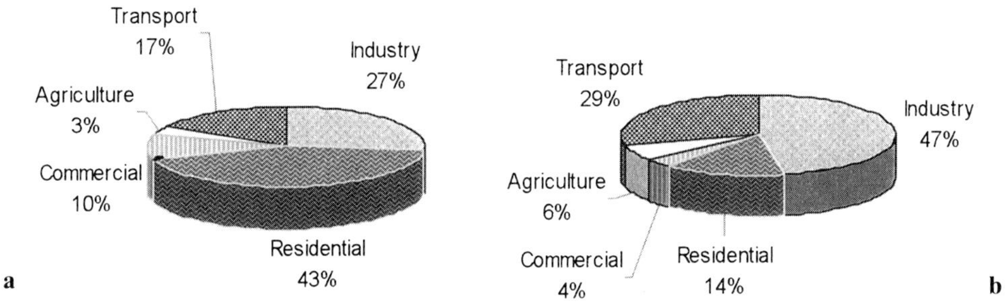

FIG. 3. Sectoral shares in **a** total and **b** commercial energy (1996–1997)

quirement would increase at a rate of 4.6% per annum from 10714 PJ in 1996–1997 to 65979 PJ in 2036–2037. Primary energy requirements over the same period were expected to increase from 14982 PJ to 96368 PJ. This indicates that the economy is far from being able to stabilize its energy requirements in the next few decades despite current efforts to decrease energy intensities and enhance the adoption of efficient fuels and technologies in all the energy-consuming sectors.

It is interesting to observe that although the overall efficiency of the system is expected to increase from 38% in 1996–1997 to 44% by 2036–2037, it will fall as low as 35% in the interim period. Such a situation is quite probable, since technological improvements and changes in processes cannot be expected to occur immediately, and the demand–supply gap will be bridged by using current inefficient technologies and fuels in the short-term. However, in the long-term, there will be a wider choice of efficient fuels and technologies, which will allow improvements in the efficiency of the system.

Sectoral Energy Requirements

The use of traditional fuels in the residential and commercial sectors results in a large variation in sectoral energy shares when examined in terms of total and commercial energy. As shown in Fig. 3, the residential and commercial sectors accounted for 43% and 10%, respectively, of total energy consumption in 1996–1997, but in terms of commercial energy, their respective shares were only 14% and 4% in the same year. The industrial sector had a 27% share in terms of total energy use, but had the largest share (47%) when compared with commercial energy use only. Similarly, the transport and agriculture sectors, which use only commercial fuels, have shares of 17% and 3%, respectively, of overall energy use, as compared with shares of 29% and 6%, respectively, of commercial energy use.

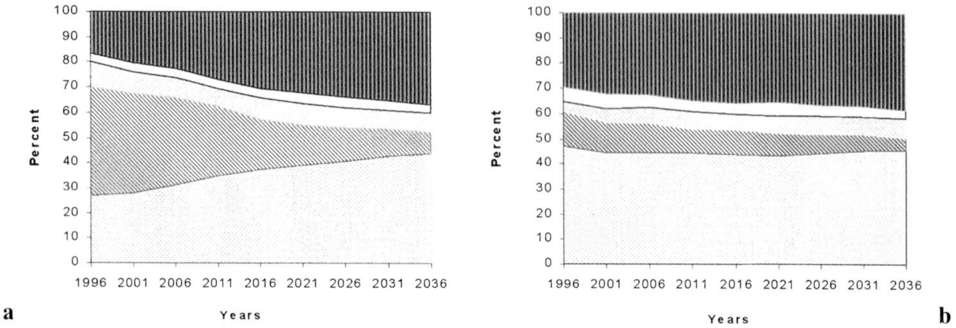

FIG. 4. Sectoral shares in **a** total and **b** commercial energy, business as usual (BAU) (1996–1997). From bottom upwards, the sectors are industry, residential, commercial, agriculture, and transport

TABLE 2. Comparison of primary energy, emissions, and costs—demand-side policy

			Change (%)
Primary energy in 2036–2037 (PJ)	BAU	96368	—
	Alt 1	94219	2.2
	Alt 2	94239	2.2
	Alt 3	92326	4.2
Cumulative CO_2 emissions (Tg)	BAU	127100	—
	Alt 1	124801	1.8
	Alt 2	124699	1.9
	Alt 3	122701	3.5
Total system costs (Rs billion)	BAU	169128	—
	Alt 1	165608	2.1
	Alt 2	168914	0.1
	Alt 3	165424	2.2

BAU, business as usual

The two parts of Fig. 4 show that sectoral shares have a tendency to converge over time. This is attributed to implicit assumptions of a decreasing prominence of traditional fuels in the residential and commercial sectors over time. Although commercial energy forms met only about 57% of their total energy targets in 1996–1997, it is expected that this will increase to about 95% in the next 40 years.[5]

Accordingly, by 2036–2037, the industry sector becomes the dominant energy consumer in both cases, followed by the transport sector and then the domestic, commercial, and agricultural sectors.

[5] This assumption is based on the premise of sustainable use of traditional energy forms. Therefore, the model assumes a constant supply of dung and crop residue fore energy use, and a decrease in fuelwood availability based on a sustainable supply from forests

TABLE 3. Comparison of energy, emissions, and costs—BAU vs. Alt 4

	Scenario	Value	Reduction (%)
Primary energy in 2036–2037 (PJ)	BAU	96368	—
	Alt 4	93209	3
Cumulative CO_2 emissions (Tg)	BAU	127100	—
	Alt 4	116945	8
Total system costs (Rs Billion)	BAU	169128	—
	Alt 4	169559	−0.3

Adopting a Multidimensional Approach Toward India's Energy–Environment Problems

A comparison of the results of Alt 1, Alt 2, and Alt 3 with BAU (Table 2) showed that the roles of energy conservation and efficiency improvements are rather limited, and cannot be expected to provide a quick and easy solution to India's energy problems.

Discussions with experts about imposing constraints to encourage the adoption of efficient technological alternatives made it clear that in many cases the changeover to efficient technologies was likely to be slow and limited owing to the numerous barriers which exist. Accordingly, the results of these three alternative scenarios indicate that reductions in primary energy use and emissions will only be about 2% of the BAU scenario by 2036–2037.

Although simulations of the supply-side scenario (Alt 4) compared with the BAU scenario in the energy sector model indicated a reduction of only about 3% in primary energy consumption by 2036–2037, reductions in cumulative levels of CO_2 were more significant at about 8% (Table 3). However, the total system costs were marginally higher than those in the BAU scenario, indicating that a higher level of investment needs to be undertaken to obtain environmental benefits of this order.

A comparison of the results of Alt 5 (integrated energy planning scenario) with BAU indicates a "win–win" situation on all counts. A multitargeted energy policy scenario could result in an 8% reduction in primary energy use by 2036–2037, leading to a 12% reduction in cumulative CO_2 emissions, as well as a 2% reduction in total system costs when compared with the BAU scenario. This model therefore indicates that a policy of simultaneous action on the demand and supply sides is the best option. Not only does this scenario result in a move toward sustainability in terms of the environment, but it would also meet the additional costs of financing clean fuels and technologies by the savings that would accrue from energy conservation measures and a greater penetration of efficient devices. This is especially important for developing countries such as India, where "paying for efficiency" is not easily understood.

The emissions abatement scenario (Alt 6) indicated that with successively greater reductions in cumulative CO_2 emissions, the system costs progressively increased over the BAU scenario (although only marginally). The major difference in the successive emissions reduction scenarios was in the share of fuels in power generation. It was found that the share of coal-based power generation was reduced with each successive emissions reduction scenario (from 63% in the BAU scenario to 56% in Alt 6, with a further 20% reduction by 2011–2012), while the shares of gas, hydroelectric power, nuclear power, and wind-based technologies increase. In the 5% emissions reduction case, the share of hydroelectric power increases from 18% to 24% by 2011–2012, while the shares of both coal- and gas-based power generation decrease in the same time-span. Although this increase is only marginal, nuclear and wind-based options also take an increased share in generating capacity.

Power Sector: An Emerging Area of Concern

With industrial progress and urbanization, the power sector will become increasingly important. The results of the energy sector model showed that electric power will increase from about 13% of total useful energy requirements in 1996–1997 to more than 18% by 2021–2022. In absolute terms, the power requirements in 2021–2022 are estimated to be 3.8 times those of 1996–1997. Further, despite an increasing share of gas and hydroelectric power, coal-based power generation is expected to continue to dominate the scene, with a 62% share in 2021–2022 compared with a 74% share in 1996–1997.

No Alternative to the Dominance of Coal

The simulation of the BAU scenario for India indicates that the share of coal in total primary energy is estimated to increase from around 38% in 1996–1997 to 41% by 2021–2022. The share of petroleum products will increase from 20% in 1996–1997 to 34% by 2021–2022, while that of gas will increase from 4% to 7% in the same period.

These results stress the need to be cognizant of the fact that coal will continue to be the dominant source of energy for India. It is therefore important to direct our efforts toward improving the efficiency of coal use, as well as limiting its environmental effects as much as possible.

Concerns of a Coal-Dominant Economy

On the environmental side, the trends described above raise concerns about the high levels of pollution that would be associated with such a pattern of energy use in the future. Apart from increases in CO_2 emissions from 787 Tg in 1996–1997 to 6957 Tg by 2036–2037, coal-based power generation would also be associated with emissions of particulates at the local level, and associated problems of ash-handling and disposal at the power plants. To date, this predicament has mainly been attributed to the deteriorating quality of non-coking coal supplied to power plants in India, but it has now been compounded by the government notification of September 1997 that requires the use of low ash coal at many power plants.

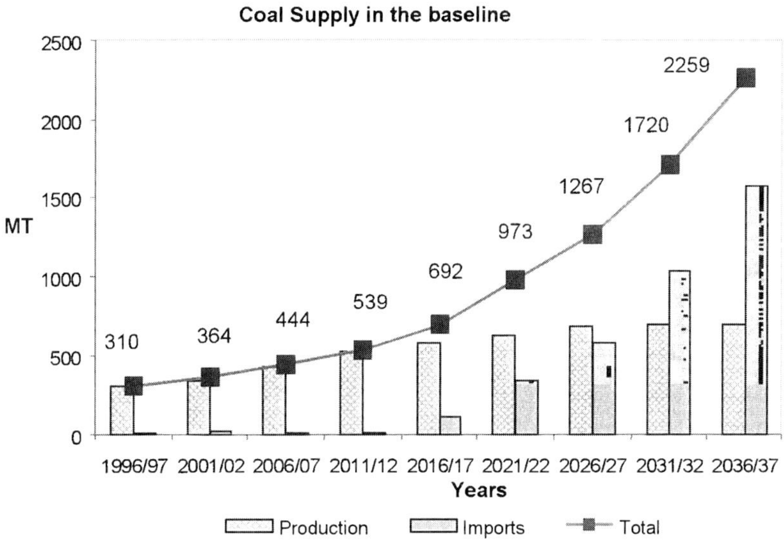

FIG. 5. Coal production and imports, 1996–1997 to 2036–2037

Based on the assumption that there will be constraints on the quality of both coking and noncoking coal used by the industry, it is likely that the country will import large quantities of coal in the future (Fig. 5).

Such high levels of coal imports in a coal-rich country such as India are completely unjustified because of the huge burden that it would impose on the economy, and also in terms of concerns about secure sources of energy in the future.

Another issue of concern is related to the vastness of the Indian subcontinent and the distribution of coal resources to the consumption centers. The geographical distribution of coal resources in India is uneven, and coal is often moved over distances of 2000 km. Historically, most of the coal was available in Eastern and Central India, and the movement of this coal to consumers was restricted to specific rail corridors running north–west or south. The Standing Committee on Coal Linkages was responsible for deciding and setting quarterly coal linkages. Linkage committees, appointed by the Government of India, apportion the coal production of each coalfield to all the major consumers based on requirements stated by them, and with the agreement of the supplying coalfields and the railway authorities. Over time, many consumers continue to be linked with the same coalfields simply because of historical associations developed many years ago. It is hypothesized that the results of an investigation into the optimality of Indian coal linkages might be similar to those associated with Chairman Mao's policy for a reversal of coal movement routes in China—a policy that was criticized for resulting in a large waste of both financial and human resources (Lu 1993).

TABLE 4. Break-up of system costs (million Rs.), 2001–2002

	BAU scenario	Free linkages scenario	Change over BAU
Freight charges for coal transportation	87624	64388	−23236 (−26.5%)
Pithead price of coal	136641	136171	−470 (−0.3%)
Power plant costs	589922	584280	−5641 (−0.9%)
Cost of ash handling and disposal	8095	8078	−16 (−0.1%)
Total system costs	822281	792917	−29364 (−3.5%)

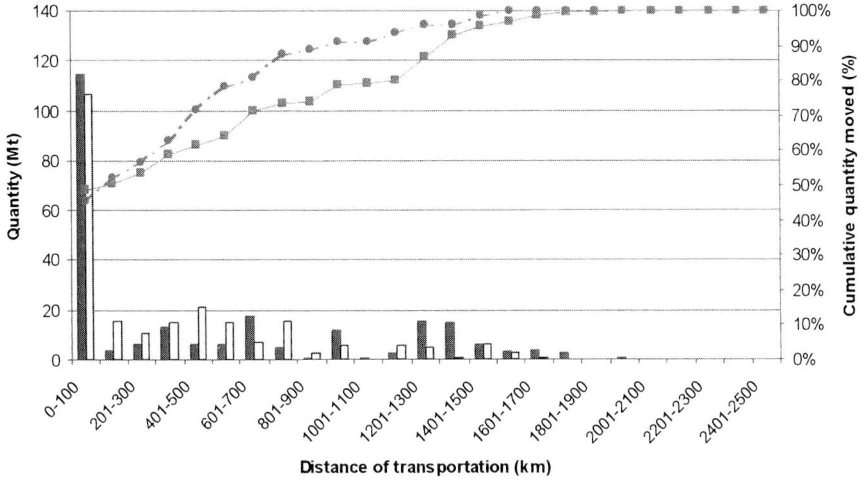

FIG. 6. Frequency distribution of coal movement. BAU vs. free linkage, 2001–2002. *Black blocks*, quantity of coal transported with BAU (Mt); *white blocks*, quantity of coal transported with free linkage (Mt); *squares*, cumulative quantity with BAU (%); *circles*, cumulative quantity with free linkage (%)

Addressing Coal Sector Issues: Results of the COLINK Model

Need to Reexamine Linkages

An analysis of the free linkage scenario for 2001–2002 using the COLINK model indicated that a reduction of as much as 27% of freight charges, as compared with BAU, could accrue merely by a reallocation of linkages (Table 4).

It was also found that coal movements to power plants would decrease from 117.2 billion tonne kilometer (Btkm) in 2001–2002 under the BAU scenario to 81.5 Btkm in the free linkage case (a reduction of approximately 30% of the BAU figure). As shown in Fig. 6, coal is transported for a maximum distance of

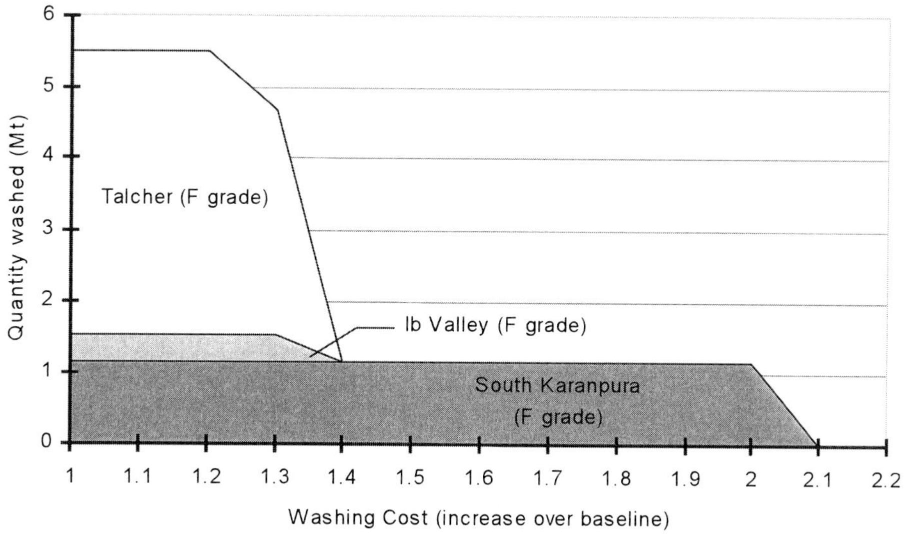

FIG. 7. Sensitivity of coal beneficiation to increases in washery costs

2000 km in the BAU scenario, while the model shows movements for a maximum distance of 1700 km in the free linkage scenario.

These results stress the need to reexamine historical linkages, and consider the possibility of changes by diverting or managing traffic on alternative routes.

Prospects for Coal Beneficiation in India

The initial runs of the COLINK model with the option of introducing washeries, using data on coal availability and quality in published documents and literature, indicated that the idea of washeries was not attractive. However, discussions with consumers revealed that the quality of coal received by the power plants was often one or two grades lower than that stated by the supplier. Consequently, the data were modified to account for grade slippages, and washery scenarios were tested again with a higher level of inferior quality coal in the system. In spite of this, it was found that although washeries became attractive, a total of only 5.5 Mt of F-grade coal from the Talcher, South Karanpura, and Ib Valley coalfields was selected for washing to a 34% ash level in 2006–2007. The reduction in system costs as compared with the BAU scenario was also marginal given the small amount of coal selected for washing. The decrease in costs can be attributed to marginal decreases in the costs of coal movement, reduced costs (O&M) at the plants, and some decreases in ash-handling and disposal costs at the power plants. However, the increases in the cost of coal and the cost of constructing washeries substantially offset the cost benefits (reductions) that accrue to the system.

Figure 7 shows the results of a sensitivity analysis conducted to examine the break-even costs for the washery process in the above scenario. For each coal-

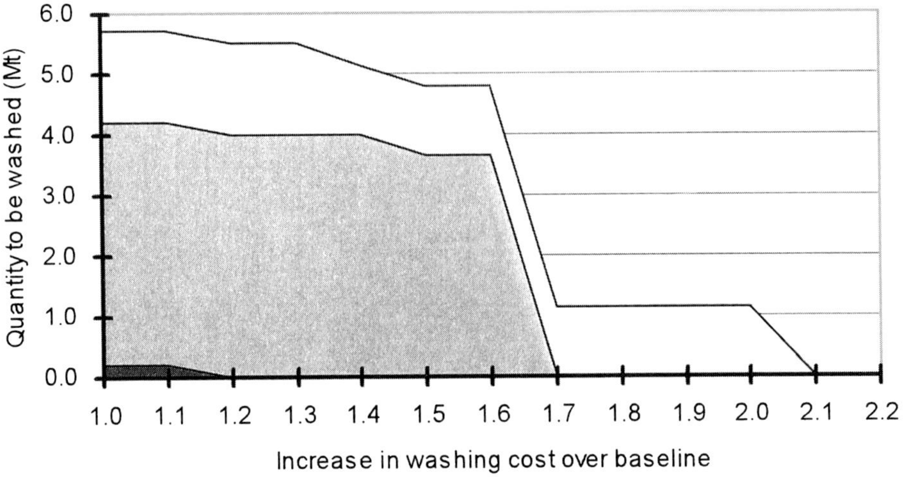

FIG. 8. Sensitivity of coal beneficiation to increasing levels of ash reduction. *White area*, 34%; *gray area*, 30%; *black area*, 25%

field, the amount of coal selected by the model for washing was recorded for every 10% increase in beneficiation costs over the baseline. It was found that the Talcher coal was the most sensitive to increases in washery costs, and although the quantity of coal washed was reduced even with a 30% increase in washery costs, no coal was taken up for washing with a 40% increase in costs over the baseline. Coal from the Ib Valley was not sensitive to changes in washing costs until there was a 40% increase over the base costs. Beyond this point, the washery was not economically attractive.

The coal from South Karanpura was extremely insensitive to changes in washery costs, and was not considered to be attractive for beneficiation until washing costs were more than twice those in the base case.

Although the model results show that low-grade coal from South Karanpura, Ib Valley, and Talcher coalfields were chosen for washing, it is clear that the economic benefits were not very substantial. Therefore, it seems unlikely that washeries will be set up in India for the beneficiation of coal supplied to thermal power plants in the absence of any stringent policy directives with regard to the environment, or to the supply or movement of high-ash coal.

A simulated scenario of introducing washeries with multiple targets of ash reduction (34%, 30%, and 25% ash levels) revealed that even with constant costs, the washeries become increasingly unattractive as higher levels of ash reduction are required (Fig. 8). With a target of 25% ash, a small quantity of coal is selected for washing, but this is highly sensitive to increases in washery costs. Increases of even 20% above existing washing costs render the washery process completely ineffective in the model for high levels of ash reduction. With a target of washing coal to 34% and 30% ash levels, a gradually increasing disincentive to wash coal

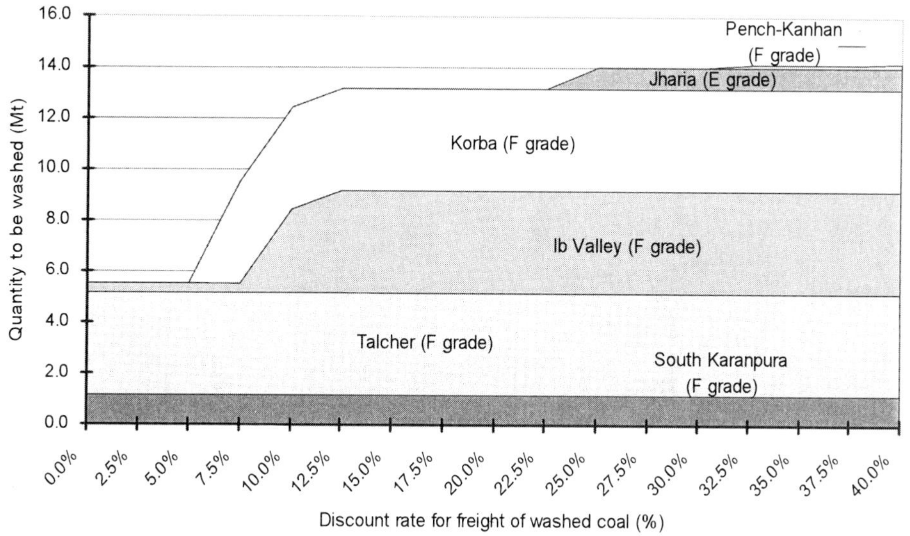

FIG. 9. Sensitivity of coal beneficiation to decreases in freight rates for washed coal

is found as washery costs increase by 60% above the existing costs. After this point, the model rejects the washing of coal.

Sensitivity of Coal Beneficiation to Freight Reductions for Washed Coal

The results of the dual-price scenario simulated by the COLINK model indicate that it becomes more economical to wash coal from different coalfields as the transportation costs for washed coal decrease (Fig. 9).

Although all the F-grade coal from the South Karanpura and Talcher coalfields is necessarily washed in the model even at existing freight rates for washed coal, it was found that higher grades of coal from other coalfields are washed when freight rates are decreased.

At existing freight rates, only 0.37 Mt of F-grade coal from the Ib Valley coalfield was selected for washing. This increased to 3.26 Mt with a 10% discount in the freight rate, and to 4 Mt when the freight rate was more than 12.5% below the base rate. F-grade coal from the Korba coalfield also becomes economical to wash and transport when freight rates are more than 5% below the base rate. Small amounts of E-grade coal from Jharia and F-grade coal from Pench–Kanhan were selected for washing when the freight rates for transporting washed coal were more than 20% below baseline rates. This clearly indicates that the quantity of coal to be washed is sensitive to freight rates.

The total quantity of coal that is washed increases from 5.5 Mt with existing freight rates for washed coal to 14.2 Mt when the rates are reduced by 40%. It was seen that even with a 10% reduction in freight rate, the quantity of

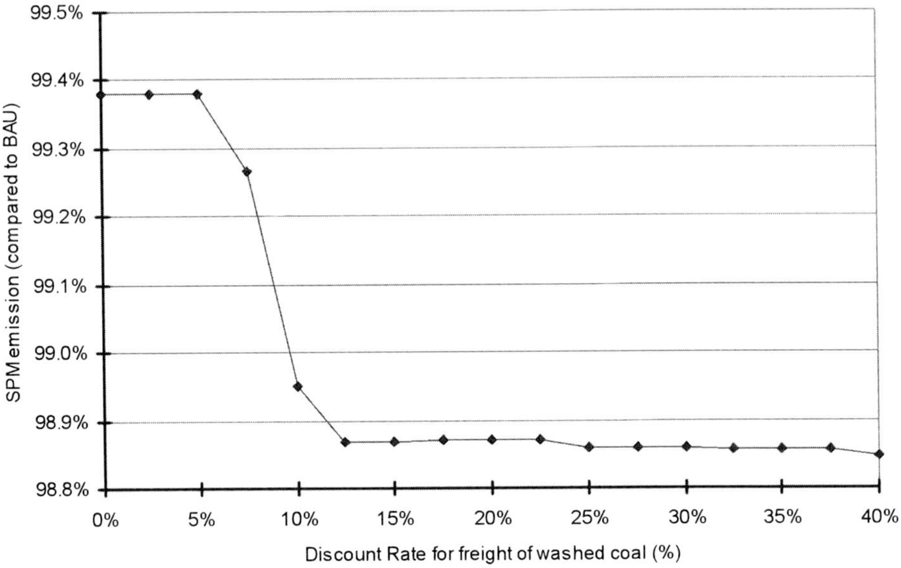

FIG. 10. Decrease in SPM emissions with reductions in freight rates for washed coal

coal selected for washing increased to about 12 Mt from the base level of about 5.5 Mt. This clearly indicates that the introduction of monetary incentives to move clean coal would provide an impetus to beneficiate thermal coals in India. Such a freight reduction policy is also recommended from the perspective of the benefits that would accrue in terms of emissions reductions after washing.

Figure 10 shows the environmental benefits of washing coal in terms of SPM emissions plotted against reductions in freight rates and the consequent amounts of coal washed.

Prospects for the Development of Alternative Coal-Based Power Generation Technologies

Figure 11 shows the results of a sensitivity run conducted with progressively increasing constraints on the levels of SPM (0%–10% of the base-level emissions) in a scenario where coal-washing technology could compete with alternative power generation technologies for the year 2006–2007. It was found that the marginal costs for reductions in SPM emissions by the washery process are higher than those achieved by adopting efficient power generation technologies. This implies that under a policy regime of environmental improvement, it is economically preferable to invest in alternative coal-based power generation technologies (supercritical or IGCC) rather than investing in coal-washing technology.

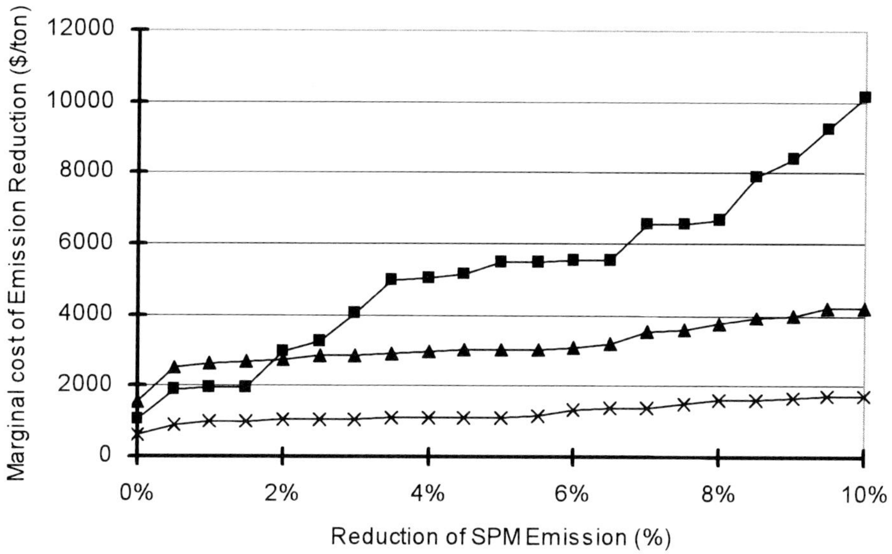

FIG. 11. Marginal costs of emissions reductions. *Squares*, subcritical technology and coal washery; *triangles*, integrated gasification combined cycle technology and no washing; *crosses*, supercritical technology and no washing

Conclusions

This study showed that energy requirements are expected to increase 6.4-fold during the period 1996–1997–2036–2037 despite current efforts toward improving efficiency as well as fuel and technology changes.

Although demand-side policy scenarios (including measures to improve efficiency and enhanced penetration of efficient technologies) indicate reductions in total system costs, primary energy requirements, and consequent CO_2 emissions as compared with the BAU scenario, the level of benefits is low. Moreover, demand-side measures are spread over numerous end-use sectors, which makes them difficult to implement.

Supply-side measures are easier to implement and monitor as they are more focused towards particular fuels or technologies. The supply-side scenario does indicate considerable potential for savings of both energy and emissions during the modeling time-frame, but the additional cost for such fuel and technological changes is likely to be an impediment for a developing country such as India, at least in the short-term. However, the introduction of policies that impose a tax on the relatively higher-polluting fuels (or a subsidy for the cleaner fuels) may be one of the ways to ensure fuel switching. Encouraging joint ventures and the participation of private parties is also important for mobilizing additional resources for fuel and technology changes on the supply side. However, for this purpose it is important to create a conducive and stable economic environment with clear and consistent policies and procedures.

The need for a simultaneous integration of environmentally compatible energy strategies on the demand as well as the supply side in order to achieve a "win–win scenario" is made clear from comparisons of energy requirements, cumulative emission levels, and total system costs with BAU levels.

This study also makes it clear that coal will continue to be the mainstay of the Indian energy sector, and therefore it is crucial to address issues related to the efficient utilization of coal in the future.

At present, the main conflict in the coal-based power generation sector relates to the requirement by many power plants for low-ash (<34%) coal, while current beneficiation technology for indigenous coal is not attractive, and a significant proportion of coal supplied to power plants has an ash content ranging from 35% to as high as 45%. In the short-term, such power plants have no options but to either import coal or shut down in order to comply with environmental regulations. Clearly these options are both undesirable. A loss of generation capacity is unwarranted in a situation where the country continues to have a large electricity demand–supply gap. Moreover, rapidly increasing coal imports in a coal-rich economy again raises concerns about the security of energy supplies and the drain on foreign exchange outflows. Although there is a continuing debate among experts about the yield of washeries for Indian thermal coal, it is felt that this yield could be even lower than that considered in our study, thereby making coal washing even more unsuitable using current techniques.

Improving the efficiency of the washery process is vital before the beneficiation of thermal coals can be exploited to a greater extent in India. There is therefore a pressing need to develop and/or adapt appropriate technologies which are suitable for beneficiating Indian noncoking coal.

Another important finding of this study is that coal washing is highly sensitive to changes in the freight rate for the transportation of washed coal. Beneficiating indigenous coal is crucial in order to shield the Indian coal market against rapidly increasing coal imports in future, as well as in terms of environmental protection. Against this background, a policy of adopting a differential freight structure that increases the attractiveness of washed coal would be valuable, at least in the short-term.

Model simulations indicate that a reallocation of coal linkages could result in savings to the power sector. In this context, it must be remembered that the model seeks to minimize total system costs, and may thereby decrease freight between some nodes by increasing freight between other nodes. In the real world, consumers would attempt to choose linkages that minimize individual costs while bearing in mind other qualitative factors such as reliability of supply and past associations. Accordingly, although the model results may not be directly applicable to actual linkages, it does seem worthwhile to assess the possibility of actual changes in linkages to ensure adequate coal movement capacity along the rail routes preferred by the model.

The study also showed that alternative technological options for power generation, such as supercritical or IGCC technology, were economically more attractive than washing coal. It is therefore becoming urgent to match technological developments and infrastructure developments to environmental policies. In this context, India must benefit from the knowledge and expertise of the developed world. Mechanisms such as the CDM could effectively facilitate the technological transfer of generation technologies such as IGCC or supercritical technologies, thereby helping sustainable development while being able to address both local and global concerns.

References

Central Electricity Authority (1997) Fourth National Power Plan 1997–2012. Government of India

Coal Directory of India (1996–1997) Calcutta: Ministry of Coal. Coal Controller's Organization, 128 p

Eighth 5-year plan (1994) Coal demand and linkages 1994/95 to 1996/97 and 2001/02: mid-term appraisal of Eighth Five Year Plan. Government of India

Hydrocarbon Vision, Government of India

Indian Market Demographics Report (1996)

Lu Y (1993) Fueling one billion—an insider's story of Chinese energy policy development. Washington Institute

Ministry of Coal (1997) Working Group Report on Coal and Lignite, Government of India

MNES (Ministry of Non-Conventional Energy Sources) (1999) Annual Report 1998–99. New Delhi

Mukhopadhay S, et al. (1999) Washing of non-coking coal in India. Indian Mining Eng J

Narasimhan KS, et al. (1997) Washability characteristics of Indian coals. J Mines, Metals Fuels 45:233–237

Planning Commission (1996) Sectoral energy forecasts

WDR (World Development Report) (1999–2000) Entering the 21st century

The Commercial Viability of the Space Solar Power System Under the Kyoto Protocol

Takamitsu Sawa* and Iwao Matsuoka†

Summary. The purpose of this chapter is to examine the commercial viability of a space solar power system (SSPS). In order for an SSPS to be able to compete with a plutonium thermal nuclear power plant, the cost of power generation by the SSPS must be lowered to 16.5 yen/kWh, which is far lower than the cost estimated by the New Energy and Industrial Technology Development Organization (NEDO). One of the most promising roles expected of a SSPS is to supply electricity to a developing country which has a supply shortage. This would certainly be classified as a clean development mechanism project, and given official sanction by the Kyoto Protocol. It would greatly contribute to reducing CO_2 emissions in the developing country, the amount of which could be counted toward the exporting country's CO_2 emissions reductions. Since the price of carbon emissions is determined in the international market of emissions trading, the commercial viability of a SSPS will be much improved because of the Kyoto Protocol.

Key words. Space solar power system (SSPS), Kyoto Protocol, Clear development mechanism (CDM), CO_2, Plutonium-thermal reactor, Generation cost, Coal-fired

Introduction

The 3rd Conference of Parties (COP-3) of the United Nations Framework Convention on Climate Change (UNFCC), held in Kyoto in December 1997, adopted the Kyoto Protocol. The Protocol includes an article that requires the 41 industrialized countries to reduce their average emissions of carbon dioxide and other greenhouse gases by at least 5% below their 1990 level during the period from 2008 to 2012 as a way to prevent global warming.

Quite suddenly, in late March 2001, the American President George W. Bush announced that his administration did not support the Kyoto Protocol. One of

*Institute of Economic Research and †Graduate School of Energy Science, Kyoto University, Yoshidahonmachi, Sakyo-ku, Kyoto 606-8501, Japan

the reasons why President Bush's administration regards the Kyoto Protocol as fatally flawed may stem from the following belief. Early actions required by the Protocol are likely to function as a disincentive in the development of large-scale technological innovations that would make significant contributions to reducing CO_2 emissions, but would require a lead-time of several decades. In addition to carbon sequestration and an improvement in nuclear power generation systems, the technological innovations being anticipated by the USA administration presumably includes a space solar power system (SSPS).

A SSPS has three components: an orbiting platform carrying solar panels, a microwave transmission system sending the generated electricity to the earth, and a rectenna, i.e., a receiving antenna on the ground that collects the transmitted microwaves and converts them into a form that will suit existing electric utility grids. In the USA, the National Aeronautics and Space Administration (NASA) designed a commercial space solar power plant in their Fresh Look study (Mankins et al. 1997). In Japan, there have been many studies relating to SSPS technology over the past few decades, and the Ministry of Economy, Trade and Industry (METI) has just given official support to research concerning an SSPS. It is quite likely that both governments might have been motivated to promote SSPS technology after recognizing its potentially significant contribution to reducing CO_2 emissions.

Given this background, it might be crucially important to estimate cost of power generation by a SSPS and evaluate its commercial viability, i.e., the possibility of such systems entering the competitive market for electricity in the not-so-distant future. It is highly likely that a SSPS would be at a disadvantage in terms of cost, and hence be unable to enter the competitive market for electricity without any support by a government.

Even if so, as the SSPS is undoubtedly one of the most effective technologies that could contribute to reducing CO_2 emissions, government support for its introduction to the electricity market through subsidies or carbon taxes may well be justified. Whether or not the subsidies are reasonable depends upon how the minimum attainable cost of power generation by a SSPS is estimated.

In addition, one of the anticipated advantages of a SSPS is that simply by setting up a rectenna it is possible to transmit electricity to any place on Earth. The Kyoto Protocol permits industrialized countries to count reductions in CO_2 emissions as a result of their investment in developing countries in their own CO_2 emissions reductions. This is what is called a clean development mechanism (CDM). By setting up a rectenna in a developing country, an industrialized country is able to reduce CO_2 emissions significantly by substituting electricity generated by coal-fired power plants in the developing country by electricity transmitted from a SSPS.

In order to estimate the cost of power generation by a SSPS, we make three assumptions: (1) The orbiting platform is located on a geosynchronous earth orbit (GEO); (2) solar panels on the platform are designed so that they always face the sun, and their total generating capacity amounts to one GW; (3) the system will start operation in the year 2020 following the timetable of the Fresh Look Study by NASA.

Maximum Permissible Cost of Power Generation by a Commercially Viable SSPS

In order to assess the competitive power of a SSPS in the future electricity market, we need two estimates.

First, we have to estimate the cost of each component of the system by taking into account any possible technological developments as well as future uncertainties, and estimate the total cost by summing the results. Our final goal is to estimate the cost (yen/kWh) of power generation by the SSPS.

Second, we have to estimate the cost of power generation by the leading alternative power sources such as nuclear power plants, which may safely be regarded as the maximum permissible cost, and should be the aim of any SSPS.

Combining these two estimates, we can make a judgement about whether or not a SSPS is commercially viable in the competitive electricity market. In the early stages, however, even if the SSPS is not commercially viable, it might be desirable for it to be supported by the government. We will suggest what types of supporting measures should be implemented in order to introduce a SSPS into the future electricity market.

Many studies have been carried out into the former estimation, but few have been done for the latter. After identifying the nuclear power plant as the most likely competitor to the SSPS in the future electricity market in Japan, we will estimate the maximum permissible generating cost of a SSPS. Since SSPSs are expected to be one of the large-scale 1-GW-class power sources, its generating costs must ultimately be lowered as close as possible to the level of a nuclear power plant. Otherwise, there will need to be permanent government subsidies.

Generally speaking, subsidies are rationalized and justified only in the early stages of a project. The decision about whether or not the government should give financial support to the research and development of a SSPS should be made on the basis of its competitive power in the future electricity market.

SSPSs and nuclear power plants have several characteristics in common. First, both are large-scale concentrated power sources, and hence require huge amounts of initial investment. Second, both are expected to be basic electricity suppliers. Third, in the process of power generation, neither of them emits any CO_2. It should be noted, however, that building materials are more or less carbon-intensive, and hence it is fair to focus upon the life-cycle assessment (LCA) of CO_2 emissions in order to compare them with thermal power plants.

According to Soda (2000) and Oshima (2000), the generating costs of a nuclear power plant are estimated to be somewhere in the range from 6.8 to 14.5 yen/kWh. The estimated generating cost is very sensitive to the assumed lifetime of the plant, because the initial investment is supposed to be paid off during the plant's lifetime. Until quite recently, it had been assumed that the average lifetime of a nuclear power plant was 16 years.

The government's official estimate (MITI 1992) of the generating cost publicized in 1992 was 9.0 yen/kWh. In 1999, however, the government (MITI 2000) lowered its cost estimate to 5.9 yen/kWh. This cost reduction was mainly due to the fact that the assumed lifetime was extended to 40 years.

We can safely assume that the generating costs of a nuclear power plant will stay around the current cost for the next few decades for the following reasons.

1. The price of uranium will stay around the current level. The reason is that since the world-wide production of uranium is increasing, while the demand is stagnating, the uranium market will not be tight enough to push up the price. Moreover, it is quite likely that nuclear fuel for Russian nuclear weapons will flow into the nuclear fuel power generation market (OECD/NEA 2000).

2. It is very unlikely that during the next two decades some technological innovation will significantly reduce the generating costs of nuclear power plants. In fact, Fig. 1 shows that the generating cost (yen/kWh) has not been reduced during the past three decades. We therefore assume that the cost of generating nuclear power will stay around the current level.

3. What are called *capacity factors*, i.e., how often maintenance of the plants is legally necessary and the exchange rate (yen/dollar), are expected to be sustained at the current level.

Until the year 2020, we cannot expect that fast-breeder reactors (FBRs) will be put into general use. Instead, plutonium–thermal reactors which use MOX fuel, i.e., a mixed fuel of uranium and plutonium, will play an important role in

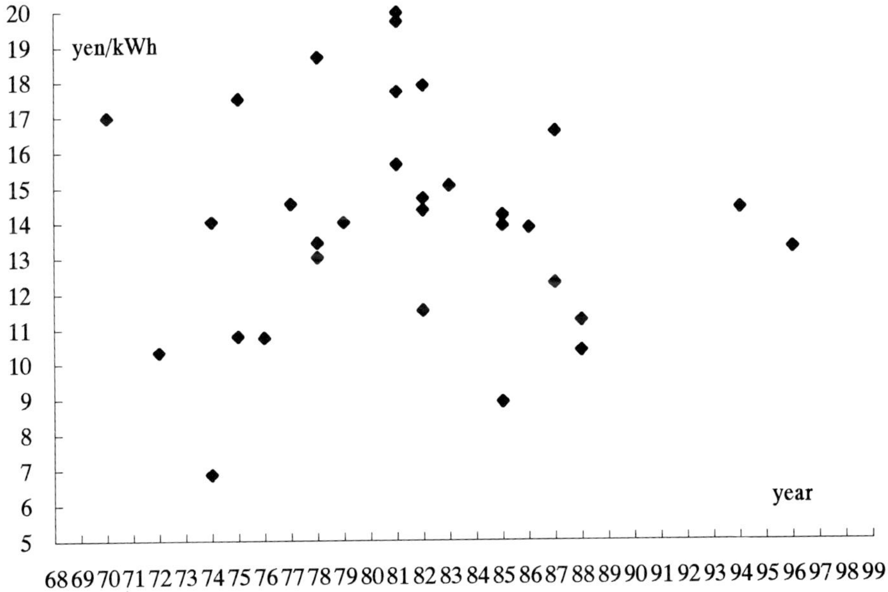

FIG. 1. The generating cost of nuclear power in Japan (from: Application for Permission for the installation of a nuclear power plant)

the nuclear fuel industry. This will push up the generating cost by 1–2 yen/kWh, mainly because the rise in the fuel costs.

Nuclear power plants are usually sited in sparsely populated areas, and hence, relatively speaking, they have more expensive transmission costs than thermal power plants. This weak point is shared by SSPSs: rectennas have to be sited well away from residential areas, because the site must be vast enough not only for the installation of a rectenna, but also to keep microwaves away from residential areas.

In conclusion, it is reasonable to assume that the main competitor to the SSPS would be the plutonium–thermal reactor, whose generating cost would be at most 16.5 yen/kWh: the highest generating cost is 14.5 yen/kWh by the nuclear power plant plus 2 yen/kWh plutonium–thermal surcharge. Hence, in order for a SSPS to be commercially viable in 2020, it is at least necessary that the generating cost be lowered to 16.5 yen/kWh.

It should be noted here that the generating cost of a nuclear power plant does not include what we call *indirect costs*: i.e., subsidies granted by the government, the costs of establishing the site, the costs of radioactive waste disposal, and so on.

Commercial Viability of an Electricity Supply to a Developing Country as a CDM Project

An platform orbiting the Earth carryies solar panels from which electricity may be transmitted to any place where a rectenna is installed. The construction costs for a rectenna are far less than those of the orbiting platform. In other words, a SSPS enables us to export electricity to a developing country in order to make up for its own shortage of electricity.

Let us suppose that an electricity supply to a developing country by a SSPS is certified as a CDM project. That is to say, if an industrialized country exports electricity generated by a SSPS to a developing country, the former country can obtain carbon credits equaling CO_2 emissions reductions because an electricity supply from the SSPS has been substituted for one from thermal power plants. The carbon credits may be counted in the developed country's emissions rights, or it can sell the credits in the emissions trading market.

The commercial viability of electricity supplied by a SSPS as a CDM project depends on the cost of the electricity supplied by the SSPS, the generating costs in the developing country, and the market price of carbon emissions. What are the costs of electricity supplied by a SSPS? The least-cost case includes only the cost of a rectenna in the developing country, while the highest-cost case includes the generating costs, most of which are the costs of equipment for the SSPS.

The economic returns on investment from installing a rectenna to transmit electricity from an orbiting platform come from two sources: one is the profit from selling electricity, and the other is carbon credits, the price of which are determined in the emissions trading market. The higher the price of carbon emissions, the higher the returns on investment.

The market price of carbon emissions is expected to be fairly high unless the USA secedes from the Kyoto Protocol. As for the selling price of electricity, it may safely be assumed that this will be equal to the cost of domestic generation in the developing country concerned, which presumably would be by coal-fired plants.

To assess the economic potentials of SSPSs quantitatively we suppose that the host country of the CDM project is China: that is to say, a rectenna is installed somewhere in China to supply electricity generated by solar panels on the platform orbiting in space. During the next few decades, it is expected that China's demands for electricity will continue to grow. Moreover, since 70% of the electricity used in China is supplied by coal-fired power plants, we can expect to obtain the maximum possible carbon credits (C-t/kWh) by exporting electricity to China.

Given the generating costs of coal-fired power plants in China and the generating costs of a SSPS, Table 1 shows how high the market price of carbon emissions must be in order to make the supply of electricity by a SSPS commercially viable as a CDM project (Yoshioka 1998). The average generating cost for coal-fired plants in China in the year 2020 may be supposed to range from 4.0 yen/kWh (the current generating cost; ZiDong 2001) to 7.0 yen/kWh (the current generating cost in coal-fired plants in Japan; MITI 2000).

We first examine the commercial viability of SSPSs in the case where the cost of electricity supplied by a SSPS is the average generating cost plus the equipment cost of a rectenna in the host country. This is the least favorable case for SSPSs.

The generating costs of SSPSs is expected to range from 7.0 yen/kWh (the minimum possible generating cost of a 1-GW nuclear power plant) to 23.0 yen/kWh (the generating cost estimated by the New Energy and Industrial Technology Development Organization (NEDO 1994/95). The figure of 16.0 yen/kWh, which lies between these limits, is the *highest permissible* generating cost which would allow SSPSs to compete with plutonium–thermal nuclear plants. In other words, even if the nuclear power plant is in the least advantageous situation, the generating cost will be less than or equal to 16.0 yen/kWh.

TABLE 1. The lowest market price of carbon emissions required to make the electricity supply to China by a SSPS commercially viable as a CDM

Coal-fired plant in China	Power generating cost		
	SSPS		
	7 yen/kWh	16 yen/kWh	23 yen/kWh
4 yen/kWh	9.1	36.5	57.8
6 yen/kWh	3.0	30.4	51.7
7 yen/kWh	0.0	27.4	48.7

SSPS, space solar power system; CDM, clean development mechanism

Suppose that the generating cost of a SSPS is reduced to the current lowest estimated cost, i.e., 7.0 yen/kWh. Then the export of electricity to China by a SSPS is commercially viable as a CDM project as long as the market price of carbon emissions is higher than or equal to 9100 yen/t-C even when the price of electricity in China is lowest. If the generating cost is as high as the NEDO estimate, i.e., 23.0 yen/kWh, the SSPS is commercially viable as a CDM project only when the market price of carbon emissions is higher than 48 700 yen/t-C even when the price of electricity in China is highest. However, such a high price for carbon emissions is unrealistic under the Kyoto Protocol, and hence it is absolutely impossible that a SSPS could be commercially viable.

According to model simulations, the market price of carbon emissions ranges from 3000 yen/t-C to 15 000 yen/t-C. Therefore, we conclude that in a business-as-usual (BAU) case, a SSPS would not be commercially viable as long as the market price of carbon emissions lies within the normal range.

Given the price of electricity in China and the market price of carbon emissions, Table 2 gives the highest permissible generating cost which would makes electricity supplied by a SSPS commercially viable as a CDM project.

We conclude that electricity supplied by a SSPS is commercially viable as a CDM project if and only if the following conditions are met: the market price of carbon emissions is at least 50 000 yen/t-C, the generating costs of coal-fired plants in China are as high as the current level in Japan, and the generating cost of the SSPS is as low as the NEDO estimate.

This estimate is based on the assumption that the orbiting platform of the SSPS is launched to generate electricity for export to China only. According to our estimate, it is safe to say that: unless an epoch-making innovation to reduce generating costs is made, it is hopeless to think that a SSPS could be commercially viable as a CDM project.

It may be more reasonable, as well as more realistic, to suppose one of two cases. First, the main purpose of constructing a SSPS is to supply electricity to Japan, while the surplus electricity, if any, is exported to developing countries on demand. In this case, the marginal cost of supplying electricity to China is quite low, and such a CDM project is fully commercially viable. Second, the generating capacity of the orbiting platform is intentionally designed to be so high that it is efficient to send the surplus electricity to China. In these cases, the marginal

TABLE 2. The highest permissible generating cost of a SSPS to make the electricity supply to China commercially viable as a CDM

Generating cost of coal-fired plant in China	Market price of carbon emissions (yen/t-C)		
	10 000	30 000	50 000
4 yen/kWh	7.3	13.9	20.4
6 yen/kWh	9.3	15.9	22.4
7 yen/kWh	10.3	16.9	23.4

generating costs should be used rather than the average generating costs when the commercial viability as a CDM project is evaluated.

We roughly estimated the marginal generating costs on the basis of data given by NEDO. We take the marginal generating costs of a SSPS as the cost of the orbiting platform without the antenna. The reason is that the size of the antenna is not influenced by the generating power, but only by the frequency and the distance between the satellite and the rectenna on Earth. We estimated that the construction costs of a rectenna in China would be about 10% of the current costs in Japan, and we assumed no land rent because the Chinese government could expropriate the land. In the second case, marginal generating cost is estimated as 20% less than the average cost, i.e., approximately 18 yen/kWh, which is still too expensive to make the electricity supplied by the SSPS commercially viable as a CDM project.

Technological innovations to improve the competitive power of SSPSs are expected in the next two decades. These should include a cost reduction for solar panels, and improvements in microwave conversion technology and earth-to-orbit (ETO) transport technology. These technological innovations will certainly contribute to a reduction in the marginal generating cost. This reduction is expected to be more than 20%.

With recent data which were discussed by the committee of the National Space Development Agency (NASDA), however, it might be expected that SSPSs will be commercially viable. We reestimated the marginal generating costs of a SSPS using data from the most inexpensive case, i.e., the cost of the frames for the solar panels on the satellites is 90% lower than the current cost, and the transportation cost is about 1% of the current cost. In this case, it is assumed that a transporter called "a 3rd generation rocket" will be used for ETO transportation, that the average generating costs of SSPSs is under 15.4 yen/kWh, and that the marginal generating cost is about 30% less than the average cost, i.e., about 10 yen/kWh. With these assumptions, a SSPS is commercially viable as a CDM project if the generating costs of coal-fired plants in China are the same as those of similar plants currently in Japan, since those are the highest costs and the market price of carbon emissions is only 10000 yen/t-C.

Conclusions

For SSPSs to be commercially viable, i.e., to be able to compete with nuclear power plants in the liberalized electricity market in the year 2020, the generating cost must be lowered to the highest current generating cost, i.e., 16.5 yen/kWh. The NEDO estimation, i.e., 23 yen/kWh, is far higher than the highest permissible current cost. To reduce the generating cost, some major technological breakthrough is essential.

As the SSPS is one of the most hopeful power sources in our efforts to reduce CO_2 emissions, governments should support the research and development of this technology in order to reduce generating costs. A carbon tax is also recommended in order to bring SSPSs into a more advantageous position.

In his speech, President George W. Bush emphasized the importance of research and development of technology that is really effective in reducing CO_2 emissions. One of the most promising candidates is certainly the SSPS. Moreover, the Kyoto Protocol encourages CDM projects. That is to say, an industrialized country is permitted to count the amount of CO_2 reductions resulting from its investment in a developing country in the total of its own CO_2 reductions. Only by installing a rectenna can we export electricity generated by SSPSs to wherever a demand for electricity exists.

The cost of supplying electricity to a developing country is ambiguous. From the NEDO's data, if the average generating cost is taken, then the export is commercially viable as a CDM project only when the market price of carbon emissions is more than or equal to 50 000 yen, which is unrealistically high. If the marginal cost is taken, the commercial viability of the SSPS as a CDM project is much improved. From recent studies, however, the SSPS might be commercially viable as a CDM project even if the market price of carbon emissions is only 20 000 yen-C/t. Considerable technological developments are therefore crucial in order to ensure the viability of the SSPS.

References

MITI (1992) Energy balance tables in Japan. Energy Data and Modelling Center (EDMC), Agency of Natural Resource and Energy

MITI (2000) Energy balance tables in Japan. EDMC, Agency of Natural Resource and Energy

NEDO (1994/95) Survey and study on the solar power satellite system. New Energy and Industrial Technology Development Organization

OECD/NuClear Energy Agency (NEA) (2000) Uranium 1999. Organization for Economic Cooperation and Development

Oshima K (2000) Estimate of generating costs from annual reports of Japanese distributors (in Japanese), Citizens' Alliance for Saving the Atmosphere and the Earth

Science Applications International Corporation/Futron Corporation/National Aeronautics and Space Administration (1997) Space solar power: a fresh look at the feasibility of generating solar power in space for use on earth. NASA

Soda Y (2000) Re-evaluation of the Japanese nuclear fuel cycle from an economic and social point of view (available in Japanese only), Japan Science-Technology Corporation, Keio University

Yoshioka K, Asakura K (1998) CO_2 load of space solar power satellite (in Japanese)

ZhiDong L (May 2001) Private interview, Mitsubishi Research Institute

Subject Index